Patent Intensity and Economic Growth

Economic growth has traditionally been attributed to the increase in national production arising from technological innovation. Using a panel of seventy-nine countries bridging the North-South divide, *Patent Intensity and Economic Growth* is an important empirical study on the uncertain relationship between patents and economic growth. It considers the impact of one-size-fits-all patent policies on developing countries and their innovation-based economic growth, including those policies originating from the World Intellectual Property Organization, the World Trade Organization and the World Health Organization, as well as initiatives derived from the TRIPS Agreement and the Washington Consensus. This book argues against patent harmonization across countries and provides an analytical framework for country group coalitioning on policy at UN level. It will appeal to scholars and students of patent law, national and international policy makers, venture capitalist investors, and research and development managers, as well as researchers in intellectual property, innovation and economic growth.

Dr. Daniel Benoliel is a law professor at the University of Haifa Faculty of Law and Haifa Center of Law and Technology (HCLT). His main fields of expertise include international intellectual property, patent law and innovation, public international law, and entrepreneurship law. Benoliel holds a Doctorate in law (J.S.D.) from UC Berkeley School of law (Boalt Hall) and is a John M. Olin Research Fellow as well as a Yale Law School Information Society Project (ISP) Visiting Fellow alumnus. Benoliel has received numerous prizes, awards, and research grants in these fields.

Cambridge Intellectual Property and Information Law

As its economic potential has rapidly expanded, intellectual property has become a subject of front-rank legal importance. *Cambridge Intellectual Property and Information Law* is a series of monograph studies of major current issues in intellectual property. Each volume contains a mix of international, European, comparative, and national law, making this a highly significant series for practitioners, judges, and academic researchers in many countries.

Series Editors

Lionel Bently
Herchel Smith Professor of Intellectual Property Law, University of Cambridge

Graeme Dinwoodie
Professor of Intellectual Property and Information Technology Law, University of Oxford

Advisory Editors

William R. Cornish, *Emeritus Herchel Smith Professor of Intellectual Property Law, University of Cambridge*

François Dessemontet, *Professor of Law, University of Lausanne*

Jane C. Ginsburg, *Morton L. Janklow Professor of Literary and Artistic Property Law, Columbia Law School*

Paul Goldstein, *Professor of Law, Stanford University*

The Rt Hon. Sir Robin Jacob, *Hugh Laddie Professor of Intellectual Property, University College, London*

Ansgar Ohly, *Professor of Intellectual Property Law, Ludwig Maximilian University of Munich, Germany*

A list of books in the series can be found at the end of this volume.

Patent Intensity and Economic Growth

Daniel Benoliel

The University of Haifa Faculty of Law

Haifa Center of Law & Technology (HCLT)

CAMBRIDGE
UNIVERSITY PRESS

University Printing House, Cambridge CB2 8BS, United Kingdom

One Liberty Plaza, 20th Floor, New York, NY 10006, USA

477 Williamstown Road, Port Melbourne, VIC 3207, Australia

314–321, 3rd Floor, Plot 3, Splendor Forum, Jasola District Centre,
New Delhi – 110025, India

79 Anson Road, #06–04/06, Singapore 079906

Cambridge University Press is part of the University of Cambridge.

It furthers the University's mission by disseminating knowledge in the pursuit of
education, learning, and research at the highest international levels of excellence.

www.cambridge.org
Information on this title: www.cambridge.org/9781107098909
DOI: 10.1017/9781316162675

First published 2017

Printed in the United Kingdom by Clays, St Ives plc

A catalogue record for this publication is available from the British Library.

Library of Congress Cataloging-in-Publication Data
Names: Benoliel, Daniel, 1972– author.
Title: Patent intensity and economic growth / Daniel Benoliel.
Description: Cambridge [UK] ; New York : Cambridge University Press,
2017. | Series: Cambridge intellectual property and information law
Identifiers: LCCN 2017012889 | ISBN 9781107098909 (hardback)
Subjects: LCSH: Patents (International law) – History. | Patent laws and
legislation. | Technological innovations – Law and legislation. | Patents –
Economic aspects. | Technological innovations – Economic aspects. |
Economic development. | BISAC: LAW / Intellectual Property / General.
Classification: LCC K1505.4 .B39 2017 | DDC 338.9–dc23
LC record available at https://lccn.loc.gov/2017012889

ISBN 978-1-107-09890-9 Hardback

To Naáma

Contents

Acknowledgments

This book is the product of numerous years of gratifying research. I would like to offer my warm thanks for the comments, advice, and support I received from the following (in alphabetical order): Katya Assaf, Eran Bareket, Uri Benoliel, Rochelle Dreyfuss, Niva Elkin-Koren, Susy Frankel, Xiaolan Fu, Michal Gal, Heinz Goddar, Stuart Graham, Dietmar Harhoff, Ralph Heinrich, Reto Hilty, Assaf Jacob, Calestous Juma, David Kaplan, Natalia Karpova, Asa Kling, Amy Landers, Keun Lee, Mark Lemley, Li Maor, Dotan Oliar, Bruno Salama, Eli Salzberger, Assaf Weiler, Peter Yu, and Lior Zemer.

I am very grateful to World Intellectual Property Organization (WIPO) members for their cooperation in the interviews I conducted at the Geneva WIPO headquarters in October 2014. These members are Dr. Francis Gurry, Director General of WIPO; Carsten Fink, Chief Economist of WIPO; Geoffrey Onyeama, former Deputy Director General of Cooperation for Development; Yoshiyuki (Yo) Takagi, Assistant Director General for Global Infrastructure; Irfan Baloch, Acting Director of the Development Agenda Coordination Division (DACD); and James Pooley, former Deputy Director General for Patents.

Earlier versions of parts of the manuscript were presented at Oxford University's Sixth Annual Conference for the Academy of Innovation and Entrepreneurship (AIE), the 12th Globelics International Conference, the Benjamin N. Cardozo School of Law, Yeshiva University 13th Annual Intellectual Property Scholars Conference; the joint Bar Ilan University Faculty of Law and University of Pennsylvania Faculty of Law Conference on Intellectual Property and Entrepreneurship; the World Intellectual Property Organization – United Nations Economic Commission for Europe (UNECE) seminar on Economic Issues of Intellectual Property Rights in Nations in Transition, in cooperation with the Israel Patent Office (ILPO), the Israeli Intellectual Property Academic Association annual workshop, and two University of Haifa Faculty of Law Caesarea Faculty Workshops.

I wish to thank Professor Ayala Cohen and Dr. Etti Dove of the Statistics Laboratory at the Faculty of Industrial Engineering and Management, the Technion – Israel Institute of Technology for their statistical support. Their input was decisive and is warmly appreciated. I would also like to acknowledge the superb technical assistance I received from Inbar Yasur regarding PatBase and Patstat data collection. Thanks, too, to my research assistants Denis Geidman, Boaz Gadot, and David Hurtado.

I wish to acknowledge the generous financial support I received from the Israel Science Foundation (ISF) as a research grant (App. 610/15), as well as supplementary funding received from the Haifa Center of Law and Technology (HCLT) at the University of Haifa's Faculty of Law. The book was further completed as part of the visiting scholar program at the Max Planck Institute for Innovation and Competition, in Munich, Germany.

I wish to thank the editorial team at the Cambridge University Press, Intellectual Property and Information Law series, and especially the wonderful Kim Hughes, Rebecca Jackaman, and Puviarassy Kalieperumal. Finally, I thank my wife Naáma and my daughters Hadar and Maáyan for the affection and support they gave me along the way.

Abbreviations

A2K	Access To Knowledge
ABS	Nagoya Protocol on Access to Genetic Resources and the Fair and Equitable Sharing of Benefits Arising from Their Utilization
ACTA	Anti-Counterfeiting Trade Agreement
AIE	Academy of Innovation and Entrepreneurship
BTAP	Beijing Treaty on Audiovisual Performances
CART	Classification and Regression Tree
CBD	Convention on Biological Diversity
CEEC	Central and Eastern European Country
CIS	Commonwealth of Independent States
CMS	Carnegie Mellon Survey
DACD	Development Agenda Coordination Division
DACD	WIPO Development Agenda Coordination Division
DLT	Design Law treaty
ENWISE	Central and Eastern European Countries and the Baltic State
EPC	European Patent Convention
EPO	European Patent Office
EU	European Union
FDI	Foreign Direct Investment
FTA	Free Trade Agreement
FTE	Full-Time Equivalent
GATT	General Agreement on Tariffs and Trade
GDP	Gross Domestic Product
GERD	Gross Domestic Expenditure on R&D
GI	Geographic Indications
GII	Global Innovation Index
GNI	Gross National Income
GSPOA	WHO Global Strategy and Plan of Action on Public Health, Innovation and Intellectual Property
GVC	Global Value Chain

HC	Head Count
HCLT	Haifa Center of Law and Technology
HDI	Human Development Index
IBSA	India, Brazil and South Africa Initiative
ICT	Information and Communications Technology
IDB	Industrial Development Board
IGWG	WHO Intergovernmental Working Group
ILO	International Labor Organization
ILPO	Israel Patent Office
IMF	International Monetary Fund
IPR	Intellectual Property Right
ISF	Israel Science Foundation
IT	Information Technology
JPO	Japanese Patent Office
LDC	Least-Developed Countries
MAR	Marshall-Arrow-Romer knowledge spillover
MDG	Millennium Development Goal
MFN	Most-Favored Nation
MNC	Multinational Corporation
MNRE	Ministry of New and Renewable Energy of India
MVT	Marrakesh Treaty to Facilitate Access to Published Works for Persons Who Are Blind, Visually Impaired, or Otherwise Print Disabled
NIC	Newly Industrialized Country
NIE	Newly Industrialized Economy
NIEO	New International Economic Order
NIH	National Institutes of Health in the United States
NIPS	National Intellectual Property Strategy
NSF	National Science Foundation
OECD	Economic Cooperation and Development
PCAST	United States President's Council of Advisors on Science and Technology
PCT	Patent Cooperation Treaty
PPP	Purchasing Power Parity
PVP	Plant Variety Protection
R&D	Research and Development
RIS	Regional Innovation Systems
RMB	Chinese Yuan Renminbi
RNA	Ribonucleic Acid
S&T	Science and Technology
SCP	Standing Committee on Patents
SIPO	Chinese State Intellectual Property Office

SPER	Second Pair of Eyes Review program
TFP	Total Factor Productivity
TPP	Trans-Pacific Strategic Economic Partnership Agreement
TRIMS	Agreement on Trade-Related Investment Measures
TRIPS	Agreement on the Trade Related Aspects of Intellectual Property
TTIP	Transatlantic Trade and Investment Partnership
TTO	Technology Transfer Office
UE	Percentage of patents granted out of patents submitted to the USPTO or EPO
UN	United Nations
UNCTAD	United Nations Conference on Trade and Development
UNDP	United Nations Development Programme
UNECE	United Nations Economic Commission for Europe
UNESCO	United Nations Educational, Scientific and Cultural Organization
UNIDO	United Nations Industrial Development Organization
USPTO	United States Patent and Trademark Office
WHA	World Health Assembly
WHO	World Health Organization
WIPO	World Intellectual Property Organization
WTO	World Trade Organization

Tables and Appendixes

Chapter III – Institutions, GERD Intensity, and Patent Clusters

Chapter IV – GERD by Type, Patenting, and Innovation

Chapter V – Patent Intensity by Employment and Human Resources

Chapter VI – Spatial Agglomeration of Innovation and Patents

Introduction

Economic growth has traditionally been attributed to the increase in national production arising from technological innovation. Nevertheless, the relationship between patents and economic growth remains uncertain. This relationship, which forms the focus of this book, is examined by means of a panel of 79 countries bridging the North–South divide for the period 1996–2013. Three groups of countries are identified by their model patent intensity as a proxy for their domestic innovation. The book's clustering empirics may ultimately question efficient growth generation by equal international patent policies.

In the past, developing countries were thought to be at an earlier phase along a linear path of historical technological catch-up by comparison with more developed countries. This course also underlined the neoclassical economics inclination toward "one-size-fits-all" patent policies for fostering innovation-based economic growth. These policies include initiatives by the World Intellectual Property Organization (WIPO), World Trade Organization (WTO) policies deriving from the Agreement on the Trade Related Aspects of Intellectual Property (TRIPS), and relevant innovation policies of the World Health Organization (WHO). This equal-country approach is also consistent with the Washington Consensus standard macroeconomic reform package for crisis-wracked developing countries promoted by the International Monetary Fund (IMF), the World Bank, and the United States Treasury Department. This neoclassical stance ultimately mirrors the constitutional legitimacy of the United Nations (UN)-level organs.

This book joins the ranks of endogenous growth dissenters who have challenged this hegemonic approach. Accordingly, it will be argued that patent harmonization across countries is not clearly necessary, empirically based, or otherwise adequate for the South. The clustering analysis presented in the book is also undertaken within the framework of "convergence" literature. This approach provides an additional and seminal insight known as "club convergence" that can help identify similarities and differences between countries based on generalized growth-related

1

hypotheses. Club convergence over patent intensity and the concept of technological catch up are, of course, interrelated. As with other methodical taxonomies, no clear line runs between the two. As a result, a serious examination of this distinction demands the contextual divorcing of the convergence premise from issues relating to any one country's productivity performance. As a result, what is important for convergence analysis, as this book shows, is how countries perform over a model patent intensity relative to each other, as opposed to how a single country performs relative to its own historical technological catch-up.

Earlier accounts of the relationship between endogenous growth theory and club convergence mostly contributed to the understanding of convergence over salaries, educational level, Gross Domestic Product (GDP), and other macroeconomic income-related indications. Within the growth economics natural flow, convergence analysis may now be expanded into other fields, including patents and innovation.

Chapter 1 sets the framework for contemporary UN-level patent and innovation-related norm-setting. It highlights the fact that much of this process is still marred by regulatory inconsistency, underlying the need for a granular empirical and conceptual approach. Chapters 2–6 explain both how and why countries across the development divide differ in terms of patent intensity as a proxy for innovation-based economic growth. The book thus elaborates on the differences between the three patent clusters based on comparisons drawn over other World Bank and IMF country-group classifications, such as income level, geographic region, and economy type. This is followed by a characterization of the three country groups by core growth indicators, such as the type of institutions performing and financing Gross Domestic expenditure on Research and Development (GERD), GERD by type of research and development (R&D), human capital and human resources indicators, and spatial growth-related indicators.

Estimates of the patent intensity of selected countries and the three designated patent clusters as proxy of their comparable domestic innovation may yield valuable information for national and international policymakers, venture capitalist investors, and R&D managers, as well as for researchers in intellectual property, innovation and economic growth, and other fields.

1   Setting the Framework: Patenting and Economic Growth Policy

Introduction

Economists have traditionally perceived the patent system as a vital lever through which policy-making affects innovation-based economic growth.[1] Yet, across different countries the precise effect of patents remains uneven, for two fundamental reasons. These two reasons relate to the ambiguous effect of national patent laws upon their enforcement and to the ambiguity associated with the impact patenting rates have on national economic growth. First, much empirical ambiguity remains regarding the legal environment shaped by the presence and enforcement of patent laws. This ambiguity relates to such aspects as the impact of a patent rule of law on the incentive to invest in research and development, their ability to increase quotas of foreign direct investment (FDI), or their ability to promote other forms of technology absorption and diffusion in different countries.[2] If

---

[1] See, e.g., Richard Gilbert and Carl Shapiro, Optimal Patent Length and Breadth, *Rand Journal of Economics*, vol. 21, 106 (1990); Paul Klemperer, How Broad Should the Scope of Patent Protection Be?, *Rand Journal of Economics*, vol. 21, 113 (1990); Nancy Gallini, Patent Policy and Costly Imitation, *Rand Journal of Economics* vol. 23, 52 (1992).

[2] The scope of the present empirical ambiguity is rather startling. See, e.g., Yi Qian, Do National Patent Laws Stimulate Domestic Innovation in a Global Patenting Environment? A Cross-Country Analysis of Pharmaceutical Patent Protection, 1978–2002, *The Review of Economics and Statistics*, vol. 89(3) 436 (2007) (evaluating the effects of patent protection on pharmaceutical innovations for 26 advanced countries that established pharmaceutical patent laws during 1978–2002. In this seminal study, Qian finds that in countries with high levels of development, education, and economic freedom, patent laws indeed stimulate innovation. Ibid., at 436; José L. Groizard, Technology Trade, 45 *Journal of Development Studies* 1526 (2009) (using panel data for 80 countries for the period 1970, the author finds that FDI is higher for countries with stronger IPRs), at 11–13. On the other hand, Groizard identifies a negative relationship between IPRs and human capital indicators. Ibid.; Sunil Kanwar and Robert Evenson, Does Intellectual Property Protection Spur Technological Change?, 55 *Oxford Economic Papers* 235 (2003) (Lower IPRs can facilitate imitation, while, on the other hand, innovation in developing countries increases in proportion to greater IPR protection), at 236; Yongmin Chen and Thitima Puttitanun, Intellectual Property Rights and Innovation in Developing Countries, 78 *Journal of Development Economics* 474 (2005), at 489.

    See also Rod Falvey, David Greenaway, and Zhihong Yu, who find evidence of a positive effect between IPR and economic growth for both low- and high-income countries, but not for middle-income ones. Extending the Melitz Model to Asymmetric Countries (University of Nottingham Research Paper Series, Research Paper 2006/07). Using panel data for 79

anything, intellectual property rights (IPRs) and patent law mostly seem to have fallen short in systematically predicting economic growth across countries, including developing ones.[3]

The second reason for this uncertainty relates to the effect of patenting rates on economic growth. This relationship, which forms the focus of this book, will be examined by means of a comparison of the impact of patent propensity rates crossed by R&D intensity across different countries (later to be defined as patent intensity) by way of an important (albeit not a sole) proxy for domestic innovation-based economic growth.[4]

Both the legal environment and patenting itself jointly affect the propensity to patent across countries. In other words, the realization that firms or countries differ in terms of their patent propensity rates acknowledges both the "capability factors" associated with the patent legal environment,[5] alongside "willingness

countries and four sub-periods (1975–1979, 1980–1984, 1985–1989, and 1990–1994), the authors conclude that the positive relationship between IPR and economic growth in low-income countries cannot be directly explained by the potential fostering of R&D and innovation.

For a negative correlation between tightening IPR and innovation, see, e.g., James Bessen and Eric Maskin, Sequential Innovation, Patents, and Imitation, Department of Economics, Massachusetts Institute of Technology working paper no. 00–01 (2000); Mariko Sakakibara and Lee Branstetter, Do Stronger Patents Induce More Innovation? Evidence from the 1988 Japanese Patent Law Reforms, NBER working paper 7066 (1999).

[3] See World Bank, Global Economic Prospects and the Developing Countries (vol. 12, 2002) ("At different times and in different regions of the world, countries have realized high rates of growth under varying degrees of IPRs protection") at 135. See also: Bruno van Pottelsberghe de la Potterie, The Quality Factor in Patent Systems, ECARES working paper 2010–027 (2010) (reviewing empirical studies which generally "lead to the conclusion that 'strong' patent systems have, at most, an ambiguous relationship with the rate of innovation"), at 7–8; Qian, Do National Patent Laws Stimulate Domestic Innovation in a Global Patenting Environment?, (note 2 above), ("The actual effect of IPR on innovation, however, remains one of the most controversial questions in the economics of technology"), at 436.

[4] The chapter focuses solely on the propensity to patent against the backdrop of other intellectual property regimes, which foster innovation, notably in developing countries. But see, e.g., Emmanuel Hassan, Ohid Yaqub and Stephanie Diepeveen, Intellectual Property and Developing Countries: A Review of the Literature, Rand Europe (2010) ("Several surveys carried out in developed countries have shown that other factors are much more effective than patents in enabling firms to profit from inventive efforts: trade secrecy, first-mover advantages and associated brand loyalty, the complexity of the learning curve and establishment of effective production, sales and marketing functions") (internal citations omitted), at 19 and sources therein.

[5] Capability factors relating to the patent legal environment include the cost of patenting: see e.g., Georg Graevenitz, Stefan Wagner and Dietmar Harhoff, Incidence and Growth of Patent Thickets: The Impact of Technological Opportunities and Complexity, *Journal of Industrial Economics* 61 (3), 521 (2013). Capability factors further include patent imitation and litigation costs. See, e.g., Hariolf Grupp and Ulrich Schmoch, Patent Statistics in the Age of Globalization: New Legal Procedures, New Analytical Methods, New Economic Interpretation, *Research Policy* 28, 377 (1999). Another capability factor is

factors"[6] associated with differing patent propensity rates even given comparable legal environments between countries.[7] This book focuses on the latter factors, assessing both how and why countries across the North–South divide differ in terms of their "patent intensity" as a proxy for innovation-based economic growth.

Our field of inquiry conveniently corresponds with the related work of economic geographer Andrés Rodríguez-Pose, explaining why, within the European Union's regional growth dynamics, peripheral and socioeconomically disadvantaged areas have consistently failed to "catch up" with the rest of the EU.[8] Rodríguez-Pose reports the presence of different "social filters" in different regions.[9] These filters provide a different "capacity to every region to assimilate and transform its own or foreign R&D related innovation into economic activity."[10] As a result, one finds "innovation

---

the lower capability to patent process innovation as opposed to product innovation: see, e.g., Erik Brouwer and Alfred Kleinknecht, Innovative Output, and a Firm's Propensity to Patent. An Exploration of CIS Microdata, *Research Policy* 28 (6), 615 (1999) (upholding that process innovations are generally less likely to be patented compared to product innovations); Anthony Arundel and Isabelle Kabla, What Percentage of Innovations are Patented? Empirical Estimates for European Firms, *Research Policy* 27(2), 127 (1988); Wesley M. Cohen, Akira Goto, Akiya Nagata, Richard R. Nelson, J. Walsh, R&D Spillovers, Patents and the Incentives to Innovate in Japan and the United States, *Research Policy* 31, 1349 (2002). See also the sources in note 6 below.

[6] The pioneering work of Guellec and van Pottelsberghe has labeled willingness to patent factors as the potential of R&D collaborations with universities, research institutions, competitors or governments; geographical specificities, namely cluster effects among firms and countries; and technological specificities, namely new-to-the-firm/new-to-the-world innovation discrepancies. See Dominique Guellec and Bruno van Pottelsberghe de la Potterie, Applications, Grants and the Value of Patent, *Economic Letters* 69 (1), 109 (2000); Dominique Guellec and Bruno van Pottelsberghe de la Potterie, The Value of Patents and Filing Strategies: Countries and Technology Areas Patterns, *Economics of Innovation and New Technology* 11 (2), 133 (2002). As said, such archetypal willingness factors will be the focus of this book.

[7] See Kuo-Feng Huang and Tsung-Chi Cheng, Determinants of Firms' Patenting or not Patenting Behaviors, *Journal of Engineering and Technology Management* vol. 36, 52 (2015) referencing business management theoretician Frederick Herzberg's Motivator-Hygiene theory on job productivity by differentiating capability factors (hygiene factors) and willingness factors (motivation factors). See Frederick Herzberg, Motivation-hygiene Theory, in Pugh, D. (ed.), Organization Theory (Penguin, 1966). Huang and Cheng distinguish capability factors to patent from willingness factors to patent (referring to patent propensity rates), which further leads them to question: why would a firm that is capable of patenting be unwilling to patent? (at 55). The intriguing interrelation between capability and willingness factors in patenting is still under-theorized and remains outside the scope of this book.

[8] More generally, see Andrés Rodríguez-Pose, Innovation Prone and Innovation Averse Societies. *Economic Performance in Europe, Growth and Change* vol. 30 75 (1999); Riccardo Crescenzi and Andrés Rodríguez-Pose, Innovation and Regional Growth in the European Union (Springer, 2011).

[9] Andrés Rodríguez-Pose, Innovation Prone and Innovation Averse Societies, Economic Performance in Europe, *Growth and Change* vol. 30 75 (1999), at 80.

[10] Ibid.

prone" and "innovation averse" societies.[11] "Innovation prone" societies are "those capable of transforming a larger share of their own R&D into innovation and economic growth."[12] Conversely, "innovation averse" societies do not manage to transform their own R&D into innovation and economic growth to the same extent.[13] In a somewhat analogous manner, the book explains how, across the entire development divide, developing countries and notably emerging economies differ in their domestic innovation – heavily proxied by their patent intensity – from advanced economies, thereby characterizing what we may term "patent averse" and "patent prone" countries, respectively, based on their relative propensity to patent rates.

Developing countries led by emerging economies clearly differ in their propensity to attract FDI, trade, and technology.[14] Arguably, they also differ in terms of their ability to innovate and patent inventions. Traditional approaches conventionally depart from the familiar North–South dichotomy, or some variant thereof.[15] The differences in the economics of developing countries highlight, in particular, innovation asymmetries between Northern countries, which are deemed to generate innovative patentable products and technologies, and Southern countries, which are generally deemed to consume them.[16] This is reflected in a lower propensity to patent rate. This chapter substantiates the book's core theoretical and empirical argument in support of measuring patent intensity among countries by comparing patent propensity and R&D intensity rates between the groups of countries adjoining the developmental divide. In so doing, it contributes toward a theory that could replace the "one-size-fits-all" innovation-based economic growth equilibrium: a theory that examines multiple tentative equilibria across the archetypical development divide, as the empirics of this book later entail.

This chapter identifies this equal-country norm-setting as a fourfold challenge. Firstly, this norm refers to institutional aspects of fragmentation among UN-level agencies, including the WTO, over innovation and patent-related policies. Secondly, the dissonance in patenting and innovation-related norm-setting is explained by the present regulatory framework, which is designed to sustain transnational bargaining over trade and IP-related minimal standards and flexibilities. This bargaining posture is excessively based on national market size approximation, to the

---

[11] Ibid.   [12] Ibid., at 82.   [13] Ibid.

[14] More generally, see Daniel Benoliel and Bruno Salama, Toward an Intellectual Property Bargaining Theory: The Post-WTO Era. 32 University of Pennsylvania Journal of International Law. 265, (2010), at 312–364.

[15] See Paul Krugman, A Model of Innovation, Technology Transfer, and the World Distribution of Income, 87 *Journal of Political Economy*, 253 (1979), at 254–55.

[16] See Carlos M. Correa, *Intellectual Property Rights, the WTO and Developing Countries: The TRIPS Agreement and Policy Options* (Zed Books, 2000), at 11.

detriment of other, more subtle development-related criteria. Thirdly, this norm-setting challenge further entails the problematic trade-orientation of the Trade-Related Aspects of Intellectual Property (TRIPS) agreement, outweighing innovation-related considerations. Fourthly, it entails the superseding role of short-termed technical assistance and capacity-building policies at UN-level agencies, at the expense of longer-term and more cumulative development strategies.

Against the backdrop of the demise of the Washington Consensus and the gradual rise of endogenous economic growth theory, WIPO's blend of somewhat exogenous economic-related trade rules, fairly equal-country proprietary policies, and a broad development agenda demands empirical and conceptual clarity. Such clarity begs answers (provided in the following chapters) as to both *how* and *why* some countries are patent prone while others are patent averse as a proxy for their comparable domestic innovation.

1.1 Economic Growth, Patent Prone, and Patent Averse Countries

1.1.1 *Patenting and Linear Innovation-Based Economic Growth*

The initial argument concerning national economic growth through innovation emerged from Cambridge University economist Nicholas Kaldor in 1957. Kaldor theorized that differences in development stages across countries could be explained by differing rates in the adoption of technology.[17] The adoption of technology is often measured through patent statistics.[18] The underlying idea was that investment and learning were related, and that the rate at which they took place determined technical progress.[19]

Against the backdrop of serious doubts concerning the impact of R&D activity on economic growth, two core findings emerged, both of which regrettably focused primarily on developed countries. Firstly, a vast body of literature published in the late 1970s, particularly by economists Zvi

---

[17] Nicholas Kaldor, A Model of Economic Growth, 67 *Economic Journal* 591 (1957), at 595.

[18] Stanford University Professors Charles Jones and Paul Romer recently exemplified the usage of patent statistics over Kaldor's growth theory. See Charles I. Jones and Paul M. Romer, The New Kaldor Facts: Ideas, Institutions, Population, and Human Capital 8 (National Bureau of Economic Research, Working Paper No. 15094, 2009) (Offering cross-country patent statistics for measuring international flows of ideas alongside trade and FDI as key facets for economic growth).

[19] Kaldor, supra note 17.

Griliches,[20] Jacques Mairesse,[21] and Bronwyn Hall,[22] established the relationship between R&D and firm-level productivity regarding the basic research.[23] Later writings corroborated these findings for firms located in high-tech industries in such advanced economies.[24] A second core finding followed. From an institutional perspective, the UN's approach has been that when there is a need for investing in R&D, this can most efficiently be made internationally, namely by Multinational Corporations (MNCs) considered best placed to orient the direction of the technological change amalgam.[25] Accordingly, it is

[20] Zvi Griliches, Issues in Assessing the Contribution of Research and Development to Productivity Growth, *Bell Journal of Economics* 10, 92 (1979); Zvi Griliches and J. Mairesse, Productivity and R&D at the Firm Level, in Zvi Griliches (ed.), *R&D, Patents and Productivity* (University of Chicago Press, 1984), 399; Zvi Griliches, Productivity, R&D and Basic Research at the Firm Level in the 1970s, *American Economic Review* 76(1), 141 (1986).

[21] See, e.g., Jacques Mairesse and Mohamed Sassenou, R&D and Productivity: A Survey of Econometric Studies at the Firm Level. *Science-Technology-Industry Review* 8, 317 (1991); Philippe Cuneo and Jacques Mairesse, Productivity and R&D at the Firm Level in French Manufacturing, in Zvi Griliches (ed.), *R&D, Patents and Productivity* (University of Chicago Press, 1984), 399.

[22] Bronwyn H. Hall and Jacques Mairesse, Exploring the Relationship between R&D and Productivity in French Manufacturing Firms, *Journal of Econometrics* 65, 263 (1995).

[23] For theoretical literature that incorporates basic research into R&D-driven growth models in closed economies, see, e.g., Lutz G. Arnold, Basic and Applied Research, *Finanzarchiv* vol. 54, 169 (1997); Guido Cozzi and Silvia Galli, Privatization of knowledge: Did the US get it right?'. MPRA Paper 29710 (2011); Guido Cozzi and Silvia Galli, Science-based R&D in Schumpeterian growth, *Scottish Journal of Political Economy* 56, 474 (2009); Guido Cozzi and Silvia Galli, Upstream innovation protection: Common law evolution and the dynamics of wage inequality, MPRA Paper 31902 (2011); Hans Gersbach, Gerhard Sorger and Christian Amon, Hierarchical growth: Basic and applied research, CER-ETH Working Papers 118, CER-ETH – Center of Economic Research at ETH Zürich (2009); Amnon J. Salter and Ben R. Martin, The Economic Benefits of Publicly Funded Basic Research: A Critical Review, *Research Policy* 30 (3), 509 (2001), at 509. For earlier discussion, see Daniel Benoliel, The International Patent Propensity Divide, *North Carolina Journal of Law and Technology* vol. 15(1) 49 (2013), at 53–60.

[24] As numerous empirical studies have shown, R&D activities are crucial to maintaining the competitiveness of firms. Additionally, within high-tech sectors corporate R&D investment may be more fruitful in terms of achieving productivity. See, e.g., Door Petra Andries, Julie Delanote, Sarah Demeulemeester, Machteld Hoskens, Nima Moshgbar, Kristof Van Criekingen and Laura Verheyden, (2009), O&O-Activiteiten van de Vlaamse bedrijven, in Koenraad Debackere and Reinhilde Veugelers (eds.), Vlaams Indicatorenboek Wetenschap, Technologie en Innovatie 2009 (Vlaamse Overheid, 2009), 53 (showing that approximately 80% of Flanders' total R&D expenditures have been conducted by firms in the high-tech segment).

[25] See, e.g., Frieder Meyer-Krahmer and Guido Reger, New Perspectives on the Innovation Strategies of Multinational Enterprises: Lessons for Technology Policy in Europe, *28 Research Policy* 751 (1999), at 752. But see Argentino Pessoa, R&D and Economic Growth: How Strong is the Link?, *Economics Letters*, vol. 107(2) 152 (May 2010) (examining the relationship between R&D outlays and economic growth in the OECD context, while doubting the effectiveness of an innovation policy that attempts to improve aggregate productivity only based on increasing R&D intensity), at 152. Pessoa explains that among

not surprising that there is a large number of scientific studies on this occurrence, or that several of these studies show an increasing internationalization of innovative activity (mainly R&D) by MNCs.[26] If meaningful patent intensity is to take place in developing countries, it will most probably be sought by the same MNCs that overwhelmingly internationalize R&D.

The growing emphasis on the internationalization of R&D by both growth theoreticians and succeeding policy-makers largely echoed another imperative theoretical breakthrough: Paul Romer's endogenous growth theory of 1990.[27] Romer found that economic growth is primarily the result of endogenous investments in industrial R&D in innovation by forward-looking, profit-seeking agents.[28]

In marked contrast to the neoclassical growth models formulated earlier by Robert Solow,[29] followed by David Cass[30] and Tjalling Koopmans,[31] whereby long-term economic growth depends on an archetypical exogenous process being a by-product of investment in machinery and equipment, Romer's hallmark economic growth insight seemingly

---

12 countries that experienced R&D intensity above the OECD average, only 3 (the United States, Finland, and South Korea) show a GDP growth rate higher than the OECD average. Pessoa further illustrates these intriguing findings for the cases of both Ireland and Sweden. He labels the Irish "Celtic Tiger" as presenting the highest rate of economic growth with low R&D intensity, and the "Swedish Paradox" illustrates an example where the highest R&D intensity coexists with a rate of output growth below the OECD average), at 153. These conclusions are, to date, still considered marginal in economic growth literature.

[26] See generally Organization for Economic Co-operation and Development, Compendium of Patent Statistics report (2008), at 28; Daniele Archibugi and Alberto Coco, The Globalization of Technology and the European Innovation System, in Manfred M. Fischer and Josef Fröhlich (eds.) Knowledge, Complexity and Innovation Systems 58 (Springer, 2001); Pari Patel and Modesto Vega, Patterns of Internationalization of Corporate Technology: Location vs. Home Country Advantages, 28 Research Policy 145 (1999); Alexander Gerybadze and Guido Reger, Globalization of R&D: Recent Changes in the Management of Innovation in Transnational Corporations, 28 Research Policy 251 (1999); Pari Patel, Localized Production of Technology for Global Markets, 19 Cambridge Journal of Economics 141 (1995) (offering evidence that there is no systematic evidence to suggest that widespread globalization of technological activities occurred in the 1980s).

[27] See Paul M. Romer, The Origins of Endogenous Growth, 8 Journal of Economic Perspectives 3, 4–10 (1994); Paul M. Romer, Endogenous Technological Change, 98 Journal of Political Economy S71, S72 (1990) ("Technological change provides the incentive for continued capital accumulation, and together, capital accumulation and technological change account for much of the increase in output per hour worked)," at 72.

[28] Ibid.

[29] Robert M. Solow, A Contribution to the Theory of Economic Growth, 70 Quarterly Journal of Economics 65, 68–73 (1956).

[30] David Cass, Optimum Growth in an Aggregate Model of Capital Accumulation, 32 The Review of Economic Studies 233 (1965), at 233–40.

[31] Tjalling Koopmans, On the Concept of Optimal Economic Growth, in (Study Week on the) Econometric Approach to Development Planning (1965), at 226–28.

prevailed.[32] Though it has been challenged by competing economic models arguing for possible inaccuracies within US patent-based innovative markets, Romer's model has survived and continues to provide the foundation for overall R&D-related growth theory.[33] Henceforth, technological change, particularly through R&D expenditures, is regarded as a sine qua non that lies at the heart of both economic growth theory and policy.[34] Be that as it may, the comparative empirics of patent propensity

[32] Romer's economic growth theory was also said to result from investment in human capital and knowledge. Soon after, Romer's insight became widely popular. See Ben Fine, Endogenous Growth Theory: A Critical Assessment, 24 *Cambridge Journal of Economics* 245 (2000) ("Over the past three years, the number of chapters explicitly drawing upon [Romer's] endogenous growth theory almost certainly borders on a thousand"), at 246.

[33] The contributions by economists Aghion and Howitt and Grossman and Helpman were particularly effective in utilizing the increasing returns to scale of innovations to explain persistent global growth of output per capita over the past two centuries. See Philippe Aghion and Peter Howitt, *A Model of Growth Through Creative Destruction*, 60(2) *Econometrica* 323 (1992), at 327–29; Gene Grossman and Elhanan Helpman, *Innovation and Growth in the Global Economy 1–6* (MIT Press, 1991). For criticism of Romer's endogenous growth model, see Paul Segerstrom, Endogenous Growth Without Scale Effects, 88(5) *American Economic Review* 1290 (1998) (arguing that data does not support the claim that the rate of growth increases with the scale of the economy because patent statistics have been roughly constant even though R&D employment as an endogenous growth indication has risen sharply between the 1970s and 2000s and because a steady increase in R&D efforts has not led to any upward trend in US economic growth rates), at 1292–95; Charles Jones, Time Series Tests of Endogenous Growth Models, 110(2) *Quarterly Journal of Economics* 495 (1995) (developing an alternative model explaining why economic growth has not accelerated despite the substantial increase in R&D efforts), at 501–2.

[34] Romer, Endogenous Technological Change (note 27), at S72; Similarly, from a policy perspective, R&D is seen as the main driver of innovation, and R&D expenditure and intensity are two of the key indicators used to monitor resources devoted to science and technology worldwide. Governments are increasingly referring to international benchmarks when defining their science polices and allocating resources. See Eurostat – Statistics explained, Glossary: R&D intensity: http://ec.europa.eu/eurostat/statistics-explained/index.php/Main_Page.

But see Nathan Rosenberg and others who argue that many process innovations involve "grubby and pedestrian" incremental processes within the firm and are not captured by figures for R&D. See Nathan Rosenberg, *Inside the Black Box: Technology and Economics* (Cambridge University Press, 1982), at 12; Edward Dennison, *Accounting for Growth* (Harvard University Press, 1985) (suggesting that R&D accounts for only 20% of all technical progress); John R. Baldwin and Moreno Da Pont, *Innovation in Canadian Manufacturing Enterprises*, Ottawa: Statistics Canada, Micro Economic Analysis Division (1996) (explaining that certain firms do not engage in any formal R&D). There is also traditional methodological critique on the usage of R&D for innovation-based growth. See, e.g., Mark Crosby, Patents, Innovation and Growth, *The Economic Record*, 76 (234), 255 (2000) (The relationship between R&D and innovation outputs is likely to be time varying, possibly nonlinear, and is also likely to occur with uncertain lags), at 256; Zvi Griliches, Productivity Puzzles and R&D: Another Non-explanation, *Journal of Economic Perspectives*, vol. 2, 9 (1988) (explaining that R&D data are problematic because of problems of definition, and the treatment of time-lags, depreciation, and inflation), at 17–19. With the introduction of the unprecedented and

rates across countries must have been assumed to flow naturally from the central role given to the measurement of R&D in relation with patenting itself.

Growth theory has traditionally gone through another innovation-related growth transformation. Prior to the recent expansion of exogenous growth theory, policy-makers implicitly adhered to a perception of innovation-based economic growth as a linear process of technological development or innovation.[35] The model of innovation classifies research activities and establishes a connection between basic and applied research and eventually commercial activities.[36] It was perceived that "[d]eveloping countries were deemed to be at an earlier stage than the more advanced economies along the linear path of historical [economic] progress."[37] Such archetypical linearity was equally said to foster growth across countries through the translation of R&D supply into "better" innovations based on a single economic equilibrium.

vast UNESCO S&T dataset on GERD-related indicators across countries, this critique remains substantively less relevant.

[35] See Data and Statistics, International Monetary Fund (2012), www.imf.org/external/data.htm. See generally United Nations Conference on Trade and Development, World Investment Report (2005) (analyzing ways to close the archetypical "technology gap" between countries with respect to innovation). In reference to the role of internationalization of R&D in closing the "technology gap," the report further reads: "Large gaps [in this area] prevail between countries – gaps that limit the ability of many of them to take part in the global networks of knowledge creation and diffusion. Addressing these gaps is a major development challenge; it is also essential to ensure that the internationalization of R&D by TNCs benefits larger parts of the world." Ibid. at 100. For a historical account of the linear theory of innovation, see Benoît Godin, The Linear Model of Innovation: The Historical Construction of an Analytical Framework, *31 Science, Technology and Human Values* 639 (2006) (adding that, historically, the linear model is not an actual scientific model of innovation or intellectual progress, but rather a variety of actors such as scientists seeking funding and economists advising government agencies, which have constructed the linear model of innovation to classify research activities and to establish a connection between basic and applied research and eventually commercial activities), at 639–41.

[36] See Godin, ibid., at 639, 657.

[37] Org. for Econ. Co-operation and Dev., Innovation and the Development Agenda (Erika Kraemer-Mbula and Watu Wamae, eds., 2010), at 40 (for a broader contextual discussion). See generally Walt Whitman Rostow, *The Stages Of Economic Growth: A Non-Communist Manifesto* (Cambridge University Press, 3rd edn., 1991), at 4 (offering a limited production function for growth, whereas "It is possible to identify all societies, in their economic dimensions, as lying within one of five categories: the traditional society, the preconditions for take-off, the take-off, the drive to maturity, and the age of high mass-consumption"); Alexander Gerschenkron, *Economic Backwardness in Historical Perspective: A Book of Essays* (Belknap Press of Harvard University Press, 1962) (introducing his theory of "Economics Backwardness" – a country undergoing industrialization will have a different experience depending on its degree of economics backwardness when industrialization begins – as a reaction to uniform stages theories such as Rostow's).

The linear theory indicated that technology is most efficiently acquired through the assimilation of the existing backlog of knowledge by developing countries by means of investment in R&D-enhancing policies. In 1962, based on his study of international aspects of the process of economic growth through innovation and learning, Harvard University economist Alexander Gerschenkron proposed a ground-breaking idea that was subsequently applied in the field. As Gerschenkron explained, "technology gaps" between technologically edged economies, essentially developed economies, and laggard developing countries offer developing countries enormous opportunities for economic growth.[38] It was not until the late 1970s that the technology gap standpoint was revived, leading to the so-called "technology gap" theory in modern innovation theory literature. In this later conceptual stage, the literature widely explored the catching-up process by lagging countries.[39] Yet, as Carlota Perez and Luc Soete remarked,[40] catching up was again perceived linearly, as a "question of relative speed in a race along a fixed track,"[41] whereas technology was perceived as a "cumulative unidirectional process."[42]

The idea of linear economic growth based on an equal-country treating through technology continued to prevail, yielding equal-country policy recommendations. The original Sussex Manifesto,[43] most notably, alongside many research contributions by scholars from developing countries, led to a stream of problematic policy recommendations. These recommendations were naturally directed at closing the widening technology gap by promoting scientific and technological outputs, while re-emphasizing the

---

[38] See Alexander Gerschenkron, *Economic Backwardness in Historical Perspective* (Belknap Press, 1962); Alexander Gerschenkron, Economic Backwardness in Historical Perspective, in *The Progress of Underdeveloped Areas* (Bert F. Hoselitz, ed.) (University of Chicago Press, 1971).

[39] See John Cornwall, *Modern Capitalism: Its Growth and Transformation* (London: Martin Robertson, 1977); Moses Abramovitz, *Rapid Growth Potential and Its Realization: The Experience of Capitalist Economics in the Postwar Period,* in 1 Economic Growth and Resources 191, Edmond Malinvaud, ed. (1979).

[40] See Carlota Perez and L. Luc Soete, Catching-Up in Technology: Entry Barriers and Windows of Opportunity, in *Technical Change and Economic Theory* 458, Giovanni Dosi, ed. (1988).

[41] Ibid., at 460. See also Nancy Birdsall and Changyong Rhee, Does R&D Contribute to Economic Growth in Developing Countries?, Mimeo, The World Bank, Washington, DC (1993) (Using UNESCO data for research and development between 1970 and 1985, the authors show that even for OECD countries, catching up technologically, not advancing R&D affects economic growth).

[42] Perez and Soete, supra note 40, at 460.

[43] See generally Hans Wolfgang Singer, The Sussex Manifesto: Science and Technology for Developing Countries during the Second Development Decade, in I.D.S. Reprints no. 101 (Institute of Development Studies 1974).

weight of scientific R&D.[44] These recommendations further called for the adoption of technical manpower, incentivizing scientific publications, and the promotion of patenting of state-of-the-art technology per se, as proxies of innovation-based growth itself.[45] This tendency was so pronounced that, to this day, much of what is typically labeled in international intellectual property law as innovation policy continues to focus on improving R&D intensity as the chief growth-related indicator. This innovation policy is based to an extent on Romer's endogenous growth theory, while implicitly adhering to a linear growth model for all countries.

The theoretical focus on developing countries within innovation-based economic growth remains limited, though some examples from different periods are nevertheless telling. In the early 1990s, Columbia University economist Frank Lichtenberg worked with a cross-section of 53 countries, many of them developing, and memorably argued in favor of the impact of private return on R&D on innovation and, ultimately, patent-policy ramifications. Lichtenberg showed that the private rate of return on R&D can be up to seven times larger than that for a fixed investment.[46] Coe, Helpman and Hoffmaister,[47] and later Keller[48] and others,[49] analyzed 77 developing countries from Africa, Asia, Latin America, and the Middle East, arguing on balance that because developing countries' own R&D expenditures are so low, they can sensibly be ignored.[50] Research and development, we were told, typically generate a higher rate of private return for developing countries, provided that this is paternalistically endorsed by industrialized or developed countries.

Henceforth, robust developing countries' empirics linked R&D activities to income level, but not to growth rate. R&D activities, World Bank

---

[44] See Gregory Tassey, *The Economics of R&D Policy* (Quorum Books) 54–55, 226 (1997); see generally Patel, supra note 26; Jeffrey L. Furman, Michael E. Porter and Scott Stern, *The Determinants of National Innovative Capacity*, 31 Research Policy 899, 900 (2002).

[45] See Tassey, ibid.

[46] Frank R. Lichtenberg, R&D Investment and International Productivity Differences, NBER Working Papers 4161, National Bureau of Economic Research, Inc. (1992).

[47] See David T. Coe, Elhanan Helpman and Alexander W. Hoffmaister, North-South R&D Spillovers, *The Economic Journal*, vol. 107(440) 134 (1997) (showing that 77 developing countries that do little research and development themselves benefit substantively from R&D that is performed in 22 industrial countries).

[48] See Wolfgang Keller, International Technology Diffusion, NBER Working Paper Series 8573, Cambridge, Massachusetts (2001).

[49] See, e.g., Ahmad Jafari Samimiand and Seyede Monireh Alerasoul, *R&D and Economic Growth: New Evidence from Some Developing Countries, Australian Journal of Basic and Applied Sciences* 3(4):3464 (2009) (using a sample of 30 developing countries for the period 2000–2006, upholding that in general no significance positive impact of R&D on growth exists in these countries, per the present R&D rates thereof).

[50] Daniel Lederman and William F. Maloney, R&D and Development, Policy Research Working Paper, 3024 (2003), at 3.

researchers explain, become prominent only after a country reaches a certain stage of development.[51] This notion surely corresponded with additional evidence showing that FDI responds to intellectual property protection only in host-countries that have reached an equivalent minimum development threshold, including exemplary capacity to imitate inventions.[52]

Years later, this standpoint was adopted by UN-led agencies, as well as the WTO, toward developing countries, albeit in a somewhat approximated form.[53] In order for this to be effective, the UN recognized that such activity should be backed by a comparatively narrower form of independent technological learning by underdeveloped countries themselves.[54]

Archetypical linear innovation-based economic growth policies were ultimately incorporated, albeit implicitly, into a wider macroeconomic setting. The IMF, the World Bank, and the United States Treasury Department eventually created a standard macroeconomic reform package for crisis-wracked developing countries. Economist John Williamson referred to these jointly as "the Washington Consensus."[55] For developing countries, these policies encompassed standard "package deal" policies in areas such as macroeconomic stabilization, economic opening with respect to both trade and R&D investment, and the expansion of market forces within the domestic economy.[56] The ironclad neoliberal belief of the Washington Consensus in unregulated markets, unhindered cross-border

---

[51] Nancy Birdsall and Changyong Rhee, Does R&D Contribute to Economic Growth in Developing Countries?, supra note 41, at 1.

[52] Peter Nunnenkamp and Julius Spatz, Intellectual Property Rights and Foreign Direct Investment: The Role of Industry and Host-Country Characteristics 2 (Kiel Institute for World Economics, Working Paper No. 1167, June 2003), at 12–13.

[53] See generally United Nations Millennium Project, Calestous Juma and Lee Yee-Cheong, Innovation: Applying Knowledge in Development (2005); Commission for Africa, Our Common Interest: Report of the Commission for Africa (2005) (emphasizing the role of innovation and underlying investment needs as a basis for economic transformation). For additional literary critique on innovation policy for developing countries, see generally Andreanne Léger and Sushmita Swaminathan, Innovation Theories: Relevance and Implications for Developing Country Innovation at 4–12 (German Institute for Economic Research, Discussion Papers 743, 2007).

[54] Juma and Yee-Cheong, ibid.; Commission for Africa, supra note 53. For an equivalent standpoint set by the World Intellectual Property Organization (WIPO), see WIPO, *The Economics of Intellectual Property: Suggestions for Further Research in Developing Countries and Countries with Economies in Transition* 22 (2009) (R&D is the most important economic indicator on how effective the innovation process is).

[55] John Williamson, What Washington Means by Policy Reform, in *Latin American Adjustment: How Much Has Happened?* 7, 7 (John Williamson, ed., Institute for International Economics, 1990). See generally Dani Rodrik, Goodbye Washington Consensus, Hello Washington Confusion? A Review of the World Bank's "Economic Growth in the 1990s: Learning from a Decade of Reform, 44(4) *Journal of Economic Literature* 973 (2006).

[56] See generally Williamson, ibid.

trade, and the spread of formal property rights, including patent rights, nevertheless seemed to be stalled.[57] Against this background, WTO and WIPO's policies adhered to international intellectual property protection as a central pillar for both short- and long-term economic growth, thereby effectively ignoring country-group differences. Adopted in 1967, the Convention Establishing the World Intellectual Property Organization stated unequivocally that WIPO's central mission is "to promote the protection of intellectual property throughout the world."[58] In that vein, WIPO has long favored the "upward harmonization of intellectual property laws" based on ever-growing intellectual property protection and enforcement in developing and developed countries alike. WIPO's allegedly IP maximalist approach remained constant even after two milestone events. The first came in 1974, when WIPO became a specialized agency of the UN. In the same year, the UN stood at the center of the New International Economic Order (NIEO). This program adopted a set of proposals raised during the early 1970s by numerous developing countries through UNCTAD. On its failure, it was suggested that this order sought to endorse these countries' terms of trade based primarily on compulsory transfers of technology.[59] At least until the establishment of the WTO and the Agreement on Trade-Related Aspects of Intellectual Property Rights (TRIPS) as part of the Marrakesh Agreement Establishing the WTO in 1994,[60] WIPO failed to fill the remaining development void created with the historical collapse of NIEO. A second event that withstood WIPO's IP maximalist approach was its first report, in 1975, including recognition of its obligation toward developing countries.[61]

---

[57] See David Kennedy, The "Rule of Law," Political Choices, and Development Common Sense, in *The New Law and Economic Development: A Critical Appraisal* 95, 128–150 (David M. Trubek and Alvaro Santos, eds.) (Cambridge University Press, 2006).

[58] Convention Establishing the World Intellectual Property Organization, July 14, 1967, 21 UST. 1749, 828 U.N.T.S. 3. See also the later Agreement between the United Nations and the World Intellectual Property Organization, entered into effect December 17, 1974 (giving WIPO responsibility for "promoting creative intellectual activity and facilitating the transfer of technology"), at Art. 1.

[59] The term was derived from the Declaration for the Establishment of a New International Economic Order, adopted by the United Nations General Assembly in 1974. The Declaration referred to a broad range of trade, financial, commodity, and debt-related issues. See Resolution, UN General Assembly, Declaration for the Establishment of a New International Economic Order, UN Doc. A/RES/S-6/3201 (1974). More generally, see *The New International Economic Order: The North–South Debate* (Jagdish N. Bhagwati, ed.) (MIT Press, 1977).

[60] Agreement on Trade-Related Aspects of Intellectual Property Rights in the Marrakesh Agreement Establishing the World Trade Organization, Annex 1C, 1869 U.N.T.S. 299 (Apr. 15, 1994), available at www.wto.org/english/docs_e/legal_e/27-trips.pdf.

[61] Christopher May, The Pre-History and Establishment of the WIPO (2009), at 25. Journal 20, www.research.lancs.ac.uk/portal/en/publications/the-prehistory-and-est ablishment-of-the-wipo(4db79f65-30d9-42a3-b7ed-da285b32f77a).html.

In time, TRIPS applied a broad intellectual property policy flatly to all WTO members, subject to transition periods extended solely to Least-developed Countries (LDCs).[62] The mandatory adoption of TRIPS standards creates two imperative costs for developing countries: reduced access to new technologies and knowledge, and higher royalty payments.[63] Against this backdrop, defenders of TRIPS have tried to cast intellectual property protection as a central pillar of modern economic policy and a catalyst for development. Their argument is twofold.[64] Firstly, intellectual property protection is held to encourage domestic innovation in developing countries explicitly, similar to the development of protections that took place in the early history of the United States.[65] Secondly, intellectual property protection is thought to induce more inward technology transfer, particularly by means of enhanced FDI and trade carried out by MNCs.[66] The availability of intellectual property protection would thus be akin to a

[62] See, e.g., Michael Blakeney, The International Protection Of Industrial Property: From The Paris Convention To The TRIPS Agreement, WIPO National Seminar on Intellectual Property, 2003, WIPO/IP/CAI/1/03/2 13–22 (2003) (detailing the large number of generally applicable provisions and principles within the TRIPS Agreement).
　　As for LDCs, TRIPS endows its Council with the power to "accord extensions" upon a "duly motivated" request by an LDC member. In doing so, the agreement recognizes their "special needs and requirements," but only in a broad sense. LDCs were for all time granted a number of exemptions in adhering to TRIPS. See TRIPS Agreement, supra note 60, art. 66(1).

[63] See Christopher S. Gibson, Globalization and the Technology Standards Game: Balancing Concerns of Protectionism and Intellectual Property in International Standards, 22 Berkeley Technology Law Journal 1403 (2007), at 1404–06.

[64] See Shahid Alikhan, Socioeconomic Benefits of Intellectual Property Protection in Developing Countries 1–9 (2000) (arguing that intellectual property protection is an integral part of the technological and economic development at a national and international level); Kamil Idris, Intellectual Property: A Power Tool for Economic Growth 1 (2d ed. 2003) (introducing the argument that intellectual property is a powerful force to benefit individuals and nations); Ali Imam, How Patent Protection Helps Developing Countries, 33 American Intellectual Property Law Association Quarterly Journal (2005) (detailing the social and economic benefits of stronger patent protection in developing countries), at 379–80.

[65] See Robert M. Sherwood, Human Creativity for Economic Development: Patents Propel Technology, 33 Akron Law Review 351 (2000) (discussing the way in which communities chose to protect technology so that the entire community could profit throughout different stages of intellectual property development). But see Frederic M. Scherer, The Political Economy of Patent Policy Reform in the United States, 7 Journal on Telecommunications and High Technology Law. 167, 205 (2009) (arguing for the historical parallel to the United States' strong patent protection to domestic residents).

[66] See Keith E. Maskus, Intellectual Property Rights in the Global Economy 11 (2000) (arguing that stronger patent protection benefits international trade and FDI); Keith E. Maskus and Mohan Penubarti, How Trade-Related Are Intellectual Property Rights?, 39 Journal of International Economics. 227 (1995) (determining that intellectual property rights directly influence the flow of trade), at 229–30, 237–43. See generally, Daniel J. Gervais, Information Technology and International Trade: Intellectual Property, Trade & Development: The State Of Play, 74 Fordham Law Review. 505 (2005) (analyzing the effects of intellectual property protection on bilateral trade and "inward" FDI), at 517–21; Edmund

"passive" industrial policy. Henceforth, innovation would be stimulated without requiring large investments of public funds, often lacking in the developing world.[67]

WTO's TRIPS assertion supported the neoclassic exogenous economic incentives that had long been offered by developed nations.[68] Yet as seminal as the Washington Consensus overarching framework became, its core neoclassical economic and protectionist approach toward innovation policy remained largely uncorroborated. From a broader perspective, Joseph Stiglitz and others criticized the Consensus policies' protectionist approach, and blamed Washington for stifling innovation or otherwise not foreseeing the policies' potential growth impact on developing countries.[69]

To this day, there are no adequate empirical findings strictly correlating the ultimate demise of Washington's growth policies to innovation-led

W. Kitch, *The Patent Policy of Developing Countries*, 13 University of California Los Angeles Pacific Basin Law Journal 166 (1994) (arguing that developing countries opt into the international property system because it is in their self-interest to do so), at 174–76; Keith E. Maskus, The Role of Intellectual Property Rights in Encouraging Foreign Direct Investment and Technology Transfer, 9 *Duke Journal of Comparative & International Law* 109 (1998) (describing the various factors involved in improving a nation's FDI); Carlos A. Primo Braga and Carsten Fink, The Relationship Between Intellectual Property Rights and Foreign Direct Investment, 9 *Duke Journal of Comparative & International Law* 163 (1998) (describing the influence of strong IP protection on the levels of FDI).

[67] See, e.g., Kenneth W. Dam, The Economic Underpinnings of Patent Law, 23 *Journal of Legal Studies* 247, 271 (1994) (discussing the economic repercussions of intellectual property protection and the economic policies which influence patent law); Robert D. Cooter and Hans-Bernd Schaefer, *Solomon's Knot: How Law Can End the Poverty of Nations* (Princeton University Press, 2009) (discussing the role of innovation in the economic growth of developing nations).

[68] See, e.g., Carolyn Deere, *Developing Countries in the Global IP System, in The Implementation Game: The TRIPS Agreement and the Global Politics of Intellectual Property Reform in Developing Countries* (Oxford University Press, 2009), at 34, 51; Peter Yu, Toward a Nonzero-Sum Approach to Resolving Global Intellectual Property Disputes: What We Can Learn from Mediators, Business Strategists, and International Relations Theorists, 70 *The University of Cincinnati Law Review* 569, 635 (2001). Cf. Christine Thelen, Carrots and Sticks: Evaluating the Tools for Securing Successful TRIPs Implementation, XXIV *Temple Journal of Science, Technology & Environmental Law* 519 (2006) (discussing four incentive mechanisms tailored for developing countries within TRIPS, namely creating short- and long-term economic growth, technical assistance, and additional time to become compliant), at 528–33.

[69] See, e.g., Joseph Stiglitz, Chief Economist, World Bank, More Instruments and Broader Goals: Moving Toward the Post-Washington Consensus, address at the 1998 WIDER Annual Lecture 17 (Jan. 7, 1998), at: http://time.dufe.edu.cn/wencong/washingtonconsen sus/instrumentsbroadergoals.pdf ("The usual argument – that protectionism itself stifled innovation – was somewhat confused. Governments could have created competition among domestic firms, which would have provided incentives to import new technology."); see also Wing Thye Woo, Some Fundamental Inadequacies of the Washington Consensus: Misunderstanding the Poor by the Brightest, at 1; available at http://papers.ssrn.com/sol3/ papers.cfm?abstract_id=622322 ("The Washington Consensus is too hooked upon trade-led growth to acknowledge that science-led growth is becoming even more important.").

economic growth or theoretical variances thereof. Nonetheless, it is dis-appointing that over the past 57 years, only seven economies (Hong Kong, China, Japan, Korea, Malta, Singapore, and Taiwan) have made a substantive growth transition notwithstanding their innovation policy stance.[70] Another swansong for this overly broad "catch up" growth narration, again largely non-indicative of linear innovation policies per se, emerges from an examination of developing countries that have had at least 27 years of consecutive growth above 7 percent since 1950. The World Bank's *Growth Commission Report of 2008* identified only 13 econo-mies that had achieved such high rates of growth.[71]

While subjecting the World Bank's ubiquitous invocation of rule of law-related policies to the trenchant critique of the Washington Consensus, scholars such as Portuguese economist Alvaro Santos alle-gorically hammered the final nail into the World Bank's much criticized policy. Along with others, he labeled this policy a post-Washington Consensus development model, thus reflecting the "fall of neoliberal thinking."[72]

Over the last two decades, in the wake of the spectacular breakdown of numerous related policies, the World Bank and other international eco-nomic institutions have finally started to move away from the Consensus' sweeping free market prescriptions. They have instead embraced a new development framework consistent with Amartya Sen's influential "development as freedom" parable,[73] linking social justice and even

---

[70] See World Bank, Innovation Policy: A Guide for Developing Countries (2010), at 43, available at https://openknowledge.worldbank.org/bitstream/handle/10986/2460/54893 0PUB0EPI11C10Dislosed061312010.pdf; see also World Bank Comm'n on Growth & Dev., The Growth Report Strategies for Sustained Growth and Inclusive Development (2008), at 111, available at www.ycsg.yale.edu/center/forms/growthReport.pdf (adding that the 10 largest developing countries "account for about 70 percent of developing countries' GDP," that industrialized countries' secular growth rate is approximately 2% per capita, and that "[s]ince 1960, only 6 countries have grown faster than 3 percent in per capita terms … ").

[71] The list of 13 includes the 6 countries mentioned above that became developed countries: Hong Kong (China), Japan, Korea, Malta, Singapore, and Taiwan (China); plus 7 others that are still developing: Botswana, Brazil, China, Indonesia, Malaysia, Oman, and Thailand. Except for Botswana and China, these nations have not sustained their rate of growth, thereby preventing these other developing countries from making the transi-tion into developed ones. See World Bank, supra note 70.

[72] Alvaro Santos, The World Bank's Uses of the "Rule of Law" Promise in Economic Development, in *The New Law and Economic Development: A Critical Appraisal* 253 (David M. Trubek and Alvaro Santos, eds.) (Cambridge University Press, 2006), at 267. Neoliberalism surely incorporates free markets, the liberalization of trade and finance, and a limited role for the state in the economic and social organization of society. See David Harvey, *A Brief History of Neoliberalism* (AbeBooks, 2005), 2–4; see also Ha-Joon Chang, *Globalization, Economic Development and the Role of the State* (Zed Books, 2003), at 47–50.

[73] Amartya Sen, *Development as Freedom* (Alfred A. Knopf, 1999).

human rights with economic growth.[74] The development-as-freedom idea has also pervaded the United Nations Development Programme, applying a multifaceted assessment of development that includes not only gross national product, but also – and remarkably – education, literacy, life expectancy, gender equality, and even political participation.[75]

These developments were reflected in the Millennium Development Goals (MDGs), the international development goals adopted following the Millennium Summit of the UN in 2000. The MDGs were rooted in the United Nations Millennium Declaration and aimed at the world's poorest populations. Progress toward the goals was uneven, however:[76] While some countries achieved many of the goals, others were not on track to realize any. Although the UN's development goals[77] do not refer to intellectual property per se, they stand at odds with one-dimensional support for broad IPRs.[78] In this context, a UN-commissioned independent report on achieving the MDGs explicitly criticizes TRIPS for taking "too little account of levels of development and varying interests and priorities."[79] All UN agencies have by now joined the WTO in criticizing the uniformity of the US-led international intellectual property regime, riding the mounting gale of development dialectics. The WHO[80] and the United Nations Educational, Scientific and Cultural Organization

[74] See Kennedy, supra note 57, at 151–58. See also Joseph E. Stiglitz and Andrew Charlton, Fair Trade for All: How Trade Can Promote Development. Oxford University Press (2005) (presenting a mainstream criticism of the Washington Consensus and presenting a detailed proposal for a liberalized trade regime that is geared toward the special interests of developing countries).

[75] See United Nations Development Programme, Human Development Indices, http://hdr .undp.org/en/content/human-development-index-hdi.

[76] For critique of the MDG see, more generally, Naila Kabeer, Can the MDGs provide a Pathway to Social Justice?: The Challenge of Intersecting Inequalities, Institute of Development Studies (2010); Amir Attaran, An Immeasurable Crisis? A Criticism of the Millennium Development Goals and Why They Cannot Be Measured," *PLOS Medicine* 2 (10):318 (October 2005); Andy Haines and Andrew Cassels, Can the Millennium Development Goals Be Attained?, *British Medical Journal*, vol. 329, No. 7462 (14 August 2004) 394.

[77] UN General Assembly Resolution 55/2 – The United Nations Millennium Declaration, Sept. 18, 2000.

[78] Cf: United Nations, The Millennium Development Goals Report 2007 (2007), at 4–5 (expressing concern that the benefits of economic growth in developing countries have been unequally shared).

[79] See Millennium Project, Investing in Development: A Practical Plan to Achieve the Millennium Development Goals (2005) (criticizing WIPO's role in advising developing countries regarding TRIPS on access to essential medicines. The report concluded that TRIPS is ill suited to developing-country needs), at 219.

[80] See World Health Organization, Globalization, TRIPs and Access to Pharmaceuticals, WHO Policy Perspectives on Medicines, No. 3, WHO/EDM/2001.2 (Mar. 2001) (encouraging developing countries to take advantage of TRIPS' safeguards mechanisms concerning access to medicines based on affordable patented pharmaceuticals), at 5.

(UNESCO)[81] expressed concern that IPRs run counter to human rights, particularly with regard to access to essential medicines. Similarly, the United Nations Industrial Development Organization (UNIDO) has argued for a deal favoring the transfer of technology to least developed countries,[82] while the United Nations Conference on Trade and Development (UNCTAD) critiqued the wearing down of least developed countries' flexibilities in intellectual property norm-setting, particularly under TRIPS-plus agreements.[83]

As freedom recitation developed, it eventually stepped on WIPO's turf. After years of deliberation through the Provisional Committee on Proposals Related to a WIPO Development Agenda and the Inter-sessional Intergovernmental Meeting on a Development Agenda for WIPO, an archetypal WIPO Development Agenda was finally adopted in October 2007.[84] The Development Agenda emerged as a reaction to what Peter Drahos and John Braithwaite have aptly described in their seminal book *Information Feudalism* as an "agenda of underdevelopment" based on the upward harmonization of global IP governance.[85] The Development Agenda decisively rejects this neoclassic IP-centric view,[86] and indeed reflected developing countries' growing resistance to the upward harmonization of IP protection required by the TRIPS and subsequent "TRIPS-plus" bilateral free trade agreements.[87] As Neil Netanel concludes, WIPO's Development Agenda should be understood as part of a broad, multipronged rejection of the Washington Consensus that

---

[81] See UNESCO, Report on the Experts' Meeting on the Right to Enjoy the Benefits of Scientific Progress and its Applications (UNESCO Pub. SHS-2007/WS/13, June 7–8, 2007) (showing concern over the tension between intellectual property and the right to benefit from scientific progress), at 3–4.

[82] See United Nations Industrial Development Organization, Strategic Long-Term Vision Statement, GC11/8/Add.1, Oct. 14, 2005, ¶ 5(A)(h).

[83] See United Nations Conference on Trade and Development, Least Developed Countries Report 2007 128–29 (UNCTAD/LDC/2007 2007).

[84] For the historical evolvement of WIPO's Development Agenda, see Press Release, World Intellectual Prop. Org., Member States Agree to Further Examine Proposal on Development, WIPO/PR/2004/396 (Oct. 4, 2004), available at www.wipo.int/press room/en/prdocs/2004/wipo_pr_2004_396.html (discussing a proposal by Brazil and Argentina encouraging the inclusion of a "Development Agenda" in WIPO); Press Release, World Intellectual Prop. Org., Member States Adopt a Development Agenda for WIPO, WIPO/PR/2007/521 (Oct. 1, 2007), available at www.wipo.int/pressroom/en/articles/2007/article_0071.html (discussing the welcoming of the Development Agenda by WIPO's former Director General Dr. Kamil Idris).

[85] Drahos, Peter and John Braithwaite, *Information Feudalism: Who Owns the Knowledge Economy* (Earthscan Publications Ltd., 2003), at 12.

[86] See Neil Netanel, *Introduction: The WIPO Development Agenda and Its Development Policy Context 1*, in *The Development Agenda: Global Intellectual Property and Developing Countries* (Neil Netanel, ed.) (Oxford University Press, 2009), at 2.

[87] Ibid.

shunted aside the NIEO while dominating development policy through the 1980s and early 1990s.[88]

Be that as it may, WIPO still interprets its legislative mandate while admitting, however, that "too little is known about how innovation takes place in lesser developed economies, how it diffuses and what its impacts are."[89] This attitude is repeatedly exemplified in the organization's rather flat promotion of many of its policies. Such were the agreements on the electronic transmission of works protected by copyrights or related rights,[90] or the negotiations concerning the harmonization of patent rights.[91] As Professors Keith Maskus and Jerome Reichman sternly conclude, whether WIPO's strategy actually benefits innovation possibly counts for little consistency across the development divide.[92]

Against the backdrop of the demise of the Washington Consensus following its rejection by multiple UN organs, WIPO's blend of exogenous economical trade rules, fairly equal-country proprietary concepts, and ambitious development goals demands empirical and conceptual clarity. Such clarity will at last explain both how and why some countries are patent prone and others patent averse as a proxy for comparable domestic innovation. Against the backdrop of the under-empiricized regulatory challenge lurks the broader question as to what explains this UN-level patenting norm-setting.

1.1.2   The UN-level Patenting Norm-Setting Challenge

At first glance, WIPO's development record is well attuned with its underlying formal aspirations. Remarkably, WIPO has to date developed and executed at least 29 projects to operationalize the 45 recommendations

---

[88] Ibid.

[89] WIPO Economics & Statistics Series, *World Intellectual Property Report – The Changing Face of Innovation* 26 (2011). For a critique of the Secretariat's interpretation of WIPO's mandate, see Keith E. Maskus and Jerome H. Reichman, The Globalization of Private Knowledge Goods and the Privatization of Global Public Goods, 7 *Journal of International Economic Law* 279 (2004), at 294 & Fn. 54.

[90] See generally WIPO Copyright Treaty, Dec. 20, 1996, 2186 U.N.T.S. 121 WIPO Doc. CRNR/DC/94 (December 23, 1996), WIPO Doc. CRNR/DC/95 (December 23, 1996), available at www.wipo.int/treaties/en/ip/wct/summary_wct.html (noting that not even the TRIPS Agreement equivalent extensions for less developed countries are to be found in the Copyright Treaty of 1996).

[91] See generally WIPO Standing Committee on the Law of Patents, Tenth Session, Draft Substantive Patent Law Treaty, SCP/10/2, available at www.wipo.int/edocs/mdocs/scp/en/scp_10/scp_10_2.pdf. General background about the WIPO, Standing Committee on the Law of Patents can be found at www.wipo.int/patent-law/en/scp.htm.

[92] Cf: Keith E. Maskus and Jerome H. Reichman, The Globalization of Private Knowledge Goods and the Privatization of Global Public Goods, 7 *Journal of International Economic Law* 279 (2004), at 294.

of the Development Agenda.[93] The organization has also executed two external treaties containing references to the Development Agenda. These are the Beijing Treaty on Audiovisual Performances (BTAP)[94] and the Marrakesh Treaty to Facilitate Access to Published Works for Persons Who Are Blind, Visually Impaired, or Otherwise Print Disabled (MVT).[95]

Despite this, the developing countries are less than optimally united over WIPO's Development Agenda, and there is not enough common ground for cooperation.[96] An equivalent developing countries' bargaining situation prevailed across other UN-level agencies and the WTO as coalitions become the de facto preferred response of developing countries to imbalances in power in these international organizations.[97]

For developing countries with small markets and limited diplomatic resources, coalitions have repeatedly proved to be the only means at their disposal for advancing their bargaining positions.[98] These coalition-building efforts certainly play a role against the backdrop of much United States-led opposition. Since the failure of the fifth WTO Ministerial Conference in Cancún (Cancún Ministerial) in 2003, most noticeably, the United States has essentially adopted a divide-and-rule strategy intended to marginalize coalition building by developing

---

[93] See Director General Report on Implementation of the Development Agenda, Committee on Development and Intellectual Property (CDIP) Thirteenth Session, CDIP/13/2, Geneva, May 19 to 23, 2014 (March 3, 2014), at 3, 19 and Annex II (For the full overview of the status of all Development Agenda projects under implementation).

[94] Beijing Treaty on Audiovisual Performances (2012). Full text available at: www.wipo.int/treaties/en/text.jsp?file_id=295837.

[95] Marrakesh Treaty to Facilitate Access to Published Works for Persons Who Are Blind, Visually Impaired or Otherwise Print Disabled (2013). Full text available at: www.wipo.int/treaties/en/text.jsp?file_id=301016.

[96] Interview with Mr. Irfan Baloch, Acting Director, DACD, WIPO, in Geneva, Switzerland on October 16, 2014 (file with author).

[97] On the background for developing countries-led coalitions at the WTO and GATT, see the seminal work of Amrita Narlikar, *International Trade and Developing Countries: Bargaining Coalitions in the GATT and WTO* (Routledge, 2003) (offering an historical typology of developing country coalitions in the GATT and WTO). See also Vicente Paolo B. Yu III, Unity in Diversity: Governance Adaptation in Multilateral Trade Institutions Through South-South Coalition-building, Research papers 17 (South Centre, July 2008), at 28, 33–34. For literature concerning intellectual property-related coalitions, see also Peter K. Yu, Building Intellectual Property Coalitions for Development, in *Implementing the World Intellectual Property Organization's Development Agenda* 79, Wilfrid Laurier University Press, CIGI, IDRC (Jeremy de Beer, ed., 2009), at 84; John S. Odell and Susan K. Sell, *Reframing the Issue: The WTO Coalition on Intellectual Property and Public Health (2001) 85 In Negotiating Trade* (John S. Odell, ed.) (Cambridge University Press, 2006) (on the coalition of developing countries for the 2001 Doha Declaration on the WTO Agreement on Trade-Related Aspects of Intellectual Property Rights (TRIPS) and Public Health), at 104.

[98] Amrita Narlikar, ibid., at 3.

countries. Accordingly, the United States has rewarded countries that are willing to work with it, while undermining efforts by Brazil, India, and other G20 members to establish a united negotiating front for less developed countries.[99]

The 2008 subprime mortgage global economic crisis made it even more difficult for countries to articulate joint intellectual property policies, as "everybody are [sic] guarding themselves by making no progress" as suggested by Irfan Baloch, the Acting Director of the Development Agenda Coordination Division (DACD) at WIPO.[100] The appetite for new norm-setting across the UN remains limited.[101]

The 54th Session of the WIPO Assemblies of September 2014 constitutes a disturbing case in point. This recent event witnessed broad disagreement between developing and developed countries.[102] The scope of disagreement therein concerned the advancement of the three treaties discussed in the 54th Session, namely the Traditional Knowledge treaty, the Copyright Broadcasting treaty, and the Design Law treaty (DLT). At the heart of this dissent in previous years stood WIPO's 54th Session, the organizations' Development Agenda and its implementation.[103]

---

[99] See Yu, Building Intellectual Property Coalitions for Development, note 97, at 84; Peter K. Yu, The Middle Intellectual Property Powers, Drake University Legal Studies, Research Paper Series, Research Paper No. 12–28, at 18, referring to Former World Bank President and US Trade Representative Robert B. Zoellick's wordings on the fifth WTO Ministerial Conference in Cancún (Cancún Ministerial) in 2003 at Robert B. Zoellick, America will not wait (September 21, 2003) ("As WTO members ponder the future, the US will not wait: we will move toward free trade with can-do countries") at www.fordschool.umich.edu/rsie/acit/TopicsDocuments/Zoellick030921.pdf. In tandem, domestic US private entities threaten or use of unilateral sanctions, as part of their coalition with Europe, and Japan at the industry level. See, e.g., Ruth L. Okediji, Public Welfare and the Role of the WTO: Reconsidering the TRIPS Agreement, 17 *Emory International Law Review* 819 (2003) (referring to the case of the pharmaceutical industry), at 844–46; Susan Sell, Private Power, Public Law: The Globalization of Intellectual Property Rights (2003).

[100] Interview with Mr. Irfan Baloch, Acting Director, DACD, WIPO, in Geneva, Switzerland, on October 16, 2014 (file with author).

[101] Ibid.

[102] See: The 54th Session of the WIPO Assemblies of 22–30 September 2014, available at: www.wipo.int/meetings/en/details.jsp?meeting_id=32482.

[103] Catherine Saez, Crisis At WIPO Over Development Agenda; Overall Objectives In Question, Intellectual Property Watch (24.5.2014), at 1. For previous disagreement across the North–South divide in the CDIP meeting from May 7–11, 2012, see in particular William New, WIPO Development Agenda Implementation: The Ongoing Fight For Development In IP, Intellectual Property Watch (9.5.2012).

The core resentment by developing countries related to WIPO's reluctance to permanently incorporate development in the organizations' specialized Standing Committees. See: Interview with Mr. Geoffrey Onyeama, Deputy Director General, Cooperation for Development, WIPO, in Geneva, Switzerland on October 15, 2014 (file with author).

This UN-level challenge over patent-policy norm-setting has a fourfold explanation. Firstly, it reflects the division among UN-level agencies over innovation and patent-related policies. Secondly, it highlights the impetus of the IP bargaining situation based on national income-based approximation instead of a more subtle bargaining situation across countries. Thirdly, it exposes the problematic policy trade-orientation of the TRIPS Agreement. Fourthly, it entails the superseding role of technical assistance and capacity-building at the UN-level action relating to trade, intellectual property, and development.

1.1.2.1 UN-level Innovation Policy Fragmentation

The disintegration of a UN patenting norm-setting has many manifestations. It is witnessed, for example, in the fact that the UN MDGs made no mention of intellectual property. Innovation is mentioned only in the context of access to essential drugs and antiretrovirals.[104] If the MDGs were intended to provide an overarching United Nation development policy, they should surely have addressed these two aspects. One hypothesis that may shed light on this is that the MDGs were largely intended as an appeal to donor countries to help Sub-Saharan Africa, based on a neoclassical economic apparatus.[105] Former UN Secretary-General Kofi Annan was, of course, an African himself, and was strongly politically invested in the continent. Moreover, the focus was on governments rather than the business sector, primarily addressing the issues of food security, drinking water, health, and housing.[106]

Naturally, it is expected that WIPO will be coordinated with the MDGs. In principle, WIPO's Development Agenda does not weaken WIPO's official institutional autonomy as a specialized UN agency. It ostensibly applies the development policies of additional agencies to bear more forcefully on WIPO norm-setting. In particular, the Agenda provides that "WIPO's norm-setting activities should be supportive of the development goals agreed within the UN system, including those contained in the Millennium Declaration."[107] It also calls upon WIPO "to

---

[104] See Millennium Development Goals (MDG), MDG 8 – Access to Essential Medicines (2000), at: http://iif.un.org/content/mdg-8-access-essential-medicines. For a critique of WIPO's access to essential medicines policy in view of the Millennium Development Goals, see Millennium Project, Investing in Development: A Practical Plan to Achieve the Millennium Development Goals (2005), at 219.

[105] Interview with Mr. Yoshiyuki (Yo) Takagi, Assistant Director General, Global Infrastructure, WIPO, in Geneva, Switzerland on October 16, 2015 (file with author).

[106] Ibid.

[107] Proposed language that would have required WIPO to ensure that its norm-setting activities are "fully compatible" with other international instruments that advance development objectives, including international human rights instruments, was not adopted as part of the Development Agenda. See WIPO, Working Document for the

intensify its cooperation on IP related issues with UN agencies, according to Member States' orientation, in particular UNCTAD, UNEP, WHO, UNIDO, UNESCO, and other relevant international organizations, especially WTO."[108]

In practice, however, fragmentation prevailed at the UN level over innovation policy.[109] WIPO was not alone in adapting with the MDGs. Notwithstanding innovation policy, countless uncoordinated, blurred, and often ineffective statements, declarations, and resolutions were upheld by UN-level agencies, as each one attempted to promote its own version of the archetypal development agenda.[110] The WTO, the WIPO, and even the WHO are all hostages to fragmentation, and all have a claim on intellectual property and innovation policy. The saga of fragmentation at the UN system in promoting the economic and social goals set out in the UN Charter continued.

WHO offers an analogous example of the diffusion of innovation policy among the UN agencies, relating to WHO's reliance on R&D as a motor of innovation for pharmaceuticals. In a similar manner to the bargaining situation assumed by the UN in creating TRIPS, WHO also failed to adhere to a consistent pan-UN innovation policy. As of May 24, 2008, for example, the World Health Assembly (WHA) the decision-making body of WHO, released a document entitled Global Strategy and Plan of Action on Public Health, Innovation and Intellectual Property.[111] In the report, WHO member states suggest implementing strategies to

Provisional Committee on Proposals Related to a WIPO Development Agenda (PCDA), WIPO Doc. PCDA/3/2, Annex B, ¶ 28, pp. 14–15, Feb. 20, 2007.

[108] WIPO, The Development Agenda, Cluster E: Institutional Matters including Mandate and Governance, sec. 40.

[109] WIPO Secretariat cooperated with the MDG in line with Recommendation 22 to the Development Agenda, as requested by Member States participating in the MDG Gap Task Force. On February 7, 2013, a Task Force meeting took place considering the draft outline of the Task Force's report for 2013. WIPO, together with the WTO and WHO, contributed to this report, published in September 2013, with a section related to access to essential and affordable medicines and intellectual property rights. See Director General Report on Implementation of the Development Agenda, Committee on Development and Intellectual Property (CDIP) Thirteenth Session, CDIP/13/2, Geneva, May 19 to 23, 2014 (March 3, 2014), at 6, Sec. 17.

[110] Mr. Irfan Baloch, Acting Director, DACD, WIPO, in Geneva, Switzerland on October 16, 2014 (file with author).

[111] Sixty-First World Health Assembly (hereinafter, WHA), WHO Global Strategy and Plan of Action on Public Health, Innovation and Intellectual Property, at 1, WHA61.21, (May 24, 2008), available at http://apps.who.int/medicinedocs/documents/s21429en/s 21429en.pdf; Exec. Bd. 124th Session, WHO, Public health, innovation and intellectual property: global strategy and plan of action: Proposed time frames and estimated funding needs, at 1, EB124/16 Add.2 (Jan. 21, 2009), available at www.who.int/gb/eb wha/pdf_files/EB124/B124_16Add2-en.pdf.

promote R&D for diseases endemic in developing countries.[112] An Intergovernmental Working Group (IGWG) consisting of representatives from more than 20 countries developed these strategies.[113] One stated aim of the action plan was to "explor[e] a range of incentive mechanisms... [and] address the de-linkage of the costs of research and development and the price of health products and methods."[114] Some of the proposed strategies include open-source research, patent pools, and prizes.[115] Unsurprisingly, soon after, on January 21, 2009, WHO released a policy document entitled Proposed Time Frames and Estimated Funding Needs, with the aim of implementing the WHO IGWG plan of action.[116] The estimated total cost of implementing the plan of action was said to be $2.064 billion, with a proposed time frame of 2009 through 2015.[117]

In reflection of the disintegration between the different access-to-medicines agendas across the UN-level agencies and the WTO, WHO assumed the rather unattainable internationalized MNC-led R&D activity. That is, the plan adopted a separate innovation incentive policy, with no reference to the rather discouraging conditions in distinct developing countries, nor to the need for inter-UN agency coordination in the implementation of this policy.[118]

Since then, WHO, WIPO, and the WTO have strengthened their cooperation on issues around public health, intellectual property, and trade. Henceforth, the WHO Global Strategy and Plan of Action on Public Health, Innovation and Intellectual Property (GSPOA), the WIPO's Development Agenda, and the WTO Declaration on the TRIPS Agreement and public health have provided the broader context for trilateral cooperation. In 2013, for the first time, the three intergovernmental organizations embarked on the multifaceted path of coordinating health, intellectual property, and trade.[119] Nevertheless, patenting

---

[112] WHO, ibid., at 1, 6.
[113] See WHO, Public health, innovation and intellectual property and trade – Expert Working Group on R&D Financing, www.who.int/phi/R_Dfinancing/en.
[114] 2008 WHA Report, supra note 111, at 5.   [115] Ibid., at 10, 14, 16–17.
[116] Exec. Bd. 124th Session, supra note 111, at 1.   [117] Ibid., at 1–2.
[118] See generally 2008 WHA Report, supra note 111, at 1; WHO Exec. Bd., supra note 111, at 1–2.
[119] See WHO-WIPO-WTO, Promoting Access to Medical Technologies and Innovation: Intersections between Public Health, Intellectual Property and Trade (February 5, 2013). The three organizations have further established a joint Technical Workshop on patentability: see WHO, WIPO, WTO Joint Technical Workshop on Patentability Criteria (October 27, 2015), and henceforth consistently hold joint symposiums. See, e.g., Public Health, Intellectual Property, and TRIPS at 20: Innovation and Access to Medicines; Learning from the Past, Illuminating the Future – Joint Symposium by WHO, WIPO, WTO (October 28, 2015).

and innovation-related norm-setting remains intricate given the inadequate empirical precision concerning patent prone and patent averse economies as a proxy for their respective domestic innovation.

1.1.2.2 National Income-Based IP Bargaining Situation

The second explanation for the patenting and innovation-related norm-setting puzzle we will now examine relates to the proclivity of the World Bank, the WTO, and other UN agencies to adhere to power-based bargaining situations of nations based on comparative market power.[120] This approach effectively denies attention to a more subtle set of national characteristics, including relative propensity to patent and conduct research and domestic development. As the World Bank put it, economic growth is defined as the increase of national-level Gross Domestic Product (GDP).[121] Economists essentially measure the level of economic development national power by means of a ballpark estimate of market size based on the overall size and diversity of each country's economy.[122] Critics such as Clibert Rist (also the leader of post-development theory), in his iconic monograph *The History of Development*, argue that this approach embodies a limited understanding of development as an alternative to a multifaceted phenomenon.[123]

Instead of opting for country-specific, possibly endogenous, and nonlinear innovation-based economic growth, UN-level agencies thus adopted the alternative choice as noted. Accordingly, market power first

---

[120] The World Bank similarly puts at the center of its framework of analysis two elementary market power determinants of technology diffusion in less developed countries. The first involves the three main channels by which developing countries are exposed to external technology markets, namely trade, FDI (and licensing) and also highly skilled human capital diasporas. The second, partly related to marker power estimation, is the country's absorptive capacity, or technological adaptive capacity. It refers to governance and the business climate, human capital and basic technology literacy, or access to credit on capital markets. See Joseph Stiglitz, Social Absorption Capability and Innovation, Stanford University CERP Publication, 292 (1991). See also: Dominique Foray, Knowledge Policy for Development, In OECD, Innovation and the Development Agenda, Published by OECD and the International Development Research Centre (IDRC), Canada (Kraemer-Mbula Erika and Wamae Watu, eds.) (2010), at 93.

[121] GDP is assessed per year for any other measure of aggregate income, such as national income per capita and consumption per capita. See World Bank, Beyond Economic Growth Student Book (2004), at: www.worldbank.org/depweb/english/beyond/global/glossary.html; Joseph Stiglitz, Making Globalization Work (W.W. Norton & Co., 2006), at 44–48; Gerald M. Meier and James E. Rauch (eds.), *Leading Issues in Economic Development* (Oxford: Oxford University Press, 1995), at 7.

[122] Richard H. Steinberg, In the Shadow of Law or Power? Consensus-Based Bargaining and Outcomes in the GATT/WTO, 56(2) Int'l. Org. 339, 347.

[123] More generally, see also Clibert Rist, The History of Development: From Western Origins to Global Faith (2002), at 8–25. See also: James M. Cypher and James L. Dietz, The Process of Economic Development (2009), at 30.

best approximation rooted in the perception of advanced economies treats domestic market opening as a cost, and foreign market opening and associated increases in export opportunities as domestic political benefits.[124]

On an institutional UN-level, three explanatory factors can be suggested for this inclination to market size first best approximation. Firstly, this method manifests the participatory vision of the constitutional legitimacy of the WTO,[125] or otherwise of WIPO,[126] which is consistent with the economic model preferred by the United States, the EU, and constituents within them advancing their development policy preference. This default market size power-based system is still endemic in the Free Trade Agreement (FTA) system and the TRIPS-plus agreements. It is also reflected in the effect of the TRIPS Agreement on international trade and the minimal standard inflexibility it imposes on developing countries, primarily in exploiting patents.[127] Secondly, the inclination toward a market size first best approximation of development policies is based on the consensus-based rule underlying the democratic voting procedures by these organizations' member state realpolitik.[128]

A third explanation for the market size first best approximation at the UN-level can also be proposed. In order for a more subtle comparison to emerge, it would have been necessary to provide an underlying definition for both developing and developed countries. Remarkably, however, the UN system has not adopted any definition for designating a country as

---

[124] Ibid.
[125] See Gregory Shaffer, Power, Governance and the WTO: A Comparative Institutional Approach, in *Power and Global Governance* 130, 133–40 (Michael Barnett and Raymond Duvall, eds.) (Cambridge University Press, 2004); See Peter M. Gerhart, The Two Constitutional Visions of the World Trade Organization, 24 *University of Pennsylvania Journal of International Economic Law* 1, 9 (2003) (exploring the external, participatory vision of the constitutional legitimacy of the WTO while emphasizing the democratic nature of the WTO).
[126] See, e.g., Joseph Straus, The Impact of the New World Order on Economic Development: The Role of the Intellectual Property Rights System, 6 *John Marshall Review of Intellectual Property Law*, 1 (2006) (referring to the power-based system within the FTAs and TRIPS-plus agreements), at 10; Joseph Straus, Comment, Bargaining Around the TRIPS Agreement: The Case for Ongoing Public-Private Initiatives to Facilitate Worldwide Intellectual Property Transactions, 9 *Duke Journal of Comparative & International Law*, 91 (1998) (hereafter, "Bargaining Around the TRIPS Agreement") (on the TRIPS Agreement on international trade and the flexibility it gives nations to exploit patents), at 95.
[127] Straus, Bargaining Around the TRIPS Agreement, at 95.
[128] See Richard H. Steinberg, In the Shadow of Law or Power? Consensus-Based Bargaining and Outcomes in the GATT/WTO, 56(2) *International Organization* 339 (2002) (explaining why the GATT/WTO legislative consensus rule has been maintained), at 342–43.

"developed" or "developing."[129] As a result, major international organizations apply different classifying criteria. The UN categorizes developing countries into several groups based predominantly on market power factors, such as income, education, healthcare, and life expectancy.[130] The IMF divides countries based primarily on their market power into advanced economies, emerging economies, and developed economies.[131] The World Bank classifies countries according to their gross national income (GNI) per capita,[132] referring to low-income, lower-middle-income, upper-middle-income, and high-income economies. And, of course, UNCTAD has its separate country-group classification.[133]

All three explanations set the context for the current Pax-Americana power-based bargaining situation. This further explains why equal-country patenting and innovation-related policies preside internationally.

More particularly, WIPO's explanation of the disparities over patent intensity across countries repeatedly focuses on patent counts as opposed to the relative yardstick of patent propensity rates. WIPO thus emphasizes the disparity between leading developed countries in patent counts in comparison with other high-income countries, China, and other low- and middle-income countries. To illustrate, an annual report on world intellectual property for 2015, entitled *Breakthrough Innovation and Economic Growth*, informs us that high-income countries account for more than 80 percent of filings. Even within the high-income countries, patent filings are concentrated, with the United States, Japan, Germany, France, the UK, and the Republic of Korea accounting for at least 75 percent of global first filings.[134] It is only natural that under this market size approximation, WIPO's official work program contains several broad policies that erroneously chose to focus on countries' absolute patent and R&D propensities, while overlooking the relative underpinnings of patenting

---

[129] OECD, Developed and Developing Countries (January 4, 2006); WTO, Who are the Developing Countries in the WTO?, at: www.wto.org/english/tratop_e/devel_e/d1who_e.htm.

[130] The United Nations Statistics Division (UNSTATS), Composition of Macro Geographical (continental) Regions, Geographical Sub-Regions, and Selected Economic and other Groupings (February 17, 2011).

[131] International Monetary Fund (IMF), WEO Groups Aggregates Information (April 1, 2010).

[132] World Bank, How We Classify Countries, at: https://datahelpdesk.worldbank.org/knowledgebase/topics/19280-country-classification.

[133] See UNCTAD, Trade and Development Report 2011, at: http://unctad.org/en/docs/tdr2011_en.pdf.

[134] See WIPO, World Intellectual Property Report 2015 – Breakthrough Innovation and Economic Growth (2015), at 11–12, figures 4 and 5, and table 2 (accounting for first patent filings in six industries, namely airplanes antibiotics, semiconductors, 3D printing, nanotechnology, and robotics between 1995 and 2011).

activity. WIPO's policies thus include a rather flatly modeled technical assistance for developed countries,[135] a unified treaty law norm-setting apparatus,[136] and so forth.[137]

Three reasons also explain WIPO's preference for an archetypal nation market size-based bargaining posture toward developing countries, as it derives regulatory insight from relatively fixed patent and other proprietary counts. All three may also explain why patenting and innovation-related norm-setting is conceptually indistinct with regard to possible development discrepancies across the North–South divide.

The first explanation relates to the shift by developed countries from WIPO as the main multilateral forum for bargaining over patents and innovation to rival UN agencies and to the WTO. Such regime or forum shifting can be seen as an attempt to transfer treaty negotiation, lawmaking initiatives, or standard setting activities from one UN agency to

---

[135] See World Intellectual Property Organization, WIPO Intellectual Property Handbook: Policy, Law and Use (WIPO Publication No. 489, 2nd edn., 2004), available at www.wipo.int/about-ip/en/iprm, at 196–203, 359–60; WTO–WIPO Cooperation Agreement, Art. 4 (Legal-Technical Assistance and Technical Cooperation), entered into force Jan. 1, 1996. Pursuant to their Cooperation Agreement, WIPO and the WTO launched a joint initiative in July 1998 to assist developing countries in complying with TRIPS.

Technical assistance is Item no. 1 of 19 items chosen primarily because their immediate implementation does not require the engagement of additional staff or financial resources. See Provisional Committee on Proposals Related to a WIPO Development Agenda (PCDA), WIPO Development Agenda; Preliminary Implementation Report in Respect of 19 Proposals, Feb. 28, 2008, at 1, available at http://ip-watch.org/files/WIPO%20comments%20on%20DA%20recs%20-%20part%201.pdf (stating that WIPO was already "[t]aking into account country specific needs, priorities and the level of development, particular the special needs of Least Developed Countries."), at Annex, at 1.

Regarding the centrality of technical assistance for developing countries themselves, see, e.g., Robert M. Sherwood, Global Prospects for the Role of Intellectual Property in Technology Transfer, 42 *IDEA* 27(1997) (noting that the judicial systems in perhaps 80% of the countries of the world are simply not up to the task of supporting intellectual property rights, much less dealing effectively with other matters), at 30; Robert M. Sherwood, Some Things Cannot Be Legislated, 10 *Cardozo Journal of International and Comparative Law* 37 (2002) ("[I]ntellectual property systems involve a high degree of administrative and judicial discretion. Unless those who operate these systems hold a belief that they serve local interests, international rules, however derived or enforced, are likely to achieve little"), at 37.

[136] Cluster B of the Development Agenda governs WIPO norm-setting. WIPO is actively involved in organizing negotiations and preparing drafts and working papers for new intellectual property treaties. The Development Agenda requires WIPO norm-setting to "take into account different levels of development," or "take into account flexibilities in international IP agreements." See, Ibid.

[137] For a review of WIPO and developing countries-related policies, see e.g., Rami M. Olwan, *Intellectual Property and Development: Theory and Practice* (Springer, 2013), at 52–55.

another.[138] In the WIPO context, this poses a dilemma for developed countries in their capacity as the organization's regulation-givers. The dilemma is whether to negotiate in the natural domicile of WIPO, despite their recognition that this organization offers them minimal standards of protection and lacks effective enforcement mechanisms, or whether to prefer competing UN agencies.[139] In practice, following the establishment of the WTO, developed countries shifted to the General Agreement on Tariffs and Trade's (GATT). As Professor Laurence Helfer predicted, this shift ultimately favored dominant intellectual property-based industries in these countries in the public health, biosciences, and genetic sectors.[140]

The second explanation for WIPO's inclination toward a nation market size-based bargaining posture relates to the added regime-shifting between multilateral and bilateral treaty-making settings. Given the disagreements seen to date in the Doha Round, the United States and other intellectual property-rich nations increasingly rely on bilateral "solutions" for trade issues, including intellectual property. The WTO's multilateral solutions, imperfect as they are, have been giving way to Berne-Plus, TRIPS-plus,[141] and even US-plus[142] intellectual property standards negotiated through

---

[138] Forum shifting encompasses three kinds of strategies: moving an agenda from one organization to another, abandoning an organization, and pursuing the same agenda in more than one organization. See John Braithwaite and Peter Drahos, *Global Business Regulation* (Cambridge University Press, 2000), at 564. On the forum shifting from WIPO to other agencies, see Laurence R. Helfer, Regime Shifting: The TRIPS Agreement and New Dynamics of International Intellectual Property Lawmaking, 29 *Yale Journal of International Law* 1 (2004) (describing regime-shifting from WIPO to other agencies, using examples of TRIPs and food, agriculture, public health, biodiversity, and human rights), at 42 n. 186; Peter K. Yu, Currents and Crosscurrents in the International Intellectual Property Regime, 38 *Loyola of Los Angeles Law Review* 323 (2004) (describing multilateral to bilateral regime-shifting as well as shifting between the WTO and WIPO), at 408–17. But see Peter Drahos, An Alternative Framework for the Global Regulation of Intellectual Property Rights, 21 *Austrian Journal of Development Studies* 1 (2005) (concerning the shift from UNCTAD to WIPO), at 7.

[139] See Viviana Munoz Tellez, The Changing Global Governance of Intellectual Property Enforcement: New Challenges for Developing Countries, In Xuan Li and Carlos M. Correa (eds.), *Intellectual Property and Enforcement* (Edward Elgar, 2009), at 6.

[140] See Laurence R. Helfer, Regime Shifting: The TRIPS Agreement and New Dynamics of International Intellectual Property Lawmaking, 29 *Yale Journal of International Law* 1 (2004), at 3–4.

[141] "TRIPS-Plus" refers to bilateral agreements or regional multilateral agreements, often denominated as "free trade agreements," in which minimum standards exceed the TRIPS standards. See e.g., Frederick M. Abbott, The Cycle of Action and Reaction: Developments and Trends in Intellectual Property and Health, in *Negotiating Health: Intellectual Property and Access to Medicines* (Pedro Roffe et al., eds.) (Earthscan, 2006), at 31–33.

[142] See e.g., Frederick M. Abbott, Intellectual Property Provisions of Bilateral and Regional Trade Agreements in Light of US. Federal Law 1 (*International Centre for Trade and Sustainable Development*, Issue Paper No. 12, Feb. 2006), available at www.unctad.org/

FTAs.[143] All told, and as Peter Drahos explains, the non-discrimination most-favored nation (MFN) principle of TRIPS has turned into a ratchet-upward for rights holders in the market sized bloc of developed countries.[144] As a result, patent and innovation norm setting in view of development has become more power-based, positional, and contentious than ever before.

A third explanation for WIPO's marker size proclivity concerning patents and innovation policies relates to the fact that many developing countries contend with a lack of policy coordination, growing complexity, and fragmentation of policy-making venues at the transnational and international levels. This is part of what has been termed the "knowledge trap" for poor countries. These countries are thus penalized by the knowledge-intensity demanded at the UN agency level as well as by the WTO.[145]

All three explanations support the conclusion that WIPO, as well as other UN-level agencies, form part of the patenting and innovation-related norm-setting challenge. Against this backdrop, it is only natural that the TRIPS Agreement is positioned in the nexus of a trade-off between trade-innovation and patent-related regulation, as we shall now discuss.

1.1.2.3 Trade-Innovation and Patent Regulatory Trade-Off

Prior to the establishment of the WTO and TRIPS, UN organs such as WIPO and UNCTAD systematically failed to devote substantial attention to innovation-policy discrepancies across countries. Accordingly, it was only natural that on its adoption, TRIPS was confined to a flat intellectual-property policy for all WTO members, reflecting neoclassical economic growth modeling.[146]

---

en/docs/iteipc20064_en.pdf (describing several examples of US-plus standards adopted by other countries), at 9, 11.

[143] See Daniel Gervais, *TRIPS and Development*, 95, In Sage Handbook on Intellectual Property (SAGE Publications Ltd, 2014) (Matthew David and Debora Halbert, eds.), at 107.

[144] Peter Drahos, An Alternative Framework for the Global Regulation of Intellectual Property Rights, 21 *Austrian Journal of Development Studies* 1 (2005), at 7.

[145] See Margaret Chon, Denis Borges Barbosa and Andrés Moncayo von Hase, Slouching Toward Development in International Intellectual Property, *Michigan State Law Review* 71 (2007), at 89 referring to Sylvia Ostry, After Doha: Fearful New World?, Bridges, Aug. 2006, at 3, available at www.ictsd.org/monthly/bridges/BRIDGES 10–5.pdf. For the WTO, see Gregory Shaffer, Can WTO Technical Assistance and Capacity Building Serve Developing Countries?, *Wisconsin International Law Journal*, vol.23 643 (2006) ("Implementation often requires developing countries, unlike developed countries, to create entirely new regulatory institutions and regimes."), at 645.

[146] See Benoliel and Salama, Toward an Intellectual Property Bargaining Theory: The Post-WTO Era, at 278; Michael Blakeney, *The International Protection of Industrial*

From the standpoint of developing countries, the mandatory adoption of TRIPS standards created two related exogenous costs toward the developed world. These were the cost of reduced access to new technologies and knowledge, and the cost of higher royalty payments.[147] TRIPS maintained international intellectual property protection as a central pillar for both short- and long-term economic growth, effectively ignoring country-group differences. This argument stood for two long-term neoclassic exogenous economic incentives offered by developed nations.[148] The first incentive promised to undertake positive efforts in the area of technology transfer – a classic form of reflexive innovation policy toward the developing countries as transferees.[149] The second incentive assured agricultural trade.[150] These incentives, backed by ancillary agreements, proved pivotal in the eventual acquiescence of the developing countries to the TRIPS Agreement.[151] Both incentives also implicitly adhered to Solow's neoclassical growth model, formulated earlier by economists David Cass[152] and Tjalling Koopmans, as explained earlier.[153] Neither these incentives nor the TRIPS regulatory setting in general included any substantive efforts to differentiate between country clusters or otherwise to consider a nonlinear underlying innovation-based economic growth.

Property: From the Paris Convention to the TRIPS Agreement 16 (2003), available at www.wipo.int/export/sites/www/arab/en/meetings/2003/ip_cai_1/pdf/wipo_ip_cai_1_03_2.pdf.

[147] See Christopher S. Gibson, Globalization and the Technology Standards Game: Balancing Concerns of Protectionism and Intellectual Property in International Standards, 22 *Berkeley Technology Law Journal* 1403, 1406 (2007).

[148] See, e.g., Carolyn Deere, Developing Countries in the Global IP System, in *The Implementation Game: The TRIPS Agreement and the Global Politics of Intellectual Property Reform in Developing Countries* (Oxford University Press, 2009) 34, at 51; Peter Yu, Toward a Nonzero-Sum Approach to Resolving Global Intellectual Property Disputes: What We Can Learn from Mediators, Business Strategists, and International Relations Theorists, 70 *University of Cincinnati Law Review* 569 (2001), at 635.

[149] See Laurence R. Helfer, Regime Shifting: The TRIPS Agreement and New Dynamics of International Intellectual Property Lawmaking, 29 *Yale Journal of International Law* 1 (2004), at 2; Carlos M. Correa, *Intellectual Property Rights, the WTO and Developing Countries: The TRIPS Agreement and Policy Options* 18 (2000) (focusing on developing countries' concern for increasing technological transfer as means of economic growth).

[150] Laurence R. Helfer, Regime Shifting: The TRIPS Agreement and New Dynamics of International Intellectual Property Lawmaking, 29 *Yale Journal of International Law* 1 (2004), at 22; Clete D. Johnson, A Barren Harvest for the Developing World? Presidential "Trade Promotion Authority" and the Unfulfilled Promise of Agriculture Negotiations in the Doha Round, 32 *Georgia Journal of International and Comparative Law* 437 (2004), at 464–65.

[151] Johnson, ibid., at 467–68.   [152] Cass, supra note 30, ibid.

[153] Koopmans, supra note 31, at 226–28.

As a result, notwithstanding its deep-rooted innovation implications, TRIPS was predominantly accepted as a trade-related compromise.[154] The WTO's Uruguay Round of multilateral trade negotiations thus succeeded where prior WIPO negotiations, particularly regarding developing countries, had failed.[155] This round was successful because it was presented as an economic package deal, or what Professor Donald Harris calls a "treaty of adhesion."[156] Such deals are rooted in dependency theories of development, whereby developing countries were flatly perceived to be dependent on developed ones and freer trade was said to impoverish countries in the "periphery."[157] TRIPS' trade-related stand may further explain its exogenous economic disconnect from more measurable and carefully articulated patenting and innovation-related norms.

---

[154] See Jayashree Watal, *Intellectual Property Rights in the WTO and Developing Countries* (Springer) 20 (2001) (explaining how developed countries agreed to phase out their quotas under the ATC (Agreement on Textiles and Clothing) on the most sensitive items of textiles and clothing in exchange for developing countries' acceptance to the phasing-in of product patents for pharmaceuticals which they perceived as the most important patent-related good). See also Frederick M. Abbott, The WTO TRIPS Agreement and Global Economic Development, in *Public Policy and Global Technological Integration* 39 (Frederick M. Abbott and David J. Gerber, eds.), Springer (1997), at 39–40; Carolyn Deere, Developing Countries in the Global IP System, in *The Implementation Game: The TRIPS Agreement and the Global Politics of Intellectual Property Reform in Developing Countries* 34, (Oxford University Press, 2009), at 2; Charles S. Levy, Implementing TRIPS–A Test of Political Will, 31 *Law and Policy in International Business* 789 (2000) (describing TRIPS as a historical breakthrough in the trade-related context), at 789–90; Robert Weissman, A Long, Strange TRIPS: The Pharmaceutical Industry Drive to Harmonize Global Intellectual Property Rules, and the Remaining WTO Legal Alternatives Available to Third World Countries, 17 *University of Pennsylvania Journal of International Law* 1079 (1996) (describing how intellectual property became a central component of the free trade agenda), at 1096.

[155] See, e.g., Ruth L. Gana (Okediji), The Myth of Development, The Progress of Rights: Human Rights to Intellectual Property and Development, 18 *Law and Policy* 315 (1996), at 334; Donald P. Harris, Carrying a Good Joke Too Far: TRIPS and Treaties of Adhesion, 27 *University of Pennsylvania Journal of International Law* 681 (2006), at 724–25; Jerome H. Reichman, The TRIPS Component of the GATT's Uruguay Round: Comparative Prospects for Intellectual Property Owners in an Integrated World Market, 4 *Fordham Intellectual Property, Media & Entertainment Law Journal* 171 (1993) (arguing that political developments such as the drafting of TRIPS diminished WIPO's authority), at 179–80.

[156] Harris, ibid., at 724.

[157] See, e.g., Raul Prebisch, International Trade and Payments in an Era of Coexistence: Commercial Policy in the Underdeveloped Countries, 49 *American Economic Review* 251(1959) (offering examples of reasoning used by developing "periphery" countries fostering an aversion to increasing free trade), at 251–52. For a seminal Latin American perspective, see Fernando Henrique Cardoso and Enzo Faletto, Dependency and Development in Latin America 149–71 (Marjory Mattingly Uriquidi, trans.) (University of California Press, 1979) (describing the tension between Latin American nationalist and populist political agendas and its impact on related international trade policies).

The acceptance of TRIPS reflected the unequal, market size-based bargaining power seen in developed and developing countries alike. The bargaining situation between developed and developing countries henceforth permitted rich developed countries to receive stronger protection for IPRs as well as a reduction in restrictions against FDI.[158] In return, the less developed countries enjoyed lower tariffs on textiles and agriculture and protection against unilateral sanctions.[159]

In short, given its unique mixture of exogenous economical trade rules and fairly equal-country international intellectual property concepts, TRIPS never truly adhered to an endogenous economic growth model.

Developing countries promoting TRIPS referred to "inherent asymmetries and imbalances" as a trading constraint within the WTO's trading system and the Uruguay Round Agreements, including TRIPS itself. Such a confrontational approach was almost certainly also in evidence at the Fourth WTO Ministerial Conference at Doha Qatar in 2001.[160] The resilient standoff by the Group of 77 and China regarding trading issues has since entered the annals of history,[161] but Doha should nonetheless be recalled as a reactive exception in innovation policy.

Lastly, TRIPS includes two particularly narrow exceptions. Firstly, chapter 65 of the agreement grants less developed and transitional countries a five-year transitional period.[162] Secondly, and more notably, chapter 66 provides LDCs with an eleven-year transitional period.[163] Such restricted egalitarian dialectic was seen, however, merely as a means to

---

[158] See Jayashree Watal, *Intellectual Property Rights in the WTO and Developing Countries* (Springer 2001), at 20; see also Frederick M. Abbott, *The WTO TRIPS Agreement and Global Economic Development, in Public Policy and Global Technological Integration* (Frederick M. Abbott and David J. Gerber, eds.) (Springer, 1997) 39, at 39–40, 42.

[159] See Jayashree Watal, *Intellectual Property Rights in the WTO and Developing Countries*, (Springer, 2001) (describing the sanctions imposed prior to the broad introduction of the TRIPS Agreement by the United States and other developed countries via the mandatory settlement process), at 20–22.

[160] See World Trade Organization, Declaration of the Group of 77 and China on the Fourth WTO Ministerial Conference at Doha, Qatar, WT/L/424 (Oct. 22, 2001), available at www.wto.org/english/thewto_e/minist_e/min01_ e/proposals_e/wt_l_424.pdf.

[161] See Inge Govaere and Paul Demaret, The TRIPS Agreement: A Response to Global Regulatory Competition or an Exercise in Global Regulatory Coercion?, in *Regulatory Competition and Economic Integration: Comparative Perspectives* (Oxford University Press, 2001) (Daniel C. Esty and Damien Geradin, eds.), at 364, 368–69.

[162] Agreement on Trade-Related Aspects of Intellectual Property Rights art. 65(1)–(3), Marrakesh Agreement Establishing the World Trade Organization, Annex 1C, 1869 U.N.T.S. 299 (Apr. 15, 1994), available at www.wto.org/english/docs_e/legal_e/27-trips.pdf.

[163] Ibid. at art. 66(1); Benoliel and Salama, supra note 14, at 360.

create "a sound and viable technological base" in these disadvantaged countries.[164]

1.1.2.4 The Value of Technical Assistance and Capacity-Building

Technical assistance and capacity-building leave a long intellectual trail of controversy across the entire UN-level.[165] At first glance, enhancing technical assistance and capacity-building at the UN, and particularly at WIPO, may seem resultant and procedural. Nevertheless, the first organization with which WTO signed a "Cooperation Agreement" for the provision of technical assistance was WIPO.[166] Moreover, when developing countries agreed to enter into the Doha trade negotiations they obtained a commitment from the WTO Secretariat to provide capacity-building technical assistance. This assistance was intended to facilitate both their participation in the negotiations and their eventual integration in the trading system. The Doha Ministerial Declaration consequently dedicated more text to capacity-building than to issues such as trade or technology transfer.[167] The *2003 World Trade Report* likewise declared that "[t]he Doha Declaration marked a new departure in the GATT/WTO approach to technical assistance and capacity-building."[168]

Despite this support, technical assistance remains one of the core resentments of the developing countries regarding UN agency level organizations. Against the backdrop of a patenting and innovation-related norm-setting challenge, this resentment is most importantly manifested toward WIPO.[169] Such policies surely are meant to educate and enhance

---

[164] Agreement on Trade-Related Aspects of Intellectual Property Rights, supra note 60, art. 66(2) (requiring developed countries to provide commercial incentives to encourage the transfer of technology to least developed countries).

[165] Gregory Shaffer, Can WTO Technical Assistance and Capacity Building Serve Developing Countries?, *Wisconsin International Law Journal*, vol.23 643 (2006), at 643; Peter Morgan, Technical Assistance: Correcting the Precedents, 2 Development Policy Journal 1, (2002) (arguing that since the 1940s "Technical assistance became for the first time, an issue of public policy"), at 1–2.

[166] Comm. on Trade & Dev., Note by Secretariat, A New Strategy For WTO Technical Cooperation: Technical Cooperation for Capacity-building, Growth and Integration, WT/COMTD/W/90 (Sept. 21, 2001), at para. 6 n.5. Similarly, early WTO capacity-building plans consistently referred to "technical missions" helping developing countries to adapt their policies to the WTO Agreements "in areas such as customs valuation, trade remedies and TRIPS, and transposition of tariff schedules." See WTO Comm. on Trade & Dev., Note by the Secretariat, Report on Technical Assistance 2000, WT/COMTD/W/83 (May 2, 2001), at 31.

[167] See WTO, Ministerial Declaration of 20 November 2001 WT/MIN(01)/DEC/1, 41 I.L.M. 746 (2002), para. 38–41.

[168] See WTO, World Trade Report 2003 (2003) (adding that "technology transfer had never been included explicitly on the GATT/WTO agenda before"), at 164.

[169] Mr. Geoffrey Onyeama, Deputy Director General, Cooperation for Development, interview on October 15, 2014, in reference to WIPO, Development Agenda (2007),

the capacity of mostly developing countries to use intellectual property, as in WIPO's case, and are continuously advanced by WIPO's training Academy.[170] As in the case of WTO, the provision of technical assistance is also in line with WIPO's Development Agenda.[171]

The essence of the critique of developing countries against the perceived hegemony of the advanced economies does not imply a lack of recognition of the need for technical assistance; the contrary is the case. Due to the central role technical assistance receives, WIPO commissioned an unprecedented external review team titling their seminal 2011 report *WIPO Technical Assistance in the Area of Cooperation for Development*.[172] The review team confirmed that the focus of any internal assessments that do take place is generally on the short-term results carrying no long-term or cumulative impact. In the area of training, by way of example, although WIPO's training activities appear to be highly appreciated by developing countries, the report found that the development impact of these activities was not well explained or monitored.[173] The critique concerns the disproportional focus this aspect received, to the detriment of other less short-term policies. It is thus said to serve as a model "catch up" exogenous economic policy toward developing countries perceived as regulation-takers.[174]

Recommendation 1 ("WIPO technical assistance shall be, inter alia, development-oriented, demand-driven and transparent, taking into account the priorities and the special needs of developing countries, especially LDCs ... ").

[170] See Director General Report on Implementation of the Development Agenda, Committee on Development and Intellectual Property (CDIP) Thirteenth Session, CDIP/13/2, Geneva, May 19 to 23, 2014 (March 3, 2014), at 19.

For an overview of WIPO's Academy training activities see the 2014 WIPO Academy Education and Training Programs Portfolio. Document available at: www.wipo.int/edocs/pubdocs/en/training/467/wipo_pub_467_2014.pdf. Details are also included in the Annual Statistical Report for 2013. Document available at: www.wipo.int/export/sites/www/academy/en/about/pdf/academy_statistics_2013.pdf.

[171] See WIPO, Development Agenda, Recommendation 3 (concerning the "Increase human and financial allocation for technical assistance programs in WIPO for promoting, inter alia, development-oriented intellectual property culture"). See also: Director General Report on Implementation of the Development Agenda, Committee on Development and Intellectual Property (CDIP) Thirteenth Session, CDIP/13/2, Geneva, May 19 to 23, 2014 (March 3, 2014), at 2–3, Sec. 7–9.

[172] The report included a large number of recommendations to the secretariat for improving its technical assistance. See Carolyn Deere Birkbeck and Santiago Roca, An External Review of WIPO Technical Assistance in the Area of Cooperation for Development (August 31, 2011), at: www.wipo.int/edocs/mdocs/mdocs/en/wipo_ip_dev_ge_11/wipo_ip_dev_ge_11_ref_2_deere.pdf.

[173] Ibid. (the report recommends that WIPO incorporates "a sufficiently clear and broad understanding of the overall purposes of its development cooperation activities"), at iv.

[174] Ibid. (In relation to technical assistance, Recommendation 28 states: "The Report recommends that focus should be on longer-term or cumulative impacts of development cooperation activities, rather than short-term projects"), at 11.

In the case of the WTO, criticism of technical assistance efforts encouraged the WTO to declare a new approach in 2001.[175] Its stated aims were to turn technical assistance more demand-driven, enhance more financial stability through the Doha trust fund, and to improve the capability of the WTO Secretariat to deliver products within its mandate to meet the needs of developing countries.[176] An audit of the WTO's *Technical Assistance Plan for 2003* similarly criticized the implementation of WTO's plan for a lack of coherence, arguing that the Secretariat was mainly responding to ad hoc requests.[177] In a nearly Sisyphean effort to address these criticisms, the WTO's Secretariat initiated yet another plan for 2004 that was to be more "quality-oriented, aiming at building long-term, i.e., sustainable, human and institutional capacity."[178] Across the UN, technical assistance and capacity-building continue to challenge patenting and innovation-related norm-setting. Perhaps unsurprisingly, scholars such as Keith Maskus and Jerome Reichman thus conclude, maybe harshly, that whether WIPO's strategy actually benefits innovation (and to which countries those benefits flow) seems to "count for little in implementing the mandate."[179]

1.2   Toward Nonlinear Innovation Patenting Policy

Insofar as UN agencies opt for more subtle country-specific patenting and innovation norm setting, a proper theoretical setting will be desirable. We should begin by noting that many scholars have acknowledged that an effective innovation strategy requires coordination of multiple layers of

But see WIPO's Secretariat response, at WIPO, Update on the Management Response to the External Review of WIPO Technical Assistance in the Area of Cooperation for Development, CDIP/16/6 (September 2015) ("WIPO has reoriented the focus of its technical assistance and capacity building activities toward long-term sustainable projects"), at 11.

[175] Comm. on Trade & Dev., Note by Secretariat, A New Strategy For WTO Technical Cooperation: Technical Cooperation for Capacity-building, Growth and Integration, WT/COMTD/W/90 (Sept. 21, 2001)

[176] Ibid.

[177] See Comm. on Trade and Dev, Note by Secretariat, Coordinated WTO Secretariat Annual Technical Assistance Plan 2003, WT/COMTD/W/104 (Oct. 3, 2002), at para. 16.

[178] See Comm. on Trade & Dev., Technical Assistance and Training Plan 2004, WT/COMTD/W/119/Rev.3 (Feb. 18, 2004), at para. 7. In support of the WTO's long-run technical assistance objective, see also: Henri Bernard Solignac Lecomte, Building Capacity to Trade: A Road Map for Development Partners: Insights from Africa and the Caribbean 7 (European Centre for Dev. Pol'y Mgmt Discussion Paper 33, 2001), available at www.ecdpm.org ("There can only be one ultimate objective: to empower developing countries in the multilateral trade system"), ibid.

[179] See Keith E. Maskus and Jerome H. Reichman (eds.), *International Public Goods and Transfer of Technology Under a Globalized Intellectual Property Regime* (Cambridge University Press, 2005), at 18 and Fn. 54 and sources therein.

support policies.[180] This core realization corresponds with the rate of return of research activity transforming into patent intensity. In theory, studies on the rate of return take two forms. Some focus on private rates of return, namely the return on investments in research that flow from an individual research project to the organization directly involved. Others examine the social rates of return to research, namely "the benefits which accrue to the whole society."[181]

The difference between the two arises because the benefits of a specific research project, or even a firm-based innovation, are not generally confined exclusively to a single firm. Thus, the scientific benefit of a basic research study may be appropriated by more than one firm, for example by imitators who replicate the new product without bearing the cost of the original research. By lowering the costs of developing new technologies or products through investing in basic research, publicly funded projects actually generate broader social benefits. Hence, estimates of the private rate of return on R&D tend to be much lower than those for the social rate of return.

Even from a social perspective, economic benefits from R&D differ from their social, environmental, or cultural benefits. However, a fuzzy boundary remains between the economic and non-economic benefits.[182] Innovation springs from the creative application of knowledge, but its underlying conditions are necessarily complex: many are not easily altered by policy, and some are the result of cultural evolutionary processes that extend beyond the reach of short-term policy-making.[183]

For example, if a new medical treatment improves health and reduces days of work lost to a particular illness, should these benefits be seen as

---

[180] See Sanjaya Lall and Morris Teubal, Market Stimulating Technology Policies in Developing Countries: A Framework with Examples from East Asia, *World Development*, vol. 26(8) 1369 (1998) ("the exact mix varies with country context and the capabilities of its policy makers"), at 1370; Bengt-Åke Lundvall and Susana Borrás, The Globalizing Learning Economy: Implications for Technology Policy, Final Report under the TSER Programme, EU Commission (1997); Dani Rodrik, One Economics, Many Recipes: Globalization, Institutions and Economic Growth (Princeton University Press, 2007); Isabel Maria Bodas Freitas and Nick von Tunzelmann, Mapping Public Support for Innovation: A Comparison of Policy Alignment in the UK and France, *Research Policy*, vol. 37(9) 1446 (2008).

[181] Keith Smith, Economic returns to R&D: method, results, and challenges, Science Policy Support Group Review Paper No. 3, London (1991), at. 4.

[182] See, e.g., Hiroyuki Odagiri, Akira Goto, Atsushi Sunami, and Richard R. Nelson (eds.), *Intellectual Property Rights, Development and Catch Up* (Oxford University Press, 2010), at 417–30; more generally, see David S. Landes, *The Wealth and Poverty of Nations: Why Some Are So Rich and Some So Poor* (W.W. Norton, 1998) (describing the numerous climatological, historical, and cultural circumstances which had a significant and complex influence on the economic development of various nations throughout the world).

[183] Ibid.

economic or social? Given this uncertainty, "economic" may be a rather broad term. Moreover, our examination here relates to economic benefits solely in the form of directly useful knowledge. However, other less direct economic benefits are surely manifested in terms of well-being or welfare on the aggregate economy-wide level of welfare economics.

In this context, the demise of the Washington Consensus by the mid-1990s came late by comparison to the key innovation theoretical findings of the 1970s.[184] By the mid-1990s, economists such as David Mowery and Nathan Rosenberg had, for more than a decade, predicted the death of the neoclassical linear model of innovation causality in technology and science or markets.[185] As Mowery and Rosenberg further explained in their 1991 monograph *Technology and the Pursuit of Economic Growth*, the contribution of economics to the understanding of technology and economic growth has been constrained by the theoretical framework employed within neoclassical economies.[186] Innovation causation was to be replaced with the recognition that innovation, since it is endogenous and nonlinear, involved a complex and country-specific mix of new knowledge and new demand, with the exact blend being technology, firm, and timing.[187]

Such studies, especially that of the above-mentioned Science Policy Research Unit, were primarily concerned with determinants of success

---

[184] See, e.g., John Weeks and Howard Stein, Washington Consensus, in *The Elgar Companion to Development Studies* 676 (David Alexander Clark, ed.) (Edward Elgar Publishing, 2006) ("The Consensus reigned hegemonic in international development policy from the early 1980s to the mid-1990s, when it came under sustained attack"), at 676.

[185] See, e.g., David Mowery and Nathan Rosenberg, The Influence Of Market Demand Upon Innovation: A Critical Review of Some Recent Empirical Studies, 8 *Research Policy* 102 (1979) (criticizing the imbalanced attention given to demand-side innovation policy considerations, the authors state, "Little consideration has been paid to the study, at a less aggregated level, of the specific innovative outputs of industries and firms, and the forces explaining differences among industries, firms and nations"), at 103.

[186] See David C. Mowery and Nathan Rosenberg, *Technology and the Pursuit of Economic Growth* (Cambridge University Press, 1991) ("The neoclassical economic framework for the analysis of R&D and innovation says very little if anything about the institutional structure of the research systems of advanced industrial economies"), at 4; see also ibid. (yielding a similar conclusion), at 16, 96.

[187] See generally Gerhard Mensch, *Stalemate in Technology: Innovations Overcome the Depression* (Ballinger Publishing Company, 1979) (criticizing linear innovation using the example of computers in the United Kingdom during the sixties, while arguing linear representations of innovation processes thereof are poorly explained linearly); Sci. Pol'y Research Unit, Report on Project SAPPHO 1971 (detailing a study of management of innovation in two science-based industries, chemicals and scientific instruments, identifying the factors which distinguish innovations that achieved commercial success). But see Slavo Radoševic and Esin Yoruk, SAPPHO Revisited: Factors of Innovation Success in Knowledge-Intensive Enterprises in Central and Eastern Europe (DRUID Working Paper No. 12–11), available at www3.druld.dk/wp/20120011.pdf.

and failure in industrial innovation. These studies, therefore, were far less concerned with the determinants of the rate and direction of the innovative activity per se.[188]

A novel critique has originated from evolutionary economics and presently represents a departure from earlier neoclassical theories and assumptions.[189] The theory is based on the Schumpeterian vision of the economic world as a chain of disequilibria that are plainly dynamic and evolutionary. It nevertheless regards invention as an endogenous process rather than an exogenous force acting on the economic scheme. As its two early advocates, Sidney Winter and Richard Nelson, theorized in *An Evolutionary Theory of Economic Change* (1982), criticism of neoclassical economics' focus on linear growth causality over innovation justified the adoption of a new setting for innovation theory.[190] Evolutionary economics models incorporated an interactive effect between variables, as opposed to the impact that any single variable might have in explaining the process of innovation and diffusion. Arguably, the models may ultimately withstand the dynamic impact of new UN-level innovation country clusters as opposed to a single or linear innovation model.

Notably, the US government much later also admitted in its 2008 *United States President's Council of Advisors on Science and Technology* (PCAST) report that there has been a growing need for nonlinear, or at least less linear, innovation theory and policy as a whole.[191] Such findings have recently been apparent across the ever-growing literature on nonlinear innovation causation within developing countries' scholarship.[192]

---

[188] The last straw for the linear model was its inability to explain how Japan could be so successful with technology despite lacking a world-class science base as compared to British firms. See generally Dianna Hicks, T. Ishizuka, and S. Sweet, Japanese Corporations, Scientific Research and Globalization, 23 *Research Policy* (1994) (rejecting that Japanese companies are "free riders" on world science as their science draws most heavily on Japanese, not foreign sources), at 4.

[189] See Léger and Swaminathan, supra note 53, ibid.

[190] For a critique of neoclassical growth theory as being overly generalized, as discussed, see Ricardo Hausmann, Dani Rodrik, and Andrés Velasco, Getting the Diagnosis Right: A New Approach to Economic Reform, 43 *Finance and Development* 12 (2006); Ricardo Hausmann, Dani Rodrik, and Andrés Velasco, Growth Diagnostics, in *The Washington Consensus Reconsidered: Toward a New Global Governance* (Narcís Serra and Joseph Stiglitz, eds.) (Oxford University Press, 2008); Dani Rodrik, The New Development Economics: We Shall Experiment, But How Shall We Learn?, Harvard University, John F. Kennedy School of Government Faculty Research Working Papers Series, Paper No. RWP08-055 (2008) (stating that new development economics have become country-specific), at 24–28.

[191] See President's Council of Advisors on Science and Technology, University-Private Sector Research Partnerships in the Innovation Ecosystem 1–2, 7, 31 (2008) (acknowledging the shift from a linear innovation paradigm into a nonlinear or less linear one).

[192] See, e.g., Nikos C. Varsakelis, The Impact of Patent Protection, Economy Openness and National Culture on R&D Investment: A Cross-Country Empirical Investigation 30

Following on the tails of nonlinear innovation analysis, one distinct theory has started to transform into policy within the Organization for Economic Co-operation and Development (OECD) and the European Union (EU). This is the National Systems of Innovation theory.[193] This seminal theory explains that innovation and technology development are the results of a complex set of domestic relationships among various state institutions,[194] including enterprises, universities, and government research institutes. The blend of the geographic measurement and the systemic nature of the innovation process itself led to the emergence of a new approach to regional innovation systems.[195] Such a system may be seen as an agglomeration or regional cluster, though it also encompasses the supportive institutions and organizations within those regions.[196]

Research Policy 1059 (2001) (finding national culture to be a determinant of R&D intensity, using a panel of developing and industrialized countries); see generally Oscar Alfranca and Wallace E. Huffman, Aggregate Private R&D Investments in Agriculture: The Role of Incentives, Public Policies, and Institutions, 52 *Economic Development and Cultural Change* 1 (2003) (showing that private agricultural R&D investments in EU countries also respond to the quality of the institutional environment, with an emphasis on bureaucracy, enforcement of contracts and IP protection), at 1–22; UNESCO Institute for Statistics, Measuring R&D: Challenges Faced by Developing Countries, UIS/TD/10–08 (2008), available at www.uis.unesco.org/Library/Documen ts/tech%205-eng.pdf.

[193] For major academic contributions, see generally Bengt-Åke Lundvall, Product Innovation and User-Producer Interaction, in 31 Industrial Development Research Series (1985), at 28–29. See generally *National System of Innovation: Towards a Theory of Innovation and Interactive Learning* (Bengt-Åke Lundvall, ed.) (Anthem Press, 1993); *National Innovation System: A Comparative Analysis* (Richard R. Nelson, ed.) (Oxford University Press, 1993); Pari Patel and Keith Pavitt, The Nature and Economic Importance of National Innovation Systems, *STI Review* 14, (1994). For a general overview, see OECD, Innovation and the Development Agenda, supra note 120, at 57 (adding that the elements of the National Innovation System theory (NIS) have close similarities to structuralist views, stressing that "development is neither linear nor sequential, but a unique process shaped by a specific [historical,]" cultural, and socio-economic context).

[194] See Lundvall, ibid. (incorporating elements and relationships in the national system of innovation, which interact in the production, diffusion and use of new, and economically useful, knowledge).

[195] See, e.g., Philip Cooke, *Regional Innovation Systems, Clusters, and the Knowledge Economy, Industrial and Corporate Change*, 10(4), 945 (2001) (concluding that Europe's innovative gap in comparison with the United States lies in a European regional firm-level market failure); David Doloreux, What We Should Know about Regional Systems of Innovation, *Technology in Society*, 24, 243 (2002).

For other contributions on spatial analysis and innovation theory, see also: Thomas Brenner and Tom Broekel, Methodological Issues in Measuring Innovation Performance of Spatial Units, Papers in Evolutionary Economic Geography, No. 04–2009, Urban & Regional Research Centre Utrecht, Utrecht University (2009); J. Vernon Henderson, Ari Kuncoro, and Matthew Turner, *Industrial Development in Cities, The Journal of Political Economy* 103, 1067 (1995).

[196] See Bjørn T. Asheim and Meric S. Gertler, The Geography of Innovation – Regional Innovation Systems, in The *Oxford Handbook of Innovation*, 291 (Jess Fagerberg, D. C.

Developed by the renowned Danish economist Bengt-Åke Lundvall, the National System of Innovation was initially and canonically used by Chris Freeman to explain the rise of developing countries. Freeman's initial case study focused on Japanese innovative firms in the 1970s and 1980s, when Japan was still a developing country. As noted, this theory gradually became a core policy concept of the OECD and the EU. As an archetypical nonlinear as well as endogenous innovation-based economic growth theory, it has shifted focus to concentrating on the formal R&D system and technical education thereof, bearing a country-specific policy orientation.[197]

It was now possible to proceed to the substantive empirical and conceptual corroboration of the nonlinearity between countries and group of countries, with emphasis on the gap between advanced and developing countries. Regarding patent propensity, analysis based on patent per Gross Domestic Expenditure on R&D (GERD), as presented in this book, facilitated this progress. In this context, particular mention should be made of an unprecedented and monumental country panel dataset completed by UNESCO in 2011. Developing countries that previously rarely reported on consolidated R&D growth indicators and related S&T statistics systems contributed enormously toward the preparation of highly detailed standardized country panel datasets that can be used instead of previous less R&D-related datasets and/or less developed S&T national statistics systems.[198]

Conclusion

As is apparent from a cursory review of the intellectual property, trade, and development indices in use throughout the UN and the WTO, no single innovation-based growth theory prevails. On the one hand, development settings such as WIPO's Development Agenda clearly require WIPO to work toward patenting and innovation-related norms that "take

Mowery, Richard R. Nelson, eds.) (Oxford University Press, 2006); Bjørn T. Asheim and Arne Isaksen, Regional Innovation Systems: The Integration of Local "Sticky" and Global "Ubiquitous" Knowledge, *Journal of Technology Transfer*, 27, 77 (2002).

[197] The dramatic breakthrough for the theory was a three-year work program known as the Technology-Economy Programme (TEP) leading to the TEP Report. The theory was later carried through also in subsequent OECD policy studies, such as the 1994 Jobs Study and the policy recommendations, the 1996 Technology, Productivity and Job Creation report, and the 1998 Technology, Productivity and Job Creation: Best Policy Practices. See Lynn K. Mytelka and Keith Smith, Innovation Theory and Innovation Policy: Bridging the Gap 12–17 (2001).

[198] See UNESCO Institute for Statistics, Measuring R&D: Challenges Faced by Developing Countries, UIS/TD/10–08 (2008), available at www.uis.unesco.org/Libra ry/Documents/tech%205-eng.pdf.

into account different levels of development"[199] while supporting member states in enhancing national policies to their "country specific"[200] conditions, reflecting "the priorities and special needs of developing countries."[201] On the other hand, WIPO's norm-setting preferences, much like those of the WTO, are still widely perceived to favor developed countries based on equal-country paternalism.[202] The 1996 diplomatic conference aimed at updating WIPO treaties to the digital age,[203] or the more recent Substantive Patent Law Treaty promoting patent harmonization,[204] serve only to underscore the complications apparent in WIPO's norm-setting inclination.

UN officials, national governments, nongovernmental organizations, and researchers in the field must henceforth be guided by more distinct, country-specific understandings regarding all aspects of patenting and innovation-related norms. These understandings will address such dimensions as pharmaceutical patents, plant genetics, or software protection. The methodology to be employed in classifying countries as patent prone or patent averse as a proxy for their relative domestic innovation demands granular, empirical, and conceptual scrutiny, and this is the central task this book seeks to address.

---

[199] WIPO, The Development Agenda, Cluster B: Norm-setting, flexibilities, public policy and public domain, Sec. 1 and Sec. 15.

[200] WIPO, The Development Agenda, Id, Cluster A: Technical Assistance and Capacity Building ("WIPO technical assistance shall be ... taking into account ... the different levels of development of Member States ... In this regard, design, delivery mechanisms and evaluation processes of technical assistance programs should be country specific"), at Sec. 1.

[201] Ibid.

[202] See Joseph E. Stiglitz and Andrew Charlton, *Fair Trade for All: How Trade can Promote Development* (Oxford University Press, 2005), at 82 (discussing the infamous "green room" methods of the Uruguay Round of WTO negotiations and continuing disabilities faced by developing countries within the WTO); Geoffrey Yu, The Structure and Process of Negotiations at the World Intellectual Property Organization, 82 *Chicago-Kent Law Review* 1445 (2007); Coenraad Visser, The Policy-Making Dynamics in Intergovernmental Organizations, 82 *Chicago-Kent Law Review* 1457 (2007), at 1459.

[203] See Pamela Samuelson, The US Digital Agenda at WIPO, 37 *Virginia Journal of International Law* 369 (1997) (criticizing WIPO's preparation for the December 1996 diplomatic conference aimed at updating WIPO treaties to the digital age based on the US digital copyright model), at 374.

[204] See Jerome H. Reichman and Rochelle Cooper Dreyfuss, Harmonization without Consensus: Critical Reflections on Drafting a Substantive Patent Law Treaty, 57 *Duke Law Journal* 85 (2007) (depicting the patent law harmonization efforts through WIPO's self-promoted Substantive Patent Law Treaty), at 92–103. The treaty was originally designed to restrain developing countries' flexibility to craft their domestic patent laws to narrow patentable subject matter, set a high bar to inventiveness for patentability, limit patent holders' exclusive rights, and impose compulsory licenses).

2   Convergence Clubs, Coalitions, and Innovation Gaps

Introduction

Endogenous growth economics continues to present a challenge, particularly at the country-group level.[1] Earlier accounts of the relationship between endogenous growth theory and club convergence between country groups centered primarily on the understanding of archetypal club convergence regarding salaries, GDP, and other macroeconomic income-related indications.[2] There is little understanding of the reasons for the convergence between country groups or clubs in terms of domestic technological creation, and not much more is known about how club convergence is achieved through the diffusion of technology through technological transfer to developing countries.[3] Lastly, there has been very little attention directed to a conceptual explanation of the manner in which technological creation is determined in the country-group clusters.[4]

---

[1] See Ron Martin and Peter Sunley, Slow Convergence? The New Endogenous Growth Theory and Regional Development, *Economic Geography*, vol. 74(3) 201 (1988) (commenting on the empirical level), at 220. See also discussion henceforth.

[2] See, e.g., Dan Ben-David, Convergence Clubs and Subsistence Economies, *Journal of Development Economics*, vol. 55(1) 155 (1988) (concluding that "income gaps have increased within most possible groupings of countries in the world. Where 'convergence clubs' tend to be more prevalent is at the two ends of the income spectrum"), at 167.

[3] See Ron Martin and Peter Sunley, supra note 1, at 210, referring to David M. Gould and Roy J. Ruffin, What Determines Economic Growth? *Federal Bank of Dallas Economic Review* 2:25, 40 (1993); Robert J. Barro and Xavier Sala-i-Martin, Convergence Across States and Regions, *Brookings Papers on Economic Activity* 2:107, 58 (1991). Such diffusion of technology requires accordingly that lagging emerging economies would have appropriate infrastructure or conditions to adopt or absorb technological innovations. See (for a supportive economic model) Stilianos Alexiadis, *Convergence Clubs and Spatial Externalities*, Advances in Spatial Science (Springer-Verlag, 2013), at 61 and Sec. 4.5. For two of the earliest and most influential statements of this view, see George H. Borts and Jerome L. Stein, *Economic Growth in a Free Market* (Columbia University Press, 1964) (offering a classic study of regional development in the United States); Jeffrey G. Williamson, Regional Inequalities and the Process of National Development, *Economic Development and Cultural Change Quarterly Journal of Economics*, 13(1) 84 (1965) (analyzing the evolution of regional income differences in advanced industrial countries).

[4] See Stilianos Alexiadis, ibid., at 61 and Sec. 4.5.

In practice, only a small number of official issue-based coalitions have emerged in the context of innovation and patenting-related policies.[5] Two structural institutional alternatives that lie beyond the scope of this chapter serve as exceptions. The first is the loose compilation of civil society groups and movements, including numerous governments and individuals, converging over broad egalitarian principles promoted by iconic movements, such as Access To Knowledge (A2K) or the broad-reaching Open Source movement. The second alternative to such issue-based coalitions is a plethora of overly generalized regional coalition blocs, such as the African Group or the EU. However, these all-purpose regional blocs actually mitigate against a more accurate delineation of countries converging over issue-based innovation and intellectual property-related policies, as the following empirical discussion will show.

Nonetheless, de facto heterogeneity between countries in other economic-related growth policies is commonly witnessed in a variety of WTO coalitions. In particular, country coalitions are increasingly becoming the preferred informal response of developing countries to imbalances in power in the WTO. In response to the few under-theorized innovation and intellectual property-related coalitions, this chapter offers a unique clustering analysis. It does so within the framework of endogenous growth theory, measuring optimal convergence by country coalitions into multiple innovation-based growth equilibria rather than through a single "one-size-fits-all" apparatus. The measurement is based on countries' patent propensity and GERD intensity rates, referred to as "patent intensity," by way of a proxy for their domestic innovation.

The measurement of patent intensity applied below derives from two policy and theoretical sources. The first is WIPO's discontinued yearly statistical endeavor in accounting for patent intensity as separately measured by resident patent filings to GDP, resident patent filings to

---

[5] See Peter K. Yu, Building Intellectual Property Coalitions for Development, in *Implementing the World Intellectual Property Organization's Development Agenda* 79, (Jeremy de Beer, ed., Wilfrid Laurier University Press, CIGI, IDRC, 2009), at 84; John S. Odell and Susan K. Sell, Reframing the Issue: The WTO Coalition on Intellectual Property and Public Health 85, in *Negotiating Trade* (John S. Odell, ed.) (Cambridge University Press, 2006) (on the coalition of developing countries for the 2001 Doha Declaration on the WTO Agreement on Trade-Related Aspects of Intellectual Property Rights (TRIPS) and Public Health), at 104; Peter Drahos, Developing Countries and International Intellectual Property Standards-Setting, 5 *Journal of World Intellectual Property* 765 (2002) (suggesting that India, Brazil, Nigeria, and China could form a "*Developing Country Quad*" – a leading working groups on key negotiations), at 780; Gunnar Sjostedt, Negotiating the Uruguay Round of the General Agreement on Tariffs and Trade, in *International Multilateral Negotiation: Approaches to the Management of Complexity* 44 (I. William Zartman, ed.) (Jossey-Bass Publishers, 1994) (on coalition strategies leading to the negotiation of the TRIPS Agreement during the Uruguay Round), at 44–54.

population, and ultimately by R&D expenditure.[6] It remains questionable whether WIPO's attempt to account for labeled patent intensity has become or is about to become a consolidated IPR policy level. The second source underlying this book's patent activity is theoretical and relates to Ed Mansfield's definition of the propensity to patent as the percentage of patentable inventions that are in fact patented.[7] In the early 1980s, as Scherer argued, little of a systematic nature was known about the propensity to patent.[8] Since then, a growing stream of research on patented inventions has analyzed the extent to which patents provide a reliable indicator of innovative activity. Seven recent studies have concluded that a positive correlation can be found between elasticity of patent propensity and the patent/GERD indicator.[9] However, all these studies were based on surveys performed solely on cross-sectional or panel data at the firm or industry levels within the developed world.[10]

[6] See, e.g., WIPO, World Patent Report: A Statistical Review – 2008 edition, at: www.wipo.int/ipstats/en/statistics/patents/wipo_pub_931.html; WIPO, WIPO Patent Report: Statistics on Worldwide Patent Activities (2007), at 18–23.

[7] Edwin Deering Mansfield, Patents and Innovation: An Empirical study, *Management Science* 32, 173 (1986). The definition per firm-level stands for the percentage of innovative firms in a sector that have applied for at least one patent over a defined time period. See Isabelle Kabla, The patent as indicator of innovation, *INSEE Studies Economic Statistics* 1, 56 (1996); Cf.: Georg Licht and Konrad Zoz, Patents and R&D: an econometric investigation using applications for German, European, and US patents by German companies, ZEW Discussion Paper 96–19, Zentrum fur Europaische Wirtschaftsforschung, Mannheim (1996).

[8] See Frederic M. Scherer, The Propensity to Patent, *International Journal of Industrial Organization* 1, 107 (1983) ("The quantity and quality of industry patenting may depend upon chance, how readily a technology lends itself to patent protection, and business decision-makers' varying perceptions of how much they will derive from patent rights"), at 107–8.

[9] Jérôme Danguy, Gaétan de Rassenfosse and Bruno van Pottelsberghe de la Potterie, The R&D-Patent Relationship: An Industry Perspective, ECARES working paper 2010–038 (September 2010); Emmanuel Duguet and Isabelle Kabla, Appropriation strategy and the motivations to use the patent system: an econometric analysis at the firm level, Ann. INSEE (2010); Wesley M. Cohen, Richard R. Nelson, John P. Walsh, Protecting Their Intellectual Assets: Appropriability Conditions and Why US Manufacturing Firms Patent (or Not), NBER Working Paper No. 7552 Issued in February 2000; Edwin Deering Mansfield, supra note 7; Christopher Thomas Taylor, Z. A. Silberston and Aubrey Silberston, The Economic Impact of the Patent System: a Study of the British Experience, Cambridge University Press(1973). To illustrate, the largest survey so far for the United States is by Cohen, Nelson and Walsh [1996] and offered preliminary patent propensity rates for innovations, weighted by R&D expenditures, for a survey of 1065 American research laboratories in manufacturing. They report that a patent application was made for 51.5% of product innovations and for 33.0% of process innovations between 1991 and 1993. See Wesley M. Cohen, Richard R. Nelson and John Walsh, Appropriability Conditions and Why Firms Patent and Why they do not in the American Manufacturing Sector, Paper presented to the Conference on New S and T Indicators for the Knowledge Based Economy, OECD, Paris, June 19–21 (1996).

[10] See Jérôme Danguy, Gaétan de Rassenfosse and Bruno van Pottelsberghe de la Potterie, supra note 9; Iiro Mäkinen, The Propensity to Patent: An Empirical Analysis at the

Within these narrow parameters, and largely avoiding country and country-group analysis, propensity to patent literature has focused on the following indications: differences between industries,[11] the type of institution and the type of research,[12] or the indices of "patent rights" based on the "strength" of patent systems.[13] To a limited degree, heterogeneity within propensity to patent theory led also to macroeconomics-related analysis. The latter analysis again focused almost entirely on differences across developed countries.[14]

Innovation Level, ETLA – The Research Institute of the Finnish Economy (2007) (file with author), at 2.

[11] See, e.g., Wesley M. Cohen, Richard R. Nelson and John Walsh, supra note 9; Bronwyn H. Hall and Rosemarie Ham Ziedonis, The Patent Paradox Revisited: An Empirical Study of Patenting in the US Semiconductor Industry, 1979–1995, *Rand Journal of Economics*, vol. 32(1) 101 (2001); James Bessen and Robert M. Hunt, An Empirical Look at Software Patents, *Journal of Economics & Management Strategy*, Wiley Blackwell, vol. 16(1) 157 (2007).

[12] See Carine Peeters and Bruno Van Pottelsberghe, *Economics and Management Perspectives on Innovation and Intellectual Property Rights* (Palgrave Macmillan, 2006); Michele Cincera, Firms' Productivity Growth and R&D Spillovers: An Analysis of Alternative Technological Proximity Measures, Economics of Innovation and New Technology, Taylor and Francis Journals, vol. 14(8), 657 (2005).

[13] See Walter G. Park, International Patent Protection: 1960–2005, *Research Policy*, 37, 761 (2008); Juan Carlos Ginarte and Walter Park, Determinants of Patent Rights: Cross-National Study, *Research Policy*, 26 (1997) (Ginarte and Park compute an index of patent strength, also known as the IPI index, or intellectual property index); Josh Lerner, Patent Protection and Innovation Over 150 Years, NBER Working Paper No. 8977 (2002) (examined the patent laws of a comprehensive number of countries from 1960 to 1990, considering five components of the laws, namely the duration of protection, extent of coverage, membership in international patent agreements, provisions for loss of protection, and enforcement measures). Theoretical work herein emerged with Ginarte and Park (1997), and the updated versions published by Walter G. Park (2008) for 110 countries. In 2002, Lerner expanded this approach for 60 countries, defining "strong" patent systems as those that are essentially applicant friendly.

[14] See, e.g., Jonathan Eaton and Samuel Kortum, Josh Lerner, International Patenting and the European Patent Office: A Quantitative Assessment, Patents, Innovation and Economic Performance: OECD Conference Proceedings (2004) (solely analyzing European patenting patterns by advanced economies); Laura Bottazzi and Giovanni Peri, The International Dynamics of R&D and Innovation in the Short and in the Long Run, NBER Working Paper No. 11524 (July 2005) at: www.nber.org/papers/w11524 .pdf?new_window=1 (estimating the relationship between employment in R&D and generation of knowledge as measured by patent applications across OECD countries); Jeffrey L. Furman, Michael E. Porter and Scott Stern, The Determinants of National Innovative Capacity, *Research Policy* 31: 899 (2002) (introducing a novel framework based on the concept *of national innovative capacity* as the ability of a country to produce and commercialize a flow of innovative technology over the long term. The survey uses a sample of 17 OECD countries from 1973 through 1996). Few narrow exceptions for the focus on developed countries in examining patent propensity-related indications exist. See, e.g., David Matthew Waguespack, Jóhanna Kristín Birnir, and Jeff Schroeder, Technological Development and Political Stability: Patenting in Latin America and the Caribbean, *Research Policy* 34: 1570 (2005) (accounting for political stability or lack thereof over the propensity to patent by Latin American countries), at 1572.

Convergence literature contributes an additional and seminal insight. Dubbed "club convergence,"[15] this insight argues, as its name suggests, that only countries that share similar structural characteristics and initial conditions will converge.

Thus, one potential innovation-led growth hypothesis could be that richer OECD countries may form one convergence club, the developing countries an additional club, and the underdeveloped a third. Alternatively, different club convergence groupings might suggest that countries and groups of countries converge over innovation and related patenting policies, or should do so.

This chapter first offers a positive theoretical framework based on endogenous growth theory and clustering analysis. It then presents an empirical yearly time series clustering analysis for 79 countries for the period 1996–2013. In connoting patent intensity, the analysis identifies groups of countries that show similar patent intensity as a proxy for their domestic innovation. The model delineates two large patent intensity-gaps and convergence patterns within the world economy. The first gap refers to the great distance that separates the middle group of "followers" country cluster from the stronger "leaders" in terms of patent intensity capabilities. The second gap refers to the similarly remarkable gap between the weaker "marginalized" club and the "followers." The analysis further negates any significant relationship between the clusters and numerous other policy-oriented indicators, with the exception of the relationship between the three clusters and the IMF's labeled economy category.

The last section (Section 2.3) follows with numerous theoretical ramifications. These ramifications relate to the need for additional corroborating research intended to explain remaining discrepancies regarding shifts and reversals in rates of country-group convergence. Thus far, little accounts for the slowness or nonexistence of inner club convergence, especially in advanced economies, but also in emerging and other developing countries.

2.1   Patent Club Convergence: The Positive Framework

2.1.1   Convergence over Innovation-Led Growth

Evidence increasingly shows that developing countries differ not only in their propensity to attract FDI, trade, and technology, but also in their

---

[15] See, e.g., Fabio Canova and Albert Marcet, The Poor Stay Poor: Non-Convergence Across Countries and Regions, Discussion Paper 1265, London: Centre for Economic Policy Research (1995); Oded Galor, Convergence? Inferences from Theoretical Models, *The Economic Journal* 106, 1056 (1996).

ability to innovate.[16] Moreover, growing evidence reveals differences between developing countries in their ability to make use of IPRs as a tool for fostering domestic innovation.[17] All these pieces of evidence are startling when placed against the backdrop of a traditional World Bank-led and rather inflexible North/South country-group dichotomy, or some variant thereof.[18] Such an innovation policy setting continually highlights the asymmetries between Northern countries, which are deemed to generate innovative products and technologies, and Southern countries, which are generally thought to consume them.[19]

[16] See particularly, UK Commission on Intellectual Property Rights, UK Intellectual Property Rights Report, Integrating Intellectual Property Rights and Development Policy (London September 2002) ("Thus developing countries are far from homogeneous, a fact which is self-evident but often forgotten. Not only do their scientific and technical capacities vary, but also their social and economic structures, and their inequalities of income and wealth"), at 2–3. See also: Daniel Benoliel and Bruno M. Salama, Toward an Intellectual Property Bargaining Theory: The Post-WTO Era, 32 *University of Pennsylvania Journal of International Law*, 265 (2010) (analyzing the heterogeneity among developing countries, codenamed the "Developing Inequality Principle"), at 275–90.

[17] See Jose Groizard, Technology Trade, *The Journal of Development Studies*, Taylor and Francis Journals, vol. 45(9) 1526 (2009) (using panel data of 80 countries for the period 1970–95, while finding that FDI is higher for countries with stronger IPRs. On the other hand, the author shows a negative relationship between IPR and human capital indicators that exist in tandem. Earlier findings are similarly ambiguous). While some works, for example Sunil Kanwar and Robert Evenson, Does Intellectual Property Protection Spur Technological Change?, 55 *Oxford Economic Papers* 235 (2003), generally find a positive effect, Yongmin Chen and Thitima Puttitanun, Intellectual Property Rights and Innovation in Developing Countries, 78 *Journal of Development Economics* 474 (2005) explain that lower IPR can facilitate imitation, while on the other hand, innovation in developing countries increases in proportion to greater IPR protection. Furthermore, see Rod Falvey, Neil Foster and David Greenaway, Intellectual Property Rights and Economic Growth, *Review of Development Economics*, 10(4): 700 (2006) (using panel data of 79 countries and four sub-periods: 1975–79, 1980–84, 1985–89, and 1990–94, the authors find evidence of a positive effect between IPR and economic growth for both low and high-income countries, but not for middle-income countries. According to the latter, the positive relationship between IPR and economic growth in low-income countries cannot be explained by the potential fostering of R&D and innovation, but by the idea that stronger IPR protection promotes imports and inner FDI from high-income countries without negatively affecting the national industry based on imitation). See also: Rod Falvey, Neil Foster and David Greenaway, Trade, Imitative Ability and Intellectual Property Rights, *Review World Economics*, 145(3), 373 (2009) (using panel data of 69 developed and developing countries over the period 1970–99, the author shows that the IPR-R&D relationship depends on the level of development, the imitative ability, and the market size of the importing country).

[18] See, e.g., Carlos M. Correa, *Intellectual Property Rights, the WTO and Developing Countries: The TRIPS Agreement and Policy Options* (Zed Books, 2000) (describing the asymmetrical distribution of technological innovation and consumption between Northern and Southern countries), at 5–6; Paul Krugman, A Model of Technology Transfer, and the World Distribution of Income, 87 *Journal of Political Economy* 253 (1979) (analyzing the TRIPS Agreement via the innovating North and non-innovating South), at 254–55. See also the discussion in Section 2.1.2.

[19] See ibid.

To be sure, some United Nation organs did not make a clear theoretical choice on the matter. WIPO has traditionally not adopted, or at least has not yet applied, a comprehensive innovation policy for developing countries, as depicted in the previous chapter in Section 1.1.2. This challenge was particularly evident before and during the establishment of the WTO and the TRIPS Agreement.[20] It was therefore only natural that, on its adoption, TRIPS merely consisted of a flat intellectual property policy for all WTO members, corresponding with an earlier World Bank-led Pax-Americana neoclassical economic growth model.[21] Accordingly, it should come as no surprise that, to this day, WIPO remains inconsistent on this matter, as witnessed in the organization's archetypical Development Agenda, adopted in October 2007 after years of deliberations.[22]

Over-simplifying drastically, a convenient way to distinguish the two views on innovation-led growth of developing countries is to ask, "Are poor economies catching up with those already innovatively advanced (and thus richer)? Or, instead, are they caught in some innovation-related poverty trap?"[23]

Discussion of these two capitalist spatial economy questions has traditionally been dominated by two opposing views regarding the expected long-term trajectories of regional development.[24] The first of these views,

[20] For official surveys, see World Intellectual Property Organization (WIPO, 1985), and the United Nations Department of Economic and Social Affairs (UNCTAD, 1974). For theoretical and empirical studies, see Helge E. Grundmann, Foreign Patent Monopolies in Developing Countries: An Empirical Analysis, 12 *Journal of Development Studies* 186 (1976); Jorge M. Katz, Patents, the Paris Convention and Less Developed Countries, Discussion Paper no. 190, at 24–27 (Yale Univ. Economic Growth Center, Nov. 1973); Douglas F. Greer, The Case against Patent Systems in Less-Developed Countries, 8 *Journal of International Law and Ecoomics* 223 (1973); Constantine Vaitsos, Patent Revisited: Their Function in Developing Countries, 9 *Journal of Development Studies* 71, 89–90 (1972).

[21] See Daniel Benoliel and Bruno M. Salama, supra note 16, at 278; Michael Blakeney, The International Protection of Industrial Property: From the Paris Convention to the TRIPS Agreement, WIPO National Seminar on Intellectual Property, 2003, WIPO/IP/CAI/1/ 03/2, at 16.

[22] See, e.g., WIPO, The 45 Adopted Recommendations under the WIPO Development Agenda, Cluster F: Other Issues (Recommendation 45: "To approach intellectual property enforcement in the context of broader societal interests and especially development-oriented concerns, with a view that "the protection and enforcement of intellectual property rights should contribute to the promotion of technological innovation and to the transfer and dissemination of technology") at: www.wipo.int/ip-development/en/ag enda/recommendations.html. Thus, both "neoclassical economics" "technological transfer" and competing endogenous contextual "societal interests" preside in tandem, implying much theoretical inconsistency toward the matter regarding innovation-led growth. See the discussion below.

[23] Cf.: Danny T. Quah, Empirics for Economic Growth and Convergence, LSE Economics Department and CEP – Center for Economic Performance, Discussion Paper No. 253 (July 1995) (Posing a similar question concerning economic growth broadly), at 1.

[24] Earlier accounts of convergence among countries or country groups have mostly been attributed to the understanding of convergence over salaries, GDP and other macroeconomic income-related indications. Little accounts for convergence over domestic

rooted in neoclassical equilibrium economics, argues that poor economies incipiently catch up with richer ones, provided there are no central barriers to the function of market processes. In an integrated national economy there are strong pressures leading to the general convergence of regional income-related indicators over time. Regional or other country-group discrepancies can only be a short-term state, since such disparities will instigate self-correcting movements in prices, wages, capital, and labor, thereby restoring the trend to regional convergence.

The convergence hypothesis that poor economies may "catch up" has generated a huge body of empirical literature, but one that to date has barely addressed innovation or intellectual property-related economic growth in developing countries at a more detailed country-group level.[25] Instead, the most popular examples covered in the literature relate to convergence in income between rich and poor parts of the EU; in plant and firm size in industries; in economic activity across different regions (states, provinces, districts, or cities) within the same country; in asset returns and inflation rates across countries in a common trade area; in political attitudes across different groups; and in wages across industries, professions, and geographical regions.[26]

technological creation, as considered herein. See Dan Ben-David, supra note 2, at 167; Ron Martin and Peter Sunley, supra note 1, at 201–2.

[25] See Jérôme Vandenbussche, Philippe Aghion, and Costas Meghir, Growth, Distance to Frontier and Composition of Human Capital, *Journal of Economic Growth* 11(2) 97 (2006) (Using a panel of 19 OECD countries between 1960 and 2000 while using an endogenous growth model, the authors show how as a country increasingly experiences economic growth, it relies more and more on innovation), at 21–30. For additional general discussion of the argument, see Emmanuel Hassan, Ohid Yaqub and Stephanie Diepeveen, Intellectual Property and Developing Countries: A Review of the Literature, Rand Europe (2010), at 17. See also the discussion in Section 2.3.

[26] See Dan Ben-David, supra note 2, at 167; Stilianos Alexiadis, supra note 3, at 61 and Sec. 4.5 (for a supportive economic model); Danny T. Quah, supra note 23, 1–2, referring to Zvi Eckstein and Jonathan Eaton, Cities and Growth: Theory and Evidence from France and Japan, Working paper, Economics Department, Tel-Aviv University, September (1994); Joan-María Esteban and Debraj Ray, On the Measurement of Polarization, *Econometrica*, 62(4):819 (1994); Joep Konings, Gross Job Flows and Wage Determination in the UK: Evidence from Firm-Level Data. PhD thesis, LSE, London (1994); Reinout Koopmans and Ana R. Lamo, Cross-Sectional Firm Dynamics: Theory and Empirical Results from the Chemical Sector, Working paper, Economics Department, LSE, London, April (1994); Danny T. Quah, Convergence across Europe, Working paper, Economics Department, LSE, London, June (1994); Danny T. Quah, One Business Cycle and One Trend from (Many,) Many Disaggregates, *European Economic Review*, 38(3/4):605 (1994); Ron Martin and Peter Sunley, supra note 1, at 210, referring to David M. Gould and Roy J. Ruffin, supra note 3; Robert J. Barro and Xavier Sala-i-Martin, supra note 3.

For two of the earliest and most influential statements of this view, see George H. Borts and Jerome L. Stein, supra note 3 (offering a classic study of regional development in the United States), and Jeffrey G. Williamson, supra note 3 (analyzing the evolution of regional income differences in advanced industrial countries).

The convergence hypothesis on per capita income convergence has also exposed a profound and possibly inspiring empirical finding for innovation-led growth analysis. Across geographically disaggregating poor and rich economies, such as within German reunification or the effects of regional redistribution within individual countries and across the EU, all appear to be converging toward each other at a steady and consistent rate of 2 percent per annum.[27]

In the meantime, convergence literature yielded another understanding. As its name implies, club convergence theory[28] hypothesizes that only countries that are similar in their structural characteristics, and share sufficiently similar initial conditions, will converge with one another. Thus, one potential innovation-led growth hypothesis could be that richer OECD countries may form one convergence club, developing countries will form another, and the underdeveloped yet another. Alternatively, different club convergence groupings may show how countries and groups thereof converge over innovation-led growth (or should do so).

To illustrate one seminal income-related finding of club convergence, numerous economists now reject the idea of any substantive convergence between these three above-mentioned clubs.[29] They further predict that broad inequalities between the different country groupings or clubs may persist or even increase in years to come, so that cross-country income distribution will remain polarized.[30]

Club convergence theory may therefore be attributed to the second approach, known as "regional divergence." In such cases, poor countries can be expected to remain caught in a model poverty trap. In other words, there is no reason why regional growth based on either innovation or other growth-related indicators will necessarily lead to convergence, even in the long term. On the contrary, regional divergence may be said to be the most likely outcome. As a case in point, models of regional growth advanced by writers such as Perroux,[31] followed by Myrdal[32] and Kaldor,[33] indeed predict that regional incomes will tend to diverge.

---

[27] See Xavier Sala-i-Martin, The Classical Approach to Convergence Analysis, *The Economic Journal* 106 (July 1996) 1019, at 1028; Martin Larch, Regional Cross-Section Growth Dynamics in the European Community, Working paper, European Institute, LSE, London, June (1994); Danny T. Quah, supra note 23.

[28] See, e.g., Fabio Canova and Albert Marcet, supra note 15; Oded Galor, supra note 15.

[29] Ibid.   [30] Ibid.

[31] See François Perroux, Economic Space: Theory and Applications, *Quarterly Journal of Economics*, 64, 89–104; François Perroux, Note sur la Notion des 'Poles du Croissance', Economie Appliquee, 1 and 2, 307–20.

[32] Gunnar Myrdal, *Economic Theory and Under-Developed Regions* (Taylor & Francis, 1957).

[33] Nicholas Kaldor, The Case for Regional Policies, *Scottish Journal of Political Economy*, November (1070) 337–48; Nicholas Kaldor, The Role of Increasing Returns, Technical Progress and Cumulative Causation in the Theory of International Trade and Economic

If left to their own devices, we are told, market forces will become spatially disequilibrating, and economies of scale and agglomeration will then lead to the collective concentration of capital, labor, and output in certain regions at the expense of others. Uneven regional development has thus been found to be self-correcting, but only within convergence clubs and not between them.

A further seminal theoretical reformulation is known as "conditional convergence."[34] Because convergence is conditional on the different structural characteristics of each economy – for instance, its preferences, technologies, rate of population growth or government policy – different structural characteristics imply that different countries will have varying steady-state relative incomes or innovative capacity. Hence, the prediction is that the growth of an economy will be a function of the fracture that divides it from its own stable state.[35] To test for conditional convergence, therefore, it is necessary to hold the state of each economy as a constant as well.

2.1.2   Coalitions and Convergence Clubs

Heterogeneity among countries over their economic growth is nowadays manifested in a plethora of coalitions.[36] In the absence of significant information asymmetries and substantive transaction costs in international forums, convergence clubs and coalitions thereof could efficiently correlate. In reality, country coalitions are rapidly becoming the de facto preferred response of developing countries to imbalances in power at the WTO.[37] Such coalitions consequently impact trade governance and

---

Growth, Economie Appliquee, 34 Reprinted in *The Essential Kaldor* (F. Targetti and A. Thirlwall, eds.) (Holmes and Meier, 1981), 327.

[34] More generally, see Ron Martin and Peter Sunley, supra note 1, at 207–8; Xavier Sala-i-Martin, supra note 27, at 1026–27; Robert J. Barro and Xavier Sala-i-Martin, Convergence, *Journal of Political Economy*, 100, 223 (1992); N. Gregory Mankiw, David Romer, David N. Weil, A Contribution to the Empirics of Economic Growth, *Quarterly Journal of Economics*, 107, 407 (1992).

[35] More generally, see Ron Martin and Peter Sunley, supra note 1, at 207–8.

[36] See Sonia E. Rolland, Developing Country Coalitions at the WTO: In Search of Legal Support, *Harvard International Law Journal*, vol. 48(2) 483 (2007), at 483.

[37] On the background for developing countries-led coalitions at the WTO and GATT, see the seminal work of Amrita Narlikar, *International Trade and Developing Countries: Bargaining Coalitions in the GATT and WTO* (Routledge, RIPE Studies in Global Political Economy, 2003) (offering an historical typology of developing country coalitions in the GATT and WTO). See also Vicente Paolo B. Yu III, Unity in Diversity: Governance Adaptation in Multilateral Trade Institutions Through South-South Coalition-building, Research papers 17 (South Centre, July 2008), at 28, 33–34; Sonia E. Rolland, supra note 36 (emphasizing that developing country-led coalitions are beginning to change the WTO's dynamics), at 483; *Negotiating Trade: Developing*

WTO-related institutional reforms.[38] To date, of the 112 members that define themselves as "developing countries," a remarkable group of 99 countries (87.61 percent of this category) are members of one or more group or coalition comprised solely of developing countries.[39]

For developing countries with small markets and limited diplomatic resources, coalitions repeatedly prove to be the only means at their disposal for advancing their bargaining positions.[40] The joint defense of a negotiating position is likely to improve the legitimacy of a proposal in consensus-based and majoritarian institutions. This explains why even developed countries with large markets search for allies in the WTO.

Countries in the WTO and NAFTA (John Odell, ed.) (Cambridge University Press, 2006); Jerome Prieur and Omar R. Serrano, Coalitions of Developing Countries in the WTO: Why Regionalism Matters? Paper Presented at the WTO Seminar at the Department of Political Science at the Graduate Institute of International Studies in Geneva, May 2006; Constantine Michalopoulos, The Participation of the Developing Countries in the WTO, *Policy Research Working Paper, World Bank*, Washington, DC (1999), at 17 (same).

For literature concerning intellectual property-related coalitions, see also Peter K. Yu, supra note 5, at 84; John S. Odell and Susan K. Sell, supra note 5, at 104; Peter Drahos, supra note 5, at 780; Gunnar Sjostedt, supra note 5.

During the early years of the WTO, there were initial attempts at bringing together an overarching group of developing countries to counter developed countries, and particularly the loose coalition of developing countries with G-77 during the first UNCTAD meeting, in 1964. These attempts were later abandoned as it became clear that differing interests and institutional capacities posed ever greater challenges to such a grouping. See e.g. Constantine Michalopoulos, ibid.; Matthew David and Debora Halbert, IP and Development 89, in *Sage Handbook on Intellectual Property* (Matthew David and Debora Halbert, eds.) (SAGE Publications Ltd., 2014), at 89.

In the pre-WTO era, developing country-led coalitions in the GATT received only limited academic attention and were largely considered ineffective. See Amrita Narlikar, Bargaining over the Doha Development Agenda: Coalitions in the World Trade Organization, Serie LATN Papers, Nª 34 (2005) (hereinafter, Amrita Narlikar, Bargaining over the Doha) ("Developing countries, even while operating in coalitions, had stood on the sidelines in the GATT, choosing to free-ride on the concessions that were exchanged"), at 2. Narlikar adds that this neglect lay partly in the fact that coalitions in the GATT were informal and harder to trace. Ibid., at 2, referring generally also to Diana Tussie, The Less Developed Countries and the World Trading System: A Challenge to the GATT, London: Francis Pinter, 1987; John Whalley, ed., *Developing Countries and the World Trading System*, Vols. 1 and 2, (Macmillan, 1989); Diana Tussie and David Glover, eds., *Developing Countries in World Trade: Policies and Bargaining Strategies* (Lynne Rienner, 1995).

For earlier discussion, see Daniel Benoliel, Patent Convergence Club Among Nations, *Marquette Intellectual Property Law Review* 297 vol. 18(2) 297 (2014), 303–8.

[38] See Mayur Patel, New Faces in the Green Room: Developing Country Coalitions and Decision-Making in the WTO, GEG Working Paper Series, 2007/WP33; Faizel Ismail, Reforming the World Trade Organization: Developing Countries in the Doha Round, Geneva: CUTS International and Friedrich Ebert Stifung (FES) (2009); Debra P. Steger, The Future of the WTO: The Case of Institutional Reforms, *Journal of International Economic Law* 12(4) 803 (2009).

[39] See Vicente Paolo B. Yu III, supra note 37, at 28.

[40] Amrita Narlikar, Bargaining over the Doha, supra note 37, at 3.

Novel types of coalitions led by developing countries first began to appear during the run-up to the 1999 Seattle Ministerial Conference and in its aftermath. These coalitions ranged from bloc-type groups such as the Like-Minded Group of the late 1990s[41] to issue-based groups such as the G-20 of the post-Cancún period. In other cases, coalitions appeared as region-based groups,[42] such as the African Group, or as groups that shared certain development characteristics, such as the LDCs.[43] Moreover, region-based groups, as well as LDCs, remain central for coalition-based action by many developing countries.

In the meantime, informal issue-based groups or coalitions, such as the G20, the G33 and the NAMA-11,[44] are also becoming a key means for group-based action by developing countries. Of the coalitions in place as of late 2009, some 67 developing countries (or 58.77 percent of developing WTO members) have joined one or more informal issue-based developing-country coalition, while 61 developing countries are members of a regional group.[45]

These coalition-building efforts certainly play a role against the backdrop of much US-led opposition. Since the failure of the fifth WTO Ministerial Conference in Cancún (Cancún Ministerial) in 2003, most noticeably, the United States has largely engaged in a divide-and-conquer strategy intended to marginalize coalition building by developing countries. Accordingly, the United States has rewarded countries that were

---

[41] For a discussion of the genesis, negotiating strategy, and results of the Like-Minded Group, see Amrita Narlikar and John Odell, The Strict Distributive Strategy for a Bargaining Coalition: The Like Minded Group in the World Trade Organization, 1998–2001, in *Negotiating Trade: Developing Countries in the WTO and NAFTA* (John Odell, ed.) (Cambridge University Press, 2006).

[42] Jerome Prieur and Omar R. Serrano, supra note 37, at 5–7; Sisule F. Musungu, Susan Villanueva, Roxana Blasetti, Utilizing TRIPS Flexibilities for Public Health Protection Through South-South Regional Frameworks (South Center, 2004) ("[a] regional approach to the use of TRIPS flexibilities will enable similarly situated countries to address their constraints jointly"), at xiv. Musungu et al. offer two models of regional cooperation over IP-related policies, namely (a) coordination, yet non-harmonization, has most commonly been adopted among the RECs in Latin America and the Caribbean region, and (b) harmonization without coordination as is mostly witnessed in Africa in the form of OAPI and ARIPO. Ibid., at 50–55. For probable support of a regional-type coalition within the context of Intellectual Property, see Peter K. Yu, supra note 5 (Regional or pro-development forums are particularly effective means for coordinating efforts by less developed countries in the areas of public health, IP, and international trade), at 90.

[43] Jerome Prieur and Omar R. Serrano, supra note 37, at 34.

[44] A group of 11 developing countries working toward strengthening NAMA. See Faizel Ismail, The G-20 and NAMA 11: The Role of Developing Countries in the WTO Doha Round, 1 *Indian Journal of International Economic Law* 80 (2008), at 11–14.

[45] This figure includes 35 that are also members of one or more issue-based groups, and 37 that are members of one or more common characteristic groups. See Vicente Paolo B. Yu III, supra note 37, at 28.

willing to work with it, while undermining efforts by Brazil, India, and other G20 members to establish a united negotiating front for less developed countries.[46]

The G20 is the most important example of a coalition of developing countries developed during the pre-negotiation phase in GATT.[47] The G20 is composed solely of developing countries (later referred to as the "G20+").[48] This coalition of developing countries appeared just before the WTO Cancún summit, attempting to block the joint US/EC proposals.[49] In so doing, it favored negotiating with developed countries over the issue of the inclusion of services in the agenda of the Uruguay Round. This group eventually merged with the G-9,[50] a group of nine developed countries, to form the "Café au Lait" group, from which negotiating proposals eventually emerged that provided the basis for the Punta del Este declaration and the commencement of the Uruguay Round.[51]

The example of the G20 is telling for an additional reason. It illustrates how, even in the midst of changes in the precise list of countries converging, club convergence remains intact based on its core members. To illustrate, as membership in the G20 coalition has changed at various points, it has included a core group of countries known as the G3+3.[52]

---

[46] See Peter K. Yu, supra note 5, at 84; Peter K. Yu, The Middle Intellectual Property Powers, Drake University Legal Studies, Research Paper Series, Research Paper No. 12–28, at 18, referring to Former World Bank President and US Trade Representative Robert B. Zoellick's wordings on the fifth WTO Ministerial Conference in Cancún (Cancún Ministerial) in 2003 at Robert B. Zoellick, America will not wait (September 21, 2003) ("As WTO members ponder the future, the US will not wait: we will move toward free trade with can-do countries."), at www.fordschool.umich.edu/rsie/acit/Top icsDocuments/Zoellick030921.pdf. In tandem, domestic US private entities threaten or use unilateral sanctions, as part of their coalition with Europe, and Japan at the industry level. See, e.g., Ruth L. Okediji, Public Welfare and the Role of the WTO: Reconsidering the TRIPS Agreement, 17 *Emory International Law Review* 819 (2003), at 844–46 (referring to the case of the pharmaceutical industry); Susan Sell, *Private Power, Public Law: The Globalization of Intellectual Property Rights* (Cambridge University Press, 2003).

[47] Composed of Bangladesh, Chile, Colombia, Ivory Coast, Hong Kong (China), Indonesia, Jamaica, Korea, Malaysia, Mexico, Pakistan, Philippines, Romania, Singapore, Sri Lanka, Thailand, Turkey, Uruguay, Zambia, and Zaire (now DR Congo).

[48] See Jerome Prieur and Omar R. Serrano, supra note 24 (Argentina, Bolivia, Brazil, Chile, China, Colombia, Costa Rica, Cuba, Ecuador, El Salvador, Guatemala, India, Mexico, Pakistan, Paraguay, Peru, Philippines, South Africa, Thailand, and Venezuela. With the addition of Egypt and Kenya, the group acquired the name of the G22), at 8.

[49] Ibid., at 8.

[50] Composed of Australia, Austria, Canada, Finland, Iceland, New Zealand, Norway, Sweden, and Switzerland.

[51] See Vicente Paolo B. Yu III, supra note 37, at 26 referring to World Bank, The Trading system and Developing Countries, pp. 489–90, at http://siteresources.worldbank.org/I NTTRADERESEARCH/Resources/Part_7.pdf.

[52] See Jerome Prieur and Omar R. Serrano, supra note 37, at 8.

This group includes the three largest members – Brazil, China, and India – alongside three important medium-sized powers – Chile, South Africa, and Argentina.[53]

Not all coalitions prevail. During the pre-negotiation phase of the Uruguay Round from 1982 to 1986, for example, a coalition of developing countries called the G10 was formed.[54] This coalition, led by Brazil and India, opposed the launch of a new trade round, and was even more vocal in its opposition to the inclusion of services in any trade negotiations within the GATT.[55] The G10 was equally opposed to the inclusion of TRIPS or the Agreement on Trade-Related Investment Measures (TRIMS). It refused to make a compromise on any of these issues until its demands of standstill and rollback of non-tariff barriers were met. However, the group's successes were limited.[56]

Other under-theorized issue-based coalitions of WTO members regularly emerge in the context of innovation-led growth and intellectual property policies.[57] These include the "Joint proposal (in intellectual property)" coalition,[58] which sponsored a proposal calling for the establishment of a Geographic Indications (GI) database and register,[59] and its

---

[53] Ibid.
[54] See Vicente Paolo B. Yu III, supra note 37 (adding that these included Argentina, Brazil, Cuba, Egypt, India, Nicaragua, Nigeria, Peru, Tanzania, and Yugoslavia), at 26.
[55] See Vicente Paolo B. Yu III, supra note 37, at 26, referring to Sylvia Ostry, The Uruguay Round North–South Bargain: Implications for Future Negotiations in *The Political Economy of International Trade Law: Essays in Honor of Robert E. Hudec* (Daniel L. M. Kennedy and James D. Southwick, eds.) (Cambridge University Press, 2002) 285, at 289; World Bank, The Trading system and Developing Countries, available at http://si teresources.worldbank.org/INTTRADERESEARCH/Resources/Part_7.pdf, at 489.
[56] Amrita Narlikar, Bargaining over the Doha, supra note 37, at 6–7. Adding that amidst this grand-standing, the coalition also refused to engage with any other coalitions and turned down overtures from other developing countries to engage in shared research initiatives or draft joint proposals. Ibid.
[57] There still remains the alternative analytical framework of a plethora of present-day loose gatherings of civil society groups and movements including governments and individuals converging on egalitarian principles of justice, freedom, and economic development. These notably include the Access To Knowledge movement, the Open Source movement, etc. See Jack Balkin, What is Access to Knowledge? (April 21, 2006), Balkanization, at: http://balkin.blogspot.co.il/2006/04/what-is-access-to-knowledge.ht ml (on A2K); Gaëlle Krikorian and Amy Kapczynski (eds.), *Access to Knowledge in the Age of Intellectual Property* (Zone Books, 2010) (same); R.E. Wyllys, Overview of the Open-Source Movement., The University of Texas at Austin Graduate School of Library & Information Science (2000) (on the open source movement).
[58] A group of 20 WTO members including Argentina, Australia, Canada, Chile, Costa Rica, Dominican Rep., Ecuador, El Salvador, Guatemala, Honduras, Israel, Japan, Korea, Mexico, New Zealand, Nicaragua, Paraguay, Chinese Taipei. See Groups in the WTO (updated March 2, 2013), at www.wto.org/english/tratop_e/dda_e/negotia ting_groups_e.pdf, at 6.
[59] Ibid.

neighbor the W52 coalition,[60] sponsoring a proposal for "modalities" in negotiations on geographical indications. Other than these, and a handful of similar examples, countries rarely join efforts over innovation-led growth and TRIPS-related concerns as part of their overly generalized regional groupings. Such are the 42 African Group members,[61] the 31 Asian developing members,[62] the world's 50 poorest countries, referred to as the "least-developed countries,"[63] and, of course, the 28 EU members.[64]

Lastly, coalitions are often multipurpose. In such cases, well-constructed coalitions should optimize conflicting interests or otherwise efficiently allocate political dividends among its member states. To date, however, the archetypal "one-size-fits-all" narrative continuously overshadows the prospect of proficiently tailored coalitioning. Three present-day coalitions illustrate the crudeness of innovation and patent-related interest alignment by member states.

The first case of such a contentious coalition emerged around the 2011 multinational treaty establishing international standards for IPRs enforcement, named the Anti-Counterfeiting Trade Agreement (ACTA).[65] ACTA was negotiated by Australia, Canada, the EU, Japan, Morocco, Mexico, New Zealand, Singapore, South Korea, Switzerland, and the United States. Of these 11 negotiating parties, all but Switzerland and five members of the EU (Cyprus, Estonia, Germany, the Netherlands, and Slovakia) have since signed the Agreement.[66] No documented consideration has been given to the fact that, in the context of innovation and patent-related policy, the EU is profoundly uneven over innovation and patent activity outputs, as this chapter shows. Mexico and Morocco, two of ACTA's additional signatories, are likewise inconsistently correlated with member states belonging to more innovation- and patent-intense country groups, as will be discussed below.

---

[60] A group of 109 WTO members. The list includes as groups: the EU, ACP, and African Group. See Groups in the WTO, supra note 45, at 5.

[61] See African group, www.wto.org/english/tratop_e/trips_e/trips_groups_e.htm.

[62] See Asian developing members, www.wto.org/english/tratop_e/trips_e/trips_groups_e.htm.

[63] See LDC members, www.wto.org/english/tratop_e/trips_e/trips_groups_e.htm.

[64] See EU members: www.wto.org/english/tratop_e/trips_e/trips_groups_e.htm.

[65] Cf.: Peter Yu, The ACTA/TPP Country Clubs 258, In Dana Beldiman (ed.), *Access to Information and Knowledge: 21st Century Challenges in Intellectual property and Knowledge Governance* (Edward Elgar, 2014); Daniel Gervais, Country Club, Empiricism, Blogs and Innovation: The Future of International Intellectual Property Norm Making in the Wake of ACTA 323, in *Trade Governance in the Digital Age* (Mira Burri and Thomas Cottier, eds.) (Cambridge University Press, 2012) (giving ACTA as an example to supposedly like-minded coalitions as opposed to geographical or regional ones), at 323–25.

[66] Ministry of Foreign Affairs of Japan, Signing Ceremony of the EU for the Anti-Counterfeiting Trade Agreement (ACTA) (Outline) (January 26, 2012).

A similarly controversial coalition is the Trans-Pacific Strategic Economic Partnership Agreement (TPP), led by the predating "P4" or "Pacific 4."[67] Chile, New Zealand, Singapore, and Brunei signed the TPP Agreement in 2005. Shortly before the eighth round of the ACTA negotiations in Wellington, New Zealand, in March 2010, Australia, Peru, Vietnam, the United States, and the existing TPP members began negotiations over an expanded agreement. Malaysia, Mexico, Canada, and Japan later joined the negotiations.[68] In what is yet another example of an inconsistent innovation and patent-related coalition artificially abridging the development divide, TPP member states flatly agreed to reinforce and develop existing TRIPS rights and obligations. In other words, they declined to ensure what the 2001 Office of the US Trade Representative report emphatically considers an "effective and balanced approach to intellectual property rights among the TPP countries."[69]

A third example is the proposed Transatlantic Trade and Investment Partnership (TTIP) between the European Union and the United States, which aims to encourage trade and economic growth. The US government considers the TTIP a neighboring agreement to the TPP.[70] Similar to the TPP and other bilateral and regional trade agreements, this negotiated agreement includes provisions concerning the protection and enforcement of IPRs.[71] Analogous to the TPP or ACTA, nonetheless, no recorded adherence has yet been given to the profound discrepancies within EU member states over innovation and patent-related policies.

On the whole, past and present innovation and patent-related coalitions systematically group themselves despite their probable

---

[67] Peter Yu, Déjà Vu in the International Intellectual Property Regime 113, in *Sage Handbook on Intellectual Property* (Matthew David and Debora Halbert, eds.) (2014), at 119; Meredith K. Lewis, Expanding the P-4 Trade Agreement into a Broader Trans-Pacific Partnership: Implications, Risks and Opportunities, *Asian Journal of the WTO and International Health Law and Policy*, vol. 4(2) 401 (2009), at 403–04; Deborah K. Elms, The Trans-Pacific Partnership Trade Negotiations: Some Outstanding Issues for the Final Stretch, *Asian Journal of the WTO and International Health Law and Policy*, vol. 8(2) 379 (2013), at 386–87.

[68] Peter Yu, ibid., at 119.

[69] See Office of the US Trade Representative report as of November 2011. The TPP was classified from the public with little ongoing public scrutiny until it was leaked by Wikileaks in August 2013. See James Love, KEI Analysis of Wikileaks Leak of TPP IPR Text, from August 30, 2013, available at: www.keionline.org/node/1825 (2013).

[70] Daniel R. Russel, Transatlantic Interests In Asia, United States Department of State (January 13, 2014), at: https://2009-2017.state.gov/p/eap/rls/rm/2014/01/219881.htm.

[71] See Joe Kaeser, Why a US-European Trade Deal Is a Win-Win, *The Wall Street Journal* (February 2, 2014) (for the claim put by the CEO of Siemens AG whereby the TTIP would strengthen United States and EU global competitiveness by improving intellectual property protections among other things).

contradicting interests. In theory, bargaining coalitions align along a spectrum, with bloc-type coalitions at one end and issue-based ones at the other.[72] The two differ from each other in two significant ways, which also explain why innovation- and patent-related policy coalitions could, at least in theory, be more effectual. First, bloc-type coalitions bind member countries through a set of ideas and an identity that go beyond the immediately instrumental. As their name implies, issue-based coalitions are bound together by a more focused and instrumental aim, rather than an overly generalized developmental aim.[73]

A second consideration can also be offered in support of a contextual transition to issue-based coalitions over innovation- and patent-related policies. Blocs usually bring together like-minded countries such as EUs or LDCs, which adopt joint positions across issues and over time. Issue-based coalitions, on the other hand, often dissipate after the specific goal is achieved. Bloc-type coalitions successfully address the problem of minimal external weight, but they also run the risk of fragmentation, as they lack internal coherence.[74] By contrast, issue-specific coalitions enjoy internal coherence, but are difficult to maintain when large diversified economies with multiple sectoral interest groups are involved.[75] Again, such coalitions, if focused on innovation-led growth and intellectual property policy, could at least theoretically have better chances of enduring. In short, developing country-led coalitions around innovation or patent policies are especially prone to integration. These countries might therefore do well to consider when issue-based coalitions are preferable to bloc-based ones.[76] Equally, such coalitions should not be excessively specific, in order to avoid the risk of disintegration by competing interests.[77]

2.1.3   Growth Theory and Convergence over Innovation-Led Growth

One more preliminary concern remains – namely, what is the proper theoretical setting for issue-based coalitions? In the broader context of growth theory, endogenous growth theory and the new growth empirics naturally prevail. This is so much the case that growth theory prefers multiple economic growth equilibria by numerous country groups or

---

[72] Amrita Narlikar, Bargaining over the Doha, supra note 37, at 5.   [73] Ibid.   [74] Ibid.
[75] Ibid.
[76] See Colleen Hamilton and John Whalley, Coalitions in the Uruguay Round, Weltwirtschaftliches Archiv, 125 (3), 1989, at 547–56; Narlikar, Bargaining over the Doha, note 37 above, at 5.
[77] Hamilton and Whalley, ibid.

clusters[78] over a single international equilibrium in a neoclassical economic growth setting.[79]

Recent discussion, as explained thus far, has focused mostly on long-term convergence in per capita income and output indicators between countries. Again, focusing on innovation-led economic growth, the following clustering analysis is based on yearly data series between 1996 and 2013 for 79 countries. It evaluates the linkage between national innovation as measured through the rate of issued patents by GERD with the United States Patent and Trademark Office (USPTO) and European Patent Office (EPO) patents listed by inventor country search categories as a proxy for state-of-the-art-technology. Moreover, it accounts for a formulation of the sum and rate of supply of R&D, as measured by countries' GERD intensity. The rate between both, namely the propensity to patent and GERD intensity, jointly prefigures the clusters therein. The statistical analysis has only recently been made possible with the publication of highly detailed R&D-related datasets by the UNESCO Institute for Statistics in 2011 covering all countries in full or in part.

The following clustering analysis contributes to the critique of the WTO's TRIPS Agreement as well as the WIPO's systematic evasion of intellectual property-related policy delineations between distinct country groups and clusters thereof.[80] The TRIPS Agreement notably consists merely of an almost flat intellectual property and related innovation policy for all WTO members.[81] Against that backdrop, TRIPS cast international intellectual property protection as a central pillar for both short- and long-term economic growth, effectively ignoring country-group differences over innovation-led growth and related patenting policies.

This argument stood for two long-term neoclassic exogenous economic incentives offered by developed nations.[82] The first incentive promised to

---

[78] The assumption of endogenous growth holds that diminishing returns to capital, implicit in the neoclassical production function (measuring income-based indicators among countries), lead to the prediction that the rate of return to capital (and therefore its growth rate) is very large when the stock of capital is small, and vice versa. See Xavier Sala-i-Martin, supra note 27, at 1025; Danny T. Quah, supra note 23, at 1.

[79] This mainly empirical debate has promoted the development of endogenous growth theory, which seeks to move beyond conventional neoclassical theory by treating as endogenous those factors – particularly technological change and human capital – demoted as exogenous by neoclassical growth models.

[80] For official surveys, see: World Intellectual Property Organization (WIPO, 1985), supra note 20, and the United Nations Department of Economic and Social Affairs (UNCTAD, 1974), supra note 20. For theoretical and empirical studies, See Helge E. Grundmann, supra note 20; Jorge M. Katz, supra note 20, at 24–27; Douglas F. Greer, supra note 20; Constantine Vaitsos, supra note 20, at 89–90.

[81] See Daniel Benoliel and Bruno M. Salama, supra note 16, at 278; Michael Blakeney, supra note 21, at 16.

[82] See, e.g., Carolyn Deere, *The Implementation Game: The TRIPS Agreement and the Global Politics of Intellectual Property Reform in Developing Countries* (Oxford University Press, 2009), at 5, 51; Peter Yu, Toward a Nonzero-Sum Approach to Resolving

undertake positive efforts in the area of technology transfer – an archetypical and reflexive form of innovation policy toward developing countries as a whole.[83] The second incentive assured agricultural trade.[84] These incentives, backed by ancillary agreements, were pivotal for the eventual acquiescence of developing countries to the TRIPS Agreement.[85] Both incentives also adhered implicitly to Solow's neoclassical growth model, formulated earlier by economists Cass,[86] Koopmans,[87] and other contributors.[88]

More specifically, there still remains a predicament regarding the first technological incentive of technology transfer.[89] Initially, it was meant to act as a force for convergence, because of the "advantage of backwardness" conferred on technological laggards, as was initially posited in 1962 by Harvard University economic historian Alexander Gerschenkron.[90] In

Global Intellectual Relations Theorists, 70 *University of Cincinnati Law Review* 569 (2001) 635 and sources therein; Cf.: Christine Thelen, Carrots and Sticks: Evaluating the Tools for Securing Successful TRIPs Implementation, *Temple Journal of Science, Technology & Environmental Law*, vol. XXIV, 519 (2006) (discussing four incentive mechanisms tailored for developing countries within TRIPS, namely creating short- and long-term economic growth, technical assistance, and additional time to become compliant), at 528–33.

[83] Laurence R. Helfer, Regime Shifting: The TRIPS Agreement and New Dynamics of International Intellectual Property Lawmaking, 29 *Yale Journal International Law*. 1, (2004), at 2; Carlos M. Correa, supra note 18 (focusing on developing countries' concerns over increasing technological transfer as a means of economic growth), at 18; For broader long-term economic growth concerns by developing countries, see also Christine Thelen, supra note 82, at 528–29.

[84] See Laurence R. Helfer, supra note 83, at 76; Clete D. Johnson, A Barren Harvest for the Developing World? Presidential "Trade Promotion Authority" and the Unfulfilled Promise of Agriculture Negotiations in the Doha Round, 32 *The Georgia Journal of International and Comparative Law* 437 (2004), at 464–65.

[85] Ibid.

[86] David Cass, Optimum growth in an aggregative model of capital accumulation, *Review of Economic Studies*, 32, 233 (1965), at 233–40.

[87] Tjalling Koopmans, On the concept of optimal economic growth, in (Study Week on the) Econometric Approach to Development Planning, chapter 4, 225 (1965), at 226–28.

[88] For earlier contributions, See Frank P. Ramsey, A Mathematical Theory of Saving, *Economic Journal*, vol. 38 (1928) 543–59; Robert M. Solow, A Contribution to the Theory of Economic Growth, *Quarterly Journal of Economics*, vol. 70 (1956) 65; Trevor W. Swan, Economic growth and capital accumulation, *Economic Record*, vol. 32, no. 63, 334 (1956).

[89] The WIPO Development Agenda of 2007 noticeably illustrates the organization's general yet implicit inclination toward neoclassical economics-related policies, with technology transfer being its archetypical example. See WIPO, supra note 22, The 45 Adopted Recommendations under the WIPO Development Agenda, Cluster C:, Cluster C: Technology Transfer, Information and Communication Technologies (ICT) and Access to Knowledge (Recommendation 28: "To explore supportive intellectual property-related policies and measures Member States, especially developed countries, could adopt for promoting transfer and dissemination of technology to developing countries.") (See also Recommendation 31: "31. To undertake initiatives agreed by Member States, which contribute to transfer of technology to developing countries, such as requesting WIPO to facilitate better access to publicly available patent information"), ibid.

[90] See Alexander Gerschenkon, *Economic Backwardness in Historical Perspective* (Belknap Press, 1962); Alexander Gerschenkon, Economic Backwardness in Historical

his later work, Gerschenkron offered a pioneering idea that was called into action by neoclassical economists and policy-makers, such as in the WTO's TRIPS example.

As Gerschenkon explained, "technology gaps" between technologically edged (mostly developed) economies and laggard developing countries provide the latter with immense opportunities for economic growth.[91] Since Gerschenkon, almost every theory of international income differences that has taken technology transfer into account has implied that all countries share the same long-term growth rate.[92]

Club convergence, and lack thereof, over patent intensity by countries worldwide, as a proxy for their domestic innovation, corroborates the critique of these neoclassical economics growth theoreticians. That critique, of course, concerns the difficulty of these theoreticians in explaining how growth rates by poor countries remained significantly lower than the rest of the world for almost two centuries.[93] As a policy concern, a novel clustering analysis of archetypal patent intensity may ultimately necessitate innovation and patent-related policy adaptations.

2.2   The Empirical Analysis

2.2.1   Methodology

2.2.1.1 Data Selection

The main purpose of the following analysis is to cluster countries based on their patent and GERD data. The profile of each country used for the clustering consists of two-yearly time series covering the years

Perspective, in *The Progress of Underdeveloped Areas* (Bert F. Hoselitz, ed.) (University of Chicago Press, 1971).

[91] Alexander Gerschenkon, ibid.

[92] See, e.g., Susanto Basu and David N. Weil, Appropriate Technology and Growth, *Quarterly Journal of Economics*, 113, 1025–54 (1998) (applying this notion to rich countries on the technological edge); Daron Acemoglu, and Fabrizio Zilibotti, Productivity Differences, *Quarterly Journal of Economics*, 116, 563–606 (2001) (same); Stephen L. Parente and Edward C. Prescott, Technology Adoption and Growth, *Journal of Political Economy*, 102, 298 (1994) (applying the notion for countries in which technology transfer can be blocked by local special interests); Stephen L. Parente and Edward C. Prescott, Monopoly Rights: A Barrier to Riches, *American Economic Review*, 89, 1216 (1999) (same); Daron Acemoglu, Philippe Aghion, and Fabrizio Zilibotti, Distance to Frontier, Selection and Economic Growth, unpublished, MIT (2002) (referring to countries with institutions that do not permit full advantage to be taken of technology transfer).

[93] See, e.g., Peter Howitt and David Mayer-Foulkes, R&D, Implementation and Stagnation: A Schumpeterian Theory of Convergence Clubs, *Journal of Money, Credit and Banking*, 37(1), 147–77 (2005).

1996–2013. The first time series is the ratio of USPTO- and EPO-granted patents divided by GERD for each previous year and accounting for that country's propensity to patent. This time series stands for the ratio corresponding to year t defined as the number of patents of year $t+1$ divided by GERD of t. The second time series is the yearly GERD for each country, as detailed in Appendix A: Patent and Gross Domestic Expenditure on R&D (GERD) Data. The date selection method enclosed in Appendix A, was used for all of the book's empirical analyses.

2.2.1.2 Clustering Analysis
The empirical distributions of the 80 countries for each of the two time series were plotted using box plots for each year. Based on these plots, Bermuda was detected as an outlier due to an extremely small GERD value and few patents. Consequently, the data corresponding to this country was deleted and the final data included 79 countries.[94] The detailed technicalities of the analysis are described in Appendix B: Clustering Procedure: *Technical* Description.

2.2.2   Findings

2.2.2.1 Patent Propensity by GERD Intensity Clusters
Three clusters were detected which can be identified as "Leaders," comprising 21 countries; "Followers," with 32 countries; and "Marginalized," with 26 countries.

Figure 2.1 displays for each country the values of its mean standardized log GERD and the mean of the standardized square root of the ratio of

---

[94] Since the number of patents is a count variable, its variance is not independent of its mean. Therefore, rather than apply the clustering algorithms on the ratios, the square roots of these ratio were used. Similarly, the log transformation on the GERD data was calculated. The reason for the usage of both series rather than only the ratios data was that ratios could be high due to both a large number of patents and to relative small GERD values. By using both series, the differences in those cases was taken into account. The KML library which was followed for the clustering uses jointly both trajectories. See Christophe Genolini, Xavier Alacoque, Mariane Sentenac, and Catherine Arnaud, kml and kml3d: R Packages to Cluster Longitudinal Data, *Journal of Statistical Software*, 65 (4), 1 (2015). Since these two series differ in their scale, each series was scaled separately. Thus, rather than using the original data Y_{ctj}, where c denotes the country, j denotes whether it is the square root of the ratio or the log GERD series, and t denotes the year, $Z_{ctj} = \frac{Y_{ctj} - \overline{Y}_{.j}}{s_{.j}}$ was used, where $\overline{Y}_{.j}$ is the grand mean of all the data of series j, and $s_{.j}$ is the standard deviation of all the data of series j. Then, for the distance (dissimilarity) measure, the Minkowski distance was used, defined as $d(Z_1, Z_2) = \sqrt{\sum_{j,x} |Z_{1tj} - Z_{2tj}|^2}$.

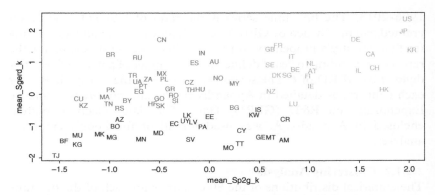

Figure 2.1 Scattered plot by Country Clusters (Mean of Standardized Square Root) (Patents per GERD) vs. Mean of Standardized Log (GERD) (1996–2013)
Legend: Light gray – Leaders cluster; Dark gray – Followers cluster; Dark – Marginalized cluster

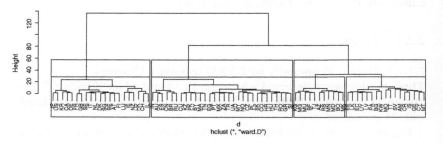

Figure 2.2 Hierarchical Clustering Dendogram

number of patents divided by the GERD of previous year. The mean is over the years. The color indicates the cluster of each country. Three such clusters were found.

As seen, the 21 countries in the cluster labeled as Leaders and colored in light gray are those that have a high number of patents by GERD as well as high GERD (e.g. the United States, Germany, Japan, and South Korea). The 32 countries in the cluster labeled as Followers, colored in dark gray, are those that have a low number of patents by GERD, though their GERD is not low (e.g. China, Romania, and Russia). The 26 countries in the cluster labeled as Marginalized and colored black are those that have both a low number of patents by GERD and low GERD (e.g. Tajikistan, Burkina Faso, and Mongolia). Figure 2.2 offers a visual representation of the hierarchical clustering dendogram and reveals the choice of three groups.

These findings show two statistical discontinuities over patent propensity and GERD intensity data: the first refers to the great distance that separates the archetypal middle cluster of followers from the stronger leaders, whereas the second refers to the gap that separates the weaker marginalized cluster from the followers.

Although empirically distinct in terms of the propensity to patent breakdown, the characteristics of these country clusters loosely resemble those of the "innovation," "imitation," and "stagnation" groups identified following Aghion, Howitt, and Mayer-Foulkes' model and the deriving empirical findings by Fulvio Castellacci offering broad technology-propensity results in a three cluster analysis.[95]

For the leaders' convergence club, the findings suggest that even among the 35 OECD countries, or the analogous 32 Advanced Economies listed by the IMF as of 16 July 2012,[96] convergence over patent propensity is not apparent. Instead, evidence shows club convergence among some OECD economies. Resembling Canova's 2004 findings on the club convergence over income-rate-related economic growth, the initially categorized follower countries in the OECD diverge from the initially remaining 20 leader countries. The latter are those which form the exclusive and enduring convergence club throughout the entire

---

[95] For later findings, see also: Fulvio Castellacci, Convergence and Divergence among Technology Clubs, DRUID Working Paper No. 06–21 1 (2006) (supporting the idea of the existence of clubs of countries characterized by different levels of technological development and different technological dynamics), at 1; Philippe Aghion, Peter Howitt, and David Mayer-Foulkes, The Effect of Financial Development on Convergence: Theory and Evidence, *The Quarterly Journal of Economics*, MIT Press, vol. 120(1) 173 2005 January (presenting evidence whereby any country with more than some critical level of financial development will converge to the growth rate of the world technology frontier, and that all other countries will have a strictly lower long-term growth rate).

[96] International Monetary Fund (IMF), Data and Statistics (2012), at: www.imf.org/exter nal/data.htm. The book's clustering analysis was unfortunately completed before the April 2016 publication of the IMF's updated list of Advanced economies. International Monetary Fund (IMF), IMF Advanced Economies List, World Economic Outlook, (April 2016), at 148.

As of July 16, 2012, Advanced Economies include: Australia, Austria, Belgium, Canada, Cyprus, Czech Republic, Denmark, Finland, France, Germany, Greece, Hong Kong, Iceland, Ireland, Israel, Italy, Japan, South Korea, Luxemburg, Malta, Netherlands, New Zealand, Norway, Portugal, Singapore, Slovak Republic, Slovenia, Spain, Sweden, Switzerland, Taiwan, the United Kingdom, and the United States. Of the six new members enlisted as Advanced economies by 2016 (Estonia, Latvia, Lithuania, Macao, Puerto Rico, San Marino), the clustering analysis ignored Macau and Latvia (hereinafter perceived as an emerging economy).

period.[97] These findings thus contradict evidence concerning income-related economic growth, especially following Barro's work, as shown in Table 2.1, listing the countries in each cluster. In his 1991 publication, he argues that over a period of almost 40 years (1950–88) convergence was restricted to OECD countries, while it was almost absent between the OECD and the less developed countries.[98]

2.2.2.2 Relationship Between Patent Activity Intensity Indicators and Clusters

Further to Table 2.1, the characterization of the patent clusters justifies the further assessment of a possible relationship between each cluster and WIPO's patent activity intensity indicators.[99] This study reviews six such indicators. The first is the relationship between this book's patent clusters and the economy categories defined by the IMF: advanced economies, emerging economies, and other developing countries (excluding emerging economies). The second relationship reviewed is that between the clusters and the countries' income group, according to the World Bank's categorization of countries into high-income countries, upper-middle-income countries, lower-middle-income countries, and low-income countries. The third possible relationship is between the clusters and geographic regions. The World Bank classifies countries into seven regions: East Asia and Pacific, Europe and Central Asia, Latin America and Caribbean, Middle East and North Africa, North America, South Asia, and Sub-Saharan Africa. A fourth indicator relates to the percentage of patents granted out of patents submitted to the USPTO or EPO (abbreviated: UE). The analysis relates only to data for the years after 2001, since the available information on the non-granted applications in the USPTO is only available for those years. The fifth indicator to be considered is the percentage of patents submitted only to PCT (and not UE) out of the USPTO, or EPO, or PCT. The sixth indicator to be reviewed is the relationship between clusters and family sizes. The set of patents filed in several countries which are related to each other by one or several common priority filings is generally known as patent family. The value

---

[97] Compare: Fabio Canova, Testing for Convergence Clubs in Income Per Capita: A Predictive Density Approach, *International Economic Review*, vol. 45(1) (2004). For a similar earlier claim, see Xavier Sala-i-Martin, supra note 27, at 1029.

[98] See, e.g., Robert J. Barro, Economic Growth in a Cross-section of Countries, *The Quarterly Journal of Economics*, vol. 106, No. 2. (May, 1991) 407. See also, Xavier Sala-i-Martin, supra note 27; Steve Dowrick and Norman Gemmell, Industrialization, Catching Up and Economic Growth: A Comparative Study Across the World's Capitalist Countries, *The Economic Journal* 101, 263–75 (1991); Steve Dowrick and Duc-Tho Nguyen, OECD Comparative Economic Growth: Catch Up and Convergence, *American Economic Review*, 79 (1989) 1010.

[99] See WIPO, World Intellectual Property Indicators 2015 (2015), at 29.

Table 2.1 *Countries by Clusters and Other Country-Group Indicators (1996–2013)*

Obs.	Country	Code	Cluster	Geographic Region (World Bank, 2016)	Income Group (World Bank, 2016)	Continent	Economy Category (IMF, 2015)
1.	AUSTRIA	AT	1	Europe & Central Asia	a	1	1
2.	BELGIUM	BE	1	Europe & Central Asia	a	1	1
3.	CANADA	CA	1	North America	a	1	1
4.	SWITZERLAND	CH	1	Europe & Central Asia	a	2	1
5.	GERMANY	DE	1	Europe & Central Asia	a	2	1
6.	DENMARK	DK	1	Europe & Central Asia	a	2	1
7.	FINLAND	FI	1	Europe & Central Asia	a	1	1
8.	FRANCE	FR	1	Europe & Central Asia	a	1	1
9.	UNITED KINGDOM	GB	1	Europe & Central Asia	a	1	1
10.	HONG KONG	HK	1	East Asia & Pacific	a	3	1
11.	IRELAND	IE	1	Europe & Central Asia	a	1	1
12.	ISRAEL	IL	1	Middle East & North Africa	a	3	1
13.	ITALY	IT	1	Europe & Central Asia	a	1	1
14.	JAPAN	JP	1	East Asia & Pacific	a	3	1
15.	REPUBLIC OF KOREA	KR	1	East Asia & Pacific	a	3	1
16.	LUXEMBOURG	LU	1	Europe & Central Asia	a	2	1
17.	NETHERLANDS	NL	1	Europe & Central Asia	a	2	1
18.	NEW ZEALAND	NZ	1	East Asia & Pacific	a	4	1
19.	SWEDEN	SE	1	Europe & Central Asia	a	2	1
20.	SINGAPORE	SG	1	East Asia & Pacific	a	3	1
21.	UNITED STATES	US	1	North America	a	1	1
22.	AUSTRALIA	AU	2	East Asia & Pacific	a	4	1
23.	CZECH REPUBLIC	CZ	2	Europe & Central Asia	a	2	1
24.	SPAIN	ES	2	Europe & Central Asia	a	1	1
25.	GREECE	GR	2	Europe & Central Asia	a	1	1
26.	NORWAY	NO	2	Europe & Central Asia	a	2	1

Table 2.1 (*cont.*)

Obs.	Country	Code	Cluster	Geographic Region (World Bank, 2016)	Income Group (World Bank, 2016)	Continent	Economy Category (IMF, 2015)
27.	PORTUGAL	PT	2	Europe & Central Asia	a	2	1
28.	SLOVENIA	SI	2	Europe & Central Asia	a	2	1
29.	SLOVAKIA	SK	2	Europe & Central Asia	a	2	1
30.	BULGARIA	BG	2	Europe & Central Asia	b	2	2
31.	BRAZIL	BR	2	Latin America & Caribbean	b	1	2
32.	CHINA	CN	2	East Asia & Pacific	b	3	2
33.	HUNGARY	HU	2	Europe & Central Asia	a	1	2
34.	INDIA	IN	2	South Asia	c	3	2
35.	MEXICO	MX	2	Latin America & Caribbean	b	1	2
36.	MALAYSIA	MY	2	East Asia & Pacific	b	3	2
37.	PAKISTAN	PK	2	South Asia	c	3	2
38.	POLAND	PL	2	Europe & Central Asia	a	2	2
39.	ROMANIA	RO	2	Europe & Central Asia	b	2	2
40.	RUSSIA	RU	2	Europe & Central Asia	b	3	2
41.	THAILAND	TH	2	East Asia & Pacific	b	3	2
42.	TURKEY	TR	2	Europe & Central Asia	b	2	2
43.	UKRAINE	UA	2	Europe & Central Asia	c	3	2
44.	SOUTH AFRICA	ZA	2	Sub-Saharan Africa	b	4	2
45.	BELARUS	BY	2	Europe & Central Asia	b	3	3
46.	COLOMBIA	CO	2	Latin America & Caribbean	b	1	2
47.	CUBA	CU	2	Latin America & Caribbean	b	1	3
48.	EGYPT	EG	2	Middle East & North Africa	c	4	3
49.	CROATIA	HR	2	Europe & Central Asia	a	1	3
50.	KAZAKHSTAN	KZ	2	Europe & Central Asia	b	3	3
51.	MOROCCO	MA	2	Middle East & North Africa	c	4	3
52.	SERBIA	RS	2	Europe & Central Asia	b	2	3
53.	TUNISIA	TN	2	Middle East & North Africa	c	4	3
54.	ARMENIA	AM	3	Europe & Central Asia	c	3	3
55.	AZERBAIJAN	AZ	3	Europe & Central Asia	b	3	3

56.	BURKINA FASO	BF	3	Sub-Saharan Africa	d	4	3
57.	BOLIVIA	BO	3	Latin America & Caribbean	c	1	3
58.	COSTA RICA	CR	3	Latin America & Caribbean	b	1	3
59.	CYPRUS	CY	3	Europe & Central Asia	a	2	1
60.	ICELAND	IS	3	Europe & Central Asia	a	1	1
61.	MACAO	MO	3	East Asia & Pacific	a	3	1
62.	MALTA	MT	3	Middle East & North Africa	a	2	1
63.	ESTONIA	EE	3	Europe & Central Asia	a	2	2
64.	LATVIA	LV	3	Europe & Central Asia	a	2	2
65.	ECUADOR	EC	3	Latin America & Caribbean	b	1	3
66.	GEORGIA	GE	3	Europe & Central Asia	b	3	3
67.	KYRGYZSTAN	KG	3	Europe & Central Asia	c	3	3
68.	KUWAIT	KW	3	Middle East & North Africa	a	3	3
69.	SRI LANKA	LK	3	South Asia	c	3	3
70.	MOLDOVA	MD	3	Europe & Central Asia	c	3	3
71.	MADAGASCAR	MG	3	Sub-Saharan Africa	d	4	3
72.	MACEDONIA	MK	3	Europe & Central Asia	b	2	3
73.	MONGOLIA	MN	3	East Asia & Pacific	C	3	3
74.	MAURITIUS	MU	3	Sub-Saharan Africa	B	4	3
75.	PANAMA	PA	3	Latin America & Caribbean	B	1	3
76.	EL SALVADOR	SV	3	Latin America & Caribbean	C	1	3
77.	TAJIKISTAN	TJ	3	Europe & Central Asia	c	3	3
78.	TRINIDAD AND TOBAGO	TT	3	Latin America & Caribbean	a	1	3
79.	URUGUAY	UY	3	Latin America & Caribbean	a	1	3

Legend:

Clusters: 1 Leaders; 2 Followers; 3 Marginalized

Continents: 1 America; 2 Europe; 3 Asia; 4 Other

Income group: a High income; b Upper middle income; c Lower middle income; d Low income

Economy category: 1 Advanced Economy; 2 Emerging Economy; 3 Other developing country excluding emerging economies

of patents is held to be associated with the number of jurisdictions in which patent protection has been sought and large international patent families have been found to be particularly valuable.

2.2.2.2.1 Economy Category

Borrowing on the IMF's economic categories, three groups are deemed relevant: advanced economies, emerging economies, and the default group of other developing countries (excluding emerging economies). Out of more than 150 developing countries, only 25 emerging economies (all but one) account for about 90 percent of the GDP of the developing countries.[100] The IMF's 23 emerging economies, as of October 2015,[101] are indeed underdeveloped economies presently perceived as hotbeds of meaningful innovation within the developing world.[102] From a political economy perspective, emerging economies are also said to possess meaningful political will to improve access to the world's intellectual output and thus lead the remaining developing countries.[103] In addition, their macroeconomics facilitates their ability to challenge the TRIPS toward developed countries.[104]

---

[100] See World Bank, The Growth Report Strategies for Sustained Growth and Inclusive Development, Commission on Growth and Development, Conference Edition (2008), at: https://openknowledge.worldbank.org/bitstream/handle/10986/6507/449860PUB0 Box3101OFFICIAL0USE0ONLY1.pdf?sequence=1&isAllowed=y (adding that the 10 largest developing countries account for about 70% of developing countries' GDP), at 111.

[101] Various sources list countries as emerging economies. While there are no agreed parameters on which the countries can be classified as such, the IMF's list is possibly the most influential one. See IMF, World Economic Outlook Adjusting to Lower Commodity Prices (October 2015) (the list includes: Bangladesh, Brazil, Bulgaria, Chile, China, Colombia, Hungary, India, Indonesia, Malaysia, Mexico, Pakistan, Peru, Philippines, Poland, Romania, Russia, South Africa, Thailand, Turkey, Ukraine, and Venezuela), at 149. Colombia is not always perceived as an emerging economy and was instead considered here as another developing country.

[102] Grace Segran, As Innovation Drives Growth in Emerging Markets, Western Economies need to Adapt (2011), at: http://knowledge.insead.edu/innovation-emerging-markets-110112.cfm?vid=515; Subhash Chandra Jain, *Emerging Economies and the Transformation of International Business* (Edward Elgar Publishing, 2006); Similarly, in her book "The Rise of 'the Rest'," Amsden identifies 12 countries that have acquired considerable manufacturing experience: China, Indonesia, India, South Korea, Malaysia, Taiwan, Thailand, Argentina, Brazil, Chile, Mexico, and Turkey. See Alice H. Amsden, The Rise of "The Rest": Challenges to the West from Late-Industrializing Economies (Oxford University Press, 2001).

[103] See, e.g., Rochelle C. Dreyfuss, The Role of India, China, Brazil and the Emerging Economies in Establishing Access Norms for Intellectual Property and Intellectual Property Lawmaking, IILJ Working Paper 2009/5, at 1.

[104] See Rochelle C. Dreyfuss, Intellectual Property Lawmaking, Global Governance and Emerging Economies, 53 in *Patent Law in Global Perspective* (Ruth L. Okediji and Margo A. Bagley, eds.) (Oxford University Press, 2014) (emphasizing that emerging economies could improve cooperation while dealing with overlapping interests also on behalf of developing countries), at 83, fn. 166, and sources therein.

Another comparable, albeit informal, country-group classification is the Newly Industrialized Countries (NICs) topping the list of developing countries by GDP.[105] Essentially, NIC's differ from the remaining developing countries in that they possess large and relatively diversified domestic economies. This fact awards them the status of strategic and fast-growing markets in and with which MNCs typically cannot refrain from investing or trading.[106] Consequently, NICs capture a disproportionally large portion of the FDI that flows to developing countries.[107] A popular method of categorization would limit the NICs to Brazil, Mexico, South Africa, China, India, Malaysia, Philippines, Thailand, and Turkey.[108] In comparison with the emerging economies' category, the latter preferably holds political will to improve access to the world's intellectual output and thus lead remaining developing countries.[109]

The clustering analysis identifies three clusters. In the leaders cluster, all 21 countries have been classified as advanced economies. In the followers cluster, comprising 32 countries, the majority (15) of the countries have been classified as emerging economies, 8 as advanced economies, and 9 as other developing countries. Finally, in the marginalized cluster, comprising 26 countries, the majority (20 countries) have been classified as other developing countries, 4 as advanced, and 2 as emerging economies.

Figure 2.3 displays for each country, in relation with Table 2.1, the values of the mean standardized log GERD and the mean of the standardized square root of the ratio: number of patents divided by the GERD of previous year. The mean is over the years. A comparison of Figure 2.1 and Table 2.1 reveals a strong association between the economy classification and the clusters.

---

[105] For a comparable analysis of nine NICs, compare: Anis Chowdhury and Iyanatul Islam, *The Newly Industrializing Economies of East Asia* (Routledge, 1997), at 4. See also Nigel Grimwade, *International Trade: New Patterns of Trade, Production and Investment* (Routledge, 1989), at 312. Accordingly, NICs tend to be more advanced than other developing countries, and less so than developed countries. There is no official or undisputed set of criteria to define an NIC, so each author sets a list of countries according to her own criteria and methods.

[106] Just consider, for instance, the fact that China already has the same number of mobile-phone users as the whole of Europe, namely 500 million. See Technology in Emerging Economies, *The Economist* (US Edition) February 9, 2008.

[107] In turn, the remaining developing countries receive proportionally much smaller shares of FDI. See, e.g., Ilene Grabel, International Private Capital Flows and Developing Countries, in *Rethinking Development Economics*, ed. Ha-Joon Chang (Anthem Press, 2003), at 327–28.

[108] David Waugh, *Geography, An Integrated Approach* (Nelson Thornes Ltd., 3rd edn., 2000); and N. Gregory Mankiw, *Principles of Economics* (South-Western College Pub., 4th edn., 2007).

[109] Rochelle C. Dreyfuss, The Role of India, China, Brazil and the Emerging Economies in Establishing Access Norms for Intellectual Property and Intellectual Property Lawmaking, IILJ Working Paper 2009/5, at 1.

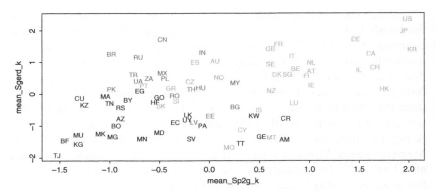

Figure 2.3 Scattered Plot by Economy Category: Mean of Standardized
Square Root (Patents to GERD) vs. Mean of Standardized Log
(GERD)
Legend: Light Gray – *Advanced economy*; Dark gray – *Emerging economy*;
Dark – *Other developing country*

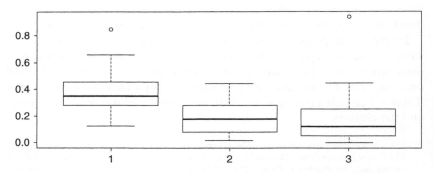

Figure 2.4 Relationship between Economy Type and UE Filings (%)
(1996–2013)

To obtain additional insight into the characteristics of the three
clusters, the classification and regression tree (CART) algorithm was
applied with the cluster identification as the dependent variable (the
classification variable) and with the two input variables, namely economy
type and the percentage of patents registered in the combined USPTO
and EPO (UE). Figure 2.4 displays the relationship between the two
input variables – the economy type and the percentage of UE filings.

Four groups of "similar" countries (leaves in the classification tree),
depicting the cluster identification and the economy relationship, were

obtained using the CART algorithm. In two leaves, CART classified 26 countries in the followers cluster (labeled 2), while in fact 22 countries were indeed in this cluster, 3 in the marginalized cluster (labeled 1), and 1 in the leaders cluster (labeled 3) (for other developing countries). CART classified 29 countries as cluster 1, while in fact 20 countries were indeed included in this cluster, while 9 were included in cluster 2. This leaf included all the other developing countries (labeled 3). CART classified 24 countries as cluster 3, while in fact 20 countries were indeed in this cluster, 3 in cluster 1, and 1 country in cluster 2. The countries in all 4 leaves are listed in Appendix C: Relationship between Patent Activity Intensity indicators and Clusters. As the Appendix further details, the economy category classification essentially resembles the patent clusters structure.

2.2.2.2.2 Income Group

The second relationship reviewed was between the clusters and the countries income group. The latter refer to the World Bank's categorization of countries as high-income, upper-middle-income, lower-middle-income, and low-income countries.[110] The following two tables display the distribution of countries according to their income classification. Table 2.2 relates to our sample of countries, while Table 2.3 relates to all countries. As we can see, our data include over-representation of the high-income category and under-representation of the low-income category.

For brevity, low income is labeled as "d"; lower middle income as "c"; upper middle income as "b"; and high income as "a." Table 2.4 reflects the strong association between income level and economy type.

The CART algorithm was again applied with the same input variables as before, namely economy type and grant rate in the UE. The income group was also added as an input variable. However, adding this variable did not change the regression tree and the same tree was obtained as when income was not included.

The plots in Figures 2.5–2.6 describe the changes in time of the two variables – the standardized log GERD and standardized square root ratio patents over GERD by income groups. Each of the four curves corresponds to countries in the same income-class classification. As we can see,

---

[110] The World Bank defines, per the 2017 fiscal year: (1) *low-income economies* with $1,025 or less in 2015 GNI per capita, calculated using the World Bank Atlas method, (2) *lower middle-income economies* with a GNI per capita between $1,026 and $4,035, (3) *upper middle-income economies* with a GNI per capita between $4,036 and $12,475, and (4) *high-income economies* with a GNI per capita of $12,476 or more. See World Bank, Country and Lending Groups, at https://datahelpdesk.worldbank.org/knowledgebase/articles/906519-world-bank-country-and-lending-groups (July 2016).

Table 2.2 *Income Group among 79 Countries*

Income group	Frequency	Percent	Cumulative Frequency	Cumulative Percent
High income	41	51.90	41	51.90
Low income	2	2.53	43	54.43
Lower middle income	14	17.72	57	72.15
Upper middle income	22	27.85	79	100.00

Table 2.3 *Income Group among All Countries*

Income Group	Frequency	Percent	Cumulative Frequency	Cumulative Percent
High income	79	36.24	79	36.24
Low income	31	14.22	110	50.46
Lower middle income	52	23.85	162	74.31
Not classified	1	0.46	163	74.77
Upper middle income	55	25.23	218	100.00

Table 2.4 *Income Group vs. Economy Category*

Income Group	Economy Category			
Frequency	1	2	3	Total
a	33	4	4	41
b	0	10	12	22
c	0	3	11	14
d	0	0	2	2
Total	33	17	29	79

with the exception of the low-income countries, all countries show a steady increase with time of their GERD, but no substantial changes in their ratio of patents over GERD.

As seen, the economy category classification discussed above resembles the patent clusters structure much more than the income group or geographic region classifications.

2.2.2.2.3 Geographic Regions

The third possible relationship is between the clusters and the geographic regions. As already noted, the World Bank categorizes countries by seven

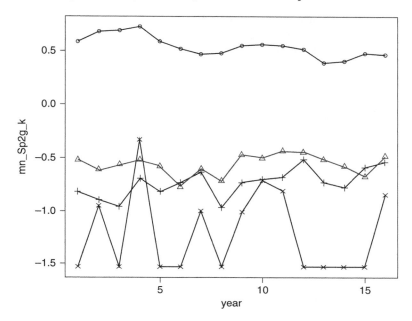

Figure 2.5 Means by Income Group Standardized Log GERD (1996–2013)

regions: East Asia and Pacific, Europe and Central Asia, Latin America and Caribbean, Middle East and North Africa, North America, South Asia, and Sub-Saharan Africa.[111] The geographic regions according to this classification include economies at all income levels. The term "country," used interchangeably with "economy," does not assume political independence, but rather refers to any territory for which authorities report separate social or economic statistics, as seen in Table 2.5.[112]

As seen, the economy category classification discussed above resembles the patent clusters structure much more than the geographic region classifications. The plots in Figures 2.7–2.8 display the changes in time of the two variables – the standardized square root ratio patents over GERD and standardized log GERD, respectively. Each of the seven curves corresponds to countries in the same region.

[111] See, World Bank, Country and Lending Groups, at https://datahelpdesk.worldbank.org/ knowledgebase/articles/906519-world-bank-country-and-lending-groups (July 2016).
[112] Ibid. Argentina, classified as high-income country in 2016, is presently unclassified and pending the expected release of revised national accounts statistics.

Table 2.5 *Clusters vs. Geographic Regions*

Cluster	East Asia & Pacific	Europe & Central Asia	Latin America & Caribbean
1	2	11	7
2	4	18	4
3	5	13	0

Cluster	Middle East & North Africa	North America	South Asia	Sub-Saharan Africa
1	2	0	1	3
2	3	0	2	1
3	1	2	0	0

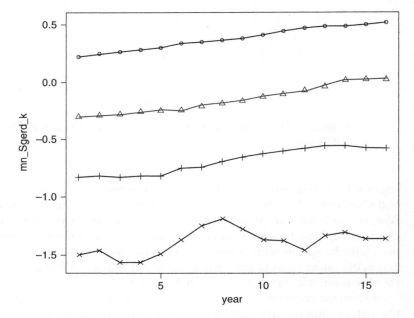

Figure 2.6 Means by Income Group Standardized Square Root Ratio
Patents over GERD (1996–2013)

As can be seen, with the exception of the Sub-Saharan Africa countries, all countries show a steady but slow increase of their GERD over time, but no substantial changes in their ratio of patents over GERD, with the exception of the countries in North America, where a decrease in the ratio of patents over GERD with time is seen.

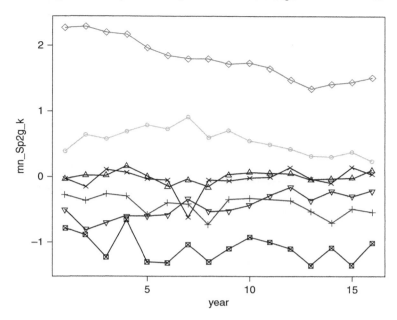

Figure 2.7 Means by Income Group Standardized Square Root Ratio Patents over GERD (1996–2013)

2.2.2.2.4 *Patent Grant Rate*

The fourth indicator assessed pertains to the percent of patents granted out of patents submitted to the USPTO or EPO (abbreviated as UE). The analysis relates solely to data for the years after 2001, since statistics for non-granted applications in the USPTO are only available after that year. Patent grant rates are measured here as the adjusted number of out of patents applied to the USPTO or the EPO in relation with the patent clusters.

Figures 2.9–2.10 below show box plots of the percent of patents granted out of patent applications to the USPTO or EPO. Figure 2.9 shows the percentage by clusters and Figure 2.10 by economy category.

The similarity between the two box plots reflects the similarity between classification by clusters and by economy type. The plots show no difference between the three clusters (or economy types) in their center of distributions (medians), though a difference in dispersion can be seen. When the Kruskal–Wallis non-parameter ANOVA was applied to test median equality for the three clusters, the p-value was 0.7 for the percent of granted patents and p-value 0.4 for the economy types.

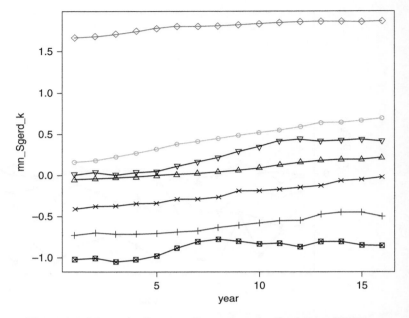

Figure 2.8 Means by Income Group Standardized Log GERD (1996–2013)

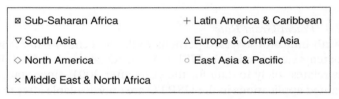

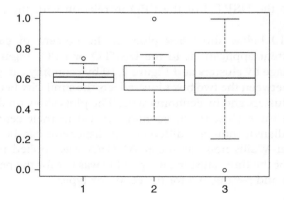

Figure 2.9 Patent Grant Rate Applied only to UE by Clusters

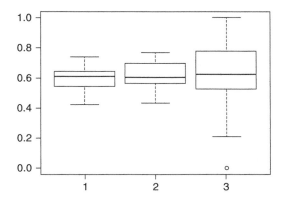

Figure 2.10 Patent Grant Rate Applied only to UE by Economy Category

2.2.2.2.5 *Patents Submitted Only to PCT*

The next aspect considered was the percentage of patent applications only through the Patent Cooperation Treaty (PCT) track (and not UE) out of all applications to the USPTO, EPO, or PCT. It should be noted, however, that "recent data" may be misleading, since there may be "good" patents that have not yet been submitted to the USPTO or EPO after being applied through the PCT system. Figures 2.11 and 2.12 show box plots for the percentage of patents submitted only to PCT (and not later submitted to UE) out of all the applications to the US, EU, or PCT, by clusters and by economy category, respectively.

As seen in the marginalized cluster (labeled 3) the values are very small, but the median ratio values do not differ substantially from those of the other two clusters. They differ in their variability.[113] The results show a significant difference in the median ratio values between the leaders cluster (labeled 1) and the followers cluster (labeled 2) (P=0.0016), and a significant difference in the median ratio values between the economy categories of advanced economies (labeled 1) and other developing countries (excluding emerging economies) (labeled 3) (P=0.01). These results, and particularly Figure 2.11 with reference to the patent clusters, corroborate the tendency of patent applicants to complete patent examination in the USPTO and EPO primarily in the marginalized and followers clusters. This signifies patent applicants' relatively lower trust in the degree of technologic and economic value of their inventions in the marginalized and followers clusters, on average.

---

[113] The Kruskal–Wallis ANOVA test was applied, followed by pairwise comparisons, using Dunn's-test for multiple comparisons of independent samples.

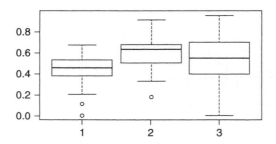

Figure 2.11 Patent Rate Submitted Only to PCT (and not UE) by Clusters

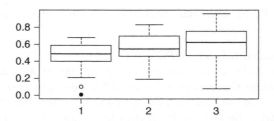

Figure 2.12 Patent Rate Submitted Only to PCT (and not UE) by Economy Category

2.2.2.2.6 *Family Size*

The sixth and final indicator to be reviewed is the relationship between clusters and family sizes. A set of patents filed in several countries which are related to each other by one or several common priority filings is generally known as a "patent family." The value of patents is believed to be associated with the number of jurisdictions in which patent protection has been sought, and large international patent families have been found to be particularly valuable.[114] In such cases, applicants might be willing to accept additional costs and delays in order to extend protection to other countries, if they deem this cost effective.

The available data are of family sizes for patents filed separately for USPTO and EPO registration. Since the same patent may be registered in both the EPO and the USPTO, it would be wrong to simply merge the two records. Therefore, the average family size during the period 2002–11 was calculated separately for the USPTO and for the EPO. Then, each

---

[114] Dietmar Harhoff, Frederic M. Scherer and Katrin Vopel, Citations, Family Size, Opposition and the Value of Patent Rights, *Research Policy*, 32(8) 1343 (2003).

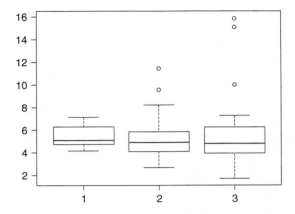

Figure 2.13 Mean of UE Family Size by Clusters

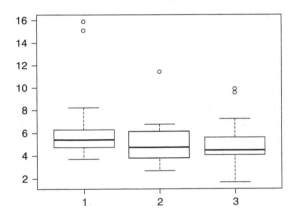

Figure 2.14 Mean of UE Family Size by Economy Category

average was weighted by the respective number of families. Figures 2.13 and 2.14 display box plots for the family sizes by clusters and by economy category, respectively.

The Kruskal–Wallis ANOVA tests were applied, followed by pairwise comparisons, using Dunn's test for multiple comparisons of independent samples. The results showed no significant difference in the median values between clusters 2 and 3 (P=0.4), and a borderline significant difference in the median ratio values between the advanced economies and other developing countries economy categories (labeled 1 and 3, respectively) (P=0.06).

2.2.2.3 Inter-Cluster and Intra-Cluster Convergence

In order to check the stability over time of the relationship between the three clusters underlying the 79 countries therein (named inter-), club convergence analysis is first required. For this purpose, all the previous clustering analyses on parts of the data were repeated. First, clustering analysis was undertaken only for the first three years (1997–99), and subsequently also for the latter three years (2008–10). The extreme ends, namely years 1996 and 2006, were not included in those three year groups since the data corresponding to those years were very sparse. The results based on all methods were identical, with one minor exception. Based on the clustering according to the last part of the series, Azerbaijan belonged to the marginalized cluster and part of the IMF's labeled emerging economies' country-group classification. In the two other cases (all years and early years only), Azerbaijan corresponded to the economy category in 1 and was classified as one of the other developing countries (excluding emerging economies).

An inner (named intra-) club convergence examination then follows. The remaining concern relates to the changes over time within each cluster according to the two decisive variables, namely the standardized square root ratio patents over GERD, accounting for patent propensity, and the standardized log GERD across clusters. As seen in Figures 2.15 and 2.16, respectively, all three clusters show an increase with time in their GERD. However, this is matched by a decrease of the ratio of patents to GERD, labeled herein as patent propensity as recalled in the leaders cluster, although no significant convergence with the other two clusters occurred.

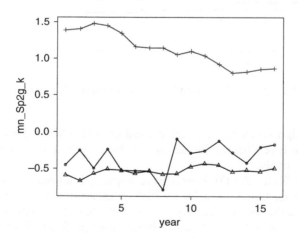

Figure 2.15 Means by Clusters Standardized Square Root Ratio Patents over GERD (1996–2013)

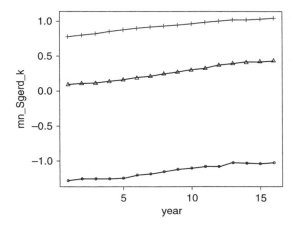

Figure 2.16 Means by Clusters Standardized Log GERD (1996–2013)

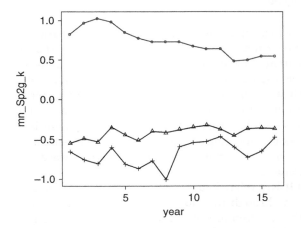

Figure 2.17 Means by Economy Category Standardized Square Root
Ratio Patents over GERD (1996–2013)

Similarly, Figures 2.17 and 2.18 display the changes over time of the
two above-mentioned aspects according to the analogous economy category country-group classification. Unsurprisingly, the two latter plots
resemble the two previous ones, due to the close similarity between the
clustering pattern and classification according to economic category.

A further detailed enhancement of these relationships is provided in the
CART output display in Appendix B. In conclusion, the gap between
the lower followers and marginalized clusters, on the one hand, and the
leaders cluster, on the other, is slowly and steadily decreasing following

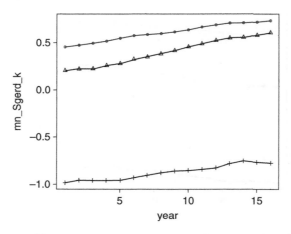

Figure 2.18 Means by Economy Category Standardized Log GERD (1996–2013)

a homogeneous marginal return of decrease in comparison with the former clusters over their slow and steady upward convergence on their propensity to patent. The archetypical form of upward convergence has been modeled as the case of poorer club members – the followers and marginalized clusters, in our case – slowly catching up in terms of their propensity to patent with wealthier members of the leaders cluster.[115] A similar pattern is seen in terms of the propensity to patent by economy category, with poorer emerging economies and other developing countries (excluding emerging economies) slowly converging upwards toward the advanced economies.

Lastly, the GERD-related findings in Figures 2.1 and 2.3 serve to demonstrate how through the global investment in R&D and possibly FDI through the intervention of MNCs and foreign governments, the marginalized and followers clusters and the respective economy groups also witness a slow but steady upward convergence toward the remaining leaders cluster or advanced economy country group.

These findings may carefully be considered to respond to Chang's seminal book *Kicking Away the Ladder*, and others similarly warning against the widening of the model North–South innovation gap presumably fostered by the TRIPS and other WTO trade-related agreements.[116] This basic

---

[115] Cf.: Dan Ben-David, supra note 2, 159.

[116] For a possible growth in the gap since the enactment of the TRIPS Agreement: Ha-Joon Chang, *Kicking Away the Ladder: Policies and Institutions for Economic Development in Historical Perspective* (Anthem Press, 2003) (arguing that TRIPS may increase in the gap

empirical finding thus corresponds in context with "catch-up" literature concerning the pulling of other countries through a "catch-up" effect. In a certain analogy to income-based growth, this finding corresponds with Baumol and Wolff's utilization of data from 72 countries demonstrating that middle-income countries (17 out of the 72 countries in the sample), mostly corresponding to the emerging economies cluster, have grown the fastest, while poor countries consistently diverge from the rest.[117]

Moreover, the present findings concerning patent propensity convergence tentatively correspond with the partial recent UN position on innovation regional convergence. In the United Nations Conference on Trade and Development (UNCTAD)'s *2005 World Investment Report*, the authors indeed warn of a widening technological gap between developing countries taking part in the global innovation network and those failing to do so in the developing world. The UNCTAD report is germane to the above finding concerning emerging economies, as it highlights the fact that developing nations, particularly in Asia, are becoming increasingly successful in attracting investment in R&D from MNCs. On balance, the UNCTAD findings tentatively uphold the assertion that developing countries that are weak in R&D, and hence presumably excluded from being emerging economies, must adopt appropriate policies if they are to benefit from this trend.[118] To illustrate, the UNCTAD report finds that more than 60 percent of the MNCs surveyed by UNCTAD plan to expand their research activities in China. For India, the figure was 29.5 percent.[119] By contrast, however, few such corporations plan to increase R&D in Africa or Latin America, with the exception of Brazil, Mexico, Morocco, and South Africa.[120] The latter findings are less relevant to our

between the most technologically advanced and the least technologically advanced nations); Christopher T. May, *The Information Society: A Skeptical View* (Polity Press, 2002) (presenting the same argument).

[117] See William J. Baumol and Edward N. Wolff, Productivity Growth, Convergence, and Welfare: Reply, *The American Economic Review*, vol. 78(5) (1988), 1155 (finding that the poorest countries have diverged from other country groups). For earlier findings concerning income-based club convergence by middle-income countries, see Hollis Chenery and Moshe Syrquin, Typical Patterns of Transformation, in (Hollis Chenery, Sherman Robinson, and Moshe Syrquin, eds.), *Industrialization and Growth: A Comparative Study* (Oxford University Press, 1986) (combining time-series and cross-sectional data for several countries while finding divergence among the poorer countries and convergence among the relatively wealthier countries); Thorkil Kristensen, *Development in Rich and Poor Countries* (Praeger, 1982) (focusing on the cross-section alone, grouped countries by their 1974 income levels and found a hump-shaped relationship between group's 1970–79 growth rates and their income levels, with the middle-income groups enjoying higher rates of growth than the wealthier and the poorer groups).

[118] United Nations Conference on Trade and Development (UNCTAD), World Investment Report 2005, New York and Geneva, United Nations (2005), at 40, 147.

[119] Ibid., at 119.   [120] Ibid., at 40, 147.

present analysis, since the marginalized cluster in our analysis excludes LDCs and most African countries. Be that as it may, these important yet incomplete UNCTAD findings are more significant concerning possible club divergence between middle-income or emerging economies and weak developing countries over innovation at large, as opposed to the narrower context concerning the propensity to patent alone.

2.3   Theoretical Ramifications

The core empirical findings correspond to a moderate extent with catch-up literature concerning the pulling of other countries through a technology "catch-up" effect. In a recently published, seminal article by Harvard University economist Jérôme Vandenbussche et al., the authors assess that the strength of this "catch-up" effect on the developing countries' frontier in fact decreases with the level of domestic technological creation.[121] As a result, it is presumed that technology creation by domestic firms becomes progressively more important as a country moves closer to the technology frontier whereby technology diffusion and absorption decline – or, in other words, as catching up possibly translates into increasingly smaller technological improvement protected through incremental patenting activity.[122]

Yet, thus far this endogenous growth analysis has remained overwhelmingly theoretical. Its validation in our case indeed is acute mostly at the country-group level.[123] As the evidence suggests, the key factors stressed by endogenous growth theory – namely, guaranteeing increasing returns, human capital and domestic technology creation – develop unevenly and could be differentiated locally and possibly also regionally.[124] However, as stated earlier, earlier accounts of endogenous growth theory's relationship to club convergence between country groups, to be sure, have mostly contributed to the understanding of archetypical club convergence over salaries, GDP, and other macroeconomic income-related indications.[125]

It is unclear why endogenous convergence between country groups or clubs over domestic state-of-the-art-technology creation exists. Similarly,

---

[121] See Jérôme Vandenbussche, Philippe Aghion and Costas Meghir, supra note 25, at 21–30. For additional general discussion of the argument, see Emmanuel Hassan, Ohid Yaqub and Stephanie Diepeveen, supra note 25, at 17.

[122] See Emmanuel Hassan, Ohid Yaqub and Stephanie Diepeveen, ibid., at 17. The present paper leaves the latter argument concerning incremental patenting outside the scope of this paper.

[123] See Ron Martin and Peter Sunley, supra note 1, at 220.

[124] See, e.g., Ron Martin and Peter Sunley, ibid.

[125] See, e.g., Dan Ben-David, supra note 2 (concluding that "income gaps have increased within most possible groupings of countries in the world. Where 'convergence clubs' tend to be more prevalent is at the two ends of the income spectrum"), at 167.

not much is known about how the latter is achieved.[126] Moreover, very little is conceptually attributed to explaining how technological creation of country-group clusters is determined.[127]

In fact, the only certain findings herein concerning club convergence negate regional divergence between advanced and emerging economies. In other words, there is slow cluster convergence between clusters over innovation, especially by developing countries which are not emerging economies toward advanced ones, measured through patent propensity and GERD intensity rates.[128]

Notwithstanding the present empirical absence concerning the exact growth model, it being exogenous or endogenous, market forces potentially have failed in disequilibrating the three patent clusters in their relative country group's progression toward a higher patent propensity and GERD intensity, as explained earlier.

Finally, one has to entertain the possibility that in the long run, comparable patent propensity rates as proxy of domestic innovation rates by advanced and emerging economies may uphold club *divergence*. Such divergence may exist instead of convergence due to possible deep international incompatibilities in economic integration.

Conclusion

Based on the examination of 79 innovating countries worldwide over the period 1996–2013, the clustering analysis offers three empirical findings. The first core finding identifies three domestic innovation-related convergence clubs (jointly referred to as patent clusters) with markedly different levels of patent propensity and GERD intensity rates (codenamed "patent intensity"). The clustering analysis shows two large gaps among the examined innovating countries and in the world innovation-based economy at

---

[126] See Ron Martin and Peter Sunley, supra note 11, at 210, referring to David M. Gould and Roy J. Ruffin, supra note 3; R. J. Barro and Xavier Sala-i-Martin, supra note 13. Such diffusion of technology requires accordingly that lagging emerging economies would have appropriate infrastructure or conditions to adopt or absorb technological innovations. See Stilianos Alexiadis, supra note 3, at 61 and Sec. 4.5 (for a supportive economic model). For two of the earliest and most influential statements of this view See George H. Borts and Jerome L. Stein, supra note 3 (offering a classic study of regional development in the United States); J. G. Williamson, supra note 3 (analyzing the evolution of regional income differences in advanced industrial countries).

[127] See Stilianos Alexiadis, supra note 3, at 61 and Sec. 4.5 (for a supportive economic model).

[128] For comparable income-related findings, see Ron Martin and Peter Sunley, supra note 1, at 210 (referring to the work of Perroux [1950, 1955], Myrdal [1957], and Kaldor [1970, 1981] predicting that regional incomes will tend to diverge, because market forces, if left to their own devices, are spatially disequilibrating).

large: the first refers to the great distance that separates the middle group of "followers" from the stronger "leaders"; the second similarly refers to the significant gap that separates the weaker "marginalized" group from the followers clubs.

The first finding offers numerous additional insights. Firstly, the relationship between each cluster and six other archetypal patent activity intensity indicators identified as relevant to patent activity offers important policy-oriented implications. As we have seen, the only significant relationship between the groupings of the three clusters and the indicators reviewed was found regarding the economy category. These are the advanced economies, emerging economies, and other developing countries (excluding emerging economies) as defined by the IMF. All the other relationships reviewed were found to be statistically insignificant. The World Bank's income groups and geographic regions were found to be irrelevant in terms of the clusters. Similarly, the percentage of patents granted, out of patents submitted to the USPTO or EPO, was found to play an insignificant role in predicting one or more cluster patenting patterns. Lastly, neither the percentage of patents submitted solely to PCT (and not UE) out of the USPTO, EPO, or PCT, nor the relationship between clusters and family sizes were found to be significant policy levers.

Moreover, the first finding shows that the leaders cluster includes 21 OECD countries out of the 35. The remainder of OECD countries belong to the followers (e.g. Australia, Norway, or Spain), and a few even belong to the marginalized cluster (Cyprus and Iceland). Thus, this analysis effectively divides the OECD countries into two approximate halves over what remains an unaccounted-for OECD patenting inner divide. Similar findings apply to other country groups, as shown, and notably the EU, whose members can be found in each of the three patent clusters.

The analysis upholds a second finding concerning convergence between (or, inter-) the three clusters. The results show that, with the exception of Azerbaijan, in the two other cases (all years, and only early years) the clustering results remain identical. These results are at odds with the notion whereby the North–South divide or variations thereof is gradually diminishing with the enactment of TRIPS.

In continuation, an inner (or intra-) club convergence examination followed. As shown, all three clusters witnessed an increase in their GERD over time. However, this is matched by a decrease in the ratio of patents to GERD, defined here as patent propensity, in the leaders cluster, although no significant convergence with the other two clusters occurred. Similar results were shown for the changes over time by the analogous economy category country-group classification. It is not surprising that the two latter plots resemble the two previous ones, due to the association between the clustering pattern and classification according to economic category, as explained. In conclusion, the gap between the lower followers and marginalized clusters on the one hand, and the leaders cluster on the other hand, is slowly and steadily decreasing following a homogeneous marginal return of decrease in comparison with the former clusters over their slow and steady upward convergence in their propensity to patent.

Lastly, the GERD-related findings further serve to demonstrate how through the global investment in R&D and possibly FDI through the intervention of MNCs and foreign governments, the marginalized and followers clusters and respective economy groups also witness a slow yet stable upward convergence toward the remaining leaders cluster or advanced economy country group. These findings correspond with recent analyses focused almost exclusively on endogenous growth-theory income-related indications. They suggest that, unlike orthodox neoclassical models by Martin and Sunley, regional convergence rates are also generally much slower.

Lastly, the analysis has numerous theoretical ramifications, relating primarily to the need for further explanation of the remaining intricacies in accounting for shifts and reversals in rates of regional convergence. Such discrepancies arise from the fact that there is still little evidence accounting for the slowness or nonexistence of inner club convergence, especially in advanced economies, but also in emerging ones. In political terms, it remains unclear to what extent mismatched countries could join new coalitions, given the conflicting interests many of them may retain within their overall WTO bargaining position.

For now, on the broad policy level, legislation to facilitate patenting entices firms either to patent a higher percentage of their innovations or even to invest more in innovation. Yet even this rather basic proposition requires further consolidation. Economists such as Falvey and others argue that strong IPRs benefit both the richest and the poorest nations, but probably not middle-income countries, such as the emerging economies.[129]

Such pro-patenting policies indeed carry complex implications that lie beyond the scope of this book.[130] As Arundel and Kabla add, policies such as those intended to reduce the cost of a patent application may

---

[129] See Falvey, Foster and Greenaway, supra note 17.
[130] Cf.: Anthony Arundel and Isabelle Kabla, What percentage of innovations are patented? Empirical estimates for European firms, *Research Policy* 27, 127 (1998), at 128.

instead increase patent propensity rates in some sectors that currently have low rates, while having little effect on firms or sectors where a majority of innovations are already patented.[131] These changes may otherwise lead to the empirically uncorroborated reduction in patent quality altogether in advanced and emerging economies alike.[132]

A final word is in order concerning the rule of law, another field that is impacted by patenting activity. The rule of law orthodoxy offered through the TRIPS Agreement and other WTO treaties is surely not the product of what Ha-Joon Chang referred to as an "innocent scholastic awakening."[133] Instead, it represents an interest in law primarily as a response to the critique and failure of earlier World Bank-led neoliberal policies. Such is the critique concerning WTO-led exogenous growth policies with their focus on technological transfer and foreign direct investment-related stand for innovation-based economic growth. As seen through the nonlinearity bestowed in the patent propensity innovative divide herein, although TRIPS require the harmonization of IPR protection, such harmonization is neither clearly necessary, empirically based, or otherwise sufficient for the South. As a result, the latter's lower patent propensity rates remain evermore at odds with the current overall "one-size-fits-all" innovation and patent-related policy.

Appendix A   Patent and Gross Domestic Expenditure on R&D (GERD) Data

A.1   Patent Data

The rationale underlying the focus on USPTO and EPO patent filing rates instead of the alternative aggregation of national patenting systems of both advanced and emerging economies is twofold. Firstly, these

---

[131] Ibid.
[132] See, noticeably, OECD Science, Technology and Industry Scoreboard 2011: Innovation and Growth in Knowledge Economies (2011) (offering a novel innovation index upholding that patent quality has declined steadily on an average of 20%), at Part 6: Competing in the Global Economy; WIPO, Standing Committee on the Law of Patents, Quality of Patents: Comments Received from Members and Observers of the Standing Committee on the Law of Patents (scp), Seventeenth Session, Geneva, December 5 to 9, 2011, SCP/17/INF/2, October 20, 2011 (concluding that patent quality varies from country to country depending on national circumstances and level of development), at 2.
[133] Compare: Ha-Joon Chang, Understanding the Relationship between Institutions and Economic Development: Some Key Theoretical Issues 1 (UN World Institute for Development Economics Research, Discussion Paper No. 2006/05, 2006), available at www.wider.unu.edu/publications/working-papers/discussion-papers/2006/en_GB/dp2006-05/ (offering skepticism concerning rule of law-related post-Washington consensus policies with emphasis on the role of the World Bank thereof).

countries, especially those in the developing world, do not have the same patentability criteria.[134] Secondly, such countries may also differ substantively in terms of their national grant rates.[135]

The available patent data were for granted patents submitted through 2013, as well as patents granted as of 2015. For example, during 2012 Kazakhstan submitted seven patents to the USPTO, only three of which were granted over the following years (through 2015). Since some of the remaining four patents may well be granted after 2015, using three to represent the number of Kazakhstan's granted patents would be an underestimation due to patent examination grant lag.

The choice of USPTO issued patents effectively serves as a proxy for R&D-related quality output assurance. The explanation for this is that patent series are by nature subject to a substantial bias, with most patents generating low or no value, and only a few patents being associated with high economic and financial value. Thus far, patent statistics studies have rarely thoroughly tested the quality sensitivity of the results of their patent count methodology or their data source.[136] The qualitative methodological improvement herein counts state-of-the-art technology that has successfully culminated in issued patents, rather than the mere filing of related patent applications. This methodological choice is related to a concern at the possibility that a quantity of innovative activity does not begin or otherwise conclude the patenting process.[137] Only technology that completes the patenting examination process is accounted for as issued patents. It is therefore a limitation of patent statistics to measure patent applications as an indication of qualitative innovation.[138]

Another approach within the patent statistics literature, irrelevant in this instance given this study's immense data coverage, has partially met this qualitative challenge. The approach proposes that instead of attempting to draw inferences about patent activity by estimating the patent

---

[134] See, e.g., Dominique Guellec and Bruno van Pottelsberghe de la Potterie, The Impact of Public R&D Expenditure on Business R&D, OECD Science, Technology and Industry Working Papers 2000/4, OECD Publishing (2000).

[135] Ibid.

[136] See OECD (2011), Science, Technology and Industry Scoreboard (September 20, 2011); Danguy, de Rassenfosse and van Pottelsberghe de la Potterie, supra note 9.

[137] See, e.g., Bronwyn H. Hall, Adam B. Jaffe, and Manuel Trajtenberg, The NBER Patent Citations Data File: Lessons, Insights and Methodological Tools, NBER Working Paper No. 849 (2001), at 4.

[138] Patent statistics literature has irregularly considered this limitation. The earliest most important contribution begins with Professor Zvi Griliches' article titled Patent statistics as economic indicators: a survey, published in the *Journal of Economic Literature* 28, 1661–707 (1990); See also Daniele Archibugi and Mario Pianta, Measuring Technological Change through Patents and Innovation Surveys, *Technovation*, 16, 451 (1996).

production function, data must be collected by asking firms directly about the fraction of innovations they generally patent.[139] This approach allows for the estimation of the propensity to patent that is closely in line with the theoretical definition of the propensity to patent as the fraction of innovations that are accounted for as issued patents.

However, two additional methodological challenges remain concerning patent filing measurement for the developing countries per se. The first regards the use of patent propensity rates to measure the percentage of innovations for which a patent application is filed.[140] In the case of developing countries in particular, many patent applications often do not lead to patent issuance, neither nationally nor at the USPTO or EPO levels. This study therefore prefers the methodological definition of the propensity to patent presented above, as a percentage of patentable inventions that are actually patented.[141]

Another challenge concerning the method for patent panel data-counting relates to the particularities of the USPTO dataset. It maintains that patents are analyzed by the inventor's country name search categories. These categories contain the country or state of residence of the inventor at the time of patent issue.[142] The inventor's country name search category indicates the inventiveness of the local laboratories and labor force of a given country. This counting method enjoys three important advantages. Firstly, it replaces the alternative search categories of "Patent Affiliate" or "Owner," which mostly represent patenting activity by multinational enterprises originating in advanced economies.[143] Secondly, the

---

[139] Van Montfort Kleinknecht and Brouwer offer to replace patent/R&D rate analysis with measuring expenditure on innovation (including non-R&D-expenditure), sales of innovative products known which may be interpreted as an indicator of imitation, or otherwise innovation not introduced earlier by competitors, which may be interpreted as an indicator of "true" innovation. See Alfred Kleinknecht, Kees van Montfort and Erik Brouwer, The Non-trivial Choice between Innovation Indicators, *Economics of Innovation and New Technology*, 11, 109 (2002) (analyzing five alternative innovation indicators: R&D, patent applications, total innovation expenditure and shares in sales taken by imitative and by innovative products measured in the Netherlands), at 113–14.

[140] See, e.g., Cohen, Nelson, and Walsh, supra note 9; Arundel and Kabla, supra note 130; Emmanuel Duguet, and Isabelle Kabla, Appropriation Strategy and the Motivations to use the Patent System: An Econometric Analysis at the Firm Level in French Manufacturing, *Annales D'Économie et de Statistique*, 49/50, 289–327 (1998); Edwin Deering Mansfield, supra note 7.

[141] Edwin Deering Mansfield, ibid.

[142] United States Patent and Trademark Office (USPTO) (2012); Patent Full-Text and Image Database – Tips on Fielded Searching (Inventor Country (ICN)), at: www.uspto .gov/patft/help/helpflds.htm#Inventor_Country.

[143] OECD, Patent Statistics Manual (2009), at: www.oecdbookshop.org/en/browse/title-detail/?ISB=9789264056442; Emmanuel Hassan, Ohid Yaqub, and Stephanie Diepeveen, Intellectual Property and Developing Countries: A review of the literature (2010), supra note 9; Anna Bergek and Maria Bruzelius (2005), Patents with Inventors

inventor's country name search category operates to minimize transaction costs associated with domestic patenting by developing countries. Thirdly, in the case of co-inventions, all co-inventions are listed.[144] Indeed, the solution presented through the inventor country search category may account for either sole or co-inventions.

Another caveat applies regarding metric issues. Patent offices rarely make public actual grant rates. To illustrate, the EPO publishes the share of patents granted for a given year as a portion of the total number of patent *actions* in the same year, namely refusals, withdrawals, and grants. Official EPO grant rates for 2007–8 were approximately 50 percent, whereas the cohort approach adopted by Lazaridis and Van Pottelsberghe[145] proposes a grant rate that ranges from 60 to 65 percent throughout the 1990s.[146] As a result, grant rate indicators are at best prejudiced approximations of patent offices' strictness in their collection procedure.[147] Moreover, there is little or no information on type I and type II errors, referring to patents mistakenly granted or patents mistakenly refused, respectively. Van Pottelsberghe adds that it is highly probable that mistakenly granted patents are more widespread than patents that are mistakenly refused.[148]

On balance, the chapter's analysis incorporates two correction methods relating to this phenomenon of underestimation. Each method will be described below, followed by a comparison between the two. As we will see, both methods yield essentially the same corrected estimates, thereby strengthening the validity of the findings.

The first method employed WIPO data, organized separately for USPTO and EPO grant lag period. Thus, for the comparison of the two methods, both methods were considered separately for each office. However, in the final stage for each year and country, the patents corresponding to both offices were pooled together.

---

from Different Countries: Exploring Some Methodological Issues through a Case Study, presented at the DRUID conference, Copenhagen, 27–29 June.

[144] OECD, Patent Statistics Manual (2009), at: www.oecdbookshop.org/en/browse/title-detail/?ISB=9789264056442.

[145] See George Lazaridis and Bruno Van Pottelsberghe, The Rigor of EPO's Patentability Criteria: An Insight into the "Induced Withdrawals," CEB Working Paper N° 07/007 (April 2007).

[146] Despite the significant increase in patent applications since the mid-1980s the EPO's grant rate remained stable at around 65%. See Bruno van Pottelsberghe de la Potterie, The Quality Factor in Patent Systems, ECARES working paper 2010–027 (2010), at 9.

[147] See Bruno van Pottelsberghe de la Potterie, ibid., at 9; George Lazaridis and Bruno van Pottelsberghe de la Potterie, supra note 145 (showing that 35% to 40% of all withdrawals take place before the request for examination).

[148] See Bruno van Pottelsberghe de la Potterie, ibid., at 9; Geroge Lazaridis and Bruno van Pottelsberghe de la Potterie, supra note 145.

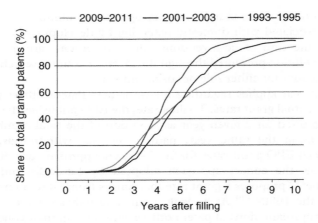

Graph A.1 EPO Pendency Time
(For years 1993–1995, 2001–2003, 2009–2011)

A.1.1 Pendency Time Estimation: The Two Methods

Method 1: Based on WIPO

WIPO pendency time – that is to say, the time it takes a patent office to process an application and decide whether to reject it or grant a patent – is represented in graphical displays for both the USPTO and the EPO offices.[149] Graphs A.1 and A.2 show the percentage of patents that were granted as a function of years after filing. These graphs are the distributions of pendency time as reported by the offices themselves. The WIPO *2013 World Intellectual Property Indicators* report explains that "Pendency time is calculated for granted patents only. Pendency time for patents that have been withdrawn, abandoned or refused are not included."[150] The three curves in each figure represent the years 1993–1995, 2001–2003, and 2009–2011, respectively.[151]

---

[149] See WIPO, World Intellectual Property Indicators (2013), at 87–88, and figure A.9.2 therein, titled "Distribution of pendency time for the top five offices."

[150] Ibid., at 87.

[151] It is assumed that patents are registered uniformly during each year. It is further assumed that the most recent curve, namely the one marked the lightest, is the most relevant as a basis for correction. Consider, for example, patents that were registered in the USPTO during 2012. Until the end of 2015, the time between registration and approval could not be longer than 3.5 years on average. According to the lightest curve, only 52% are usually granted during such a time interval. Going back to our previous example of a country that registered 7 patents in 2012, and only 3 out of them were granted during the following years up to 2015, the estimate of the correct number of final granted patents would be 6 (3/0.52= 5.77). I refer to the 52% as the relevant correction percent according to method 1. Similarly, the number of patents that were registered in the

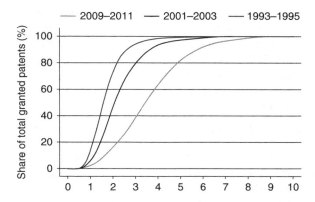

Graph A.2 USPTO Pendency Time
(for years 1993–1995, 2001–2003, 2009–2011)

The disadvantages of this method are that the same correction is applied to all countries, an approach that is reasonable only if pendency time does not differ between different countries. Moreover, variance in pendency time may also be found between different technologies. In addition, though the correction is based on the latest curve, this distribution may have changed. Nevertheless, this method is useful in order to compare its correction results and those obtained by means of the next suggested method.

Method 2: Based on Estimation
It is assumed that the percentage of patents granted out of those submitted has remained relatively stable over recent years. The USPTO granting rate for 2007 and 2008 were considered as representing the rate for the following years. Similarly, the EPO grant rate for 2004 and 2005 represents the rate for the following years.[152]

Like the former method, this one also applies equally to all countries. However, the present method can easily be modified by applying different correction factors to different countries. This flexibility is the main reason why this method is ultimately preferred.[153] The corrected factors for both

US by US inventors during 2012 and granted by 2015 was 36,354. Thus, after correction it will be 36,354/0.52 = 69911.54.

[152] These years were selected since they are far enough apart in the time series and thus could not miss patents with long pendency time. Nearly 100% of the registered patents in those years were already granted. Based on past data, it is noted that approval time was usually longer for those registered at the EPO compared with the USPTO. Therefore, different years were selected for both patent offices as representative.

[153] According to past records, the average acceptance rate for patents registered in the USPTO during 2007–2008 was: (0.64665+0.66634)/2 =0.65650, and for the EPO,

methods are quite close, as seen in Table A.1, which displays the correction percentages corresponding to each of the two methods.

The only year where a significant difference can be seen between the two methods is 2013, the last year in the series.[154] Accordingly, this year will not be included in the analyzed data set.

A.1.2   Data Pooling

Adding the numbers registered in both offices would lead to inflated numbers, since some patents were registered at both offices. The data available were the lists of each of the USPTO and EPO offices, as well as the pooled list of patents in which each patent was included only once. The following table displays the overlap of the USPTO and EPO lists (labeled *US* and *EP*, respectively). The pooled data are labeled as *UE*:

EP_UE specifies the proportion of patents that were registered in the EPO among the pooled list

EP_UE_ ONLY specifies the proportion of patents that were registered only in the EPO among the pooled list

US_UE specifies the proportion of patents that were registered in USPTO among the pooled list

US_UE_ ONLY specifies the proportion of patents that were registered only in the USPTO among the pooled list

By definition, *EP* is the proportion of patents that were registered at the EPO among the pooled list plus the proportion of patents that were registered only in the USPTO among the pooled list add up to one. Similarly, the proportion of patents that were registered only in the EPO among the pooled list plus the proportion of patents that were registered in USPTO among the pooled list total one. As Table A.2 shows, there are very minor changes during the years, and most of the patents were registered in the USPTO.

corresponding to years 2004–2005, it was (0.47549+0.46020)/2=0.46785. For each year, the ratio of the observed percent granted divided by the expected percent was derived (which is 0.65650 for those registered in the USPTO and 0.46785 for those granted at the EPO). Thus, during 2012, for example, the percent of granted patents was only 0.336 in the USPTO, and these are only 51% of the expected. I refer to 51% as the correction percent of the present method. The correction required would therefore be to divide the uncorrected granted number by 0.51. Since 0.51= observed granted rate/ expected granted rate: Corrected granted number = Uncorrected granted number x expected granted rate/ observed uncorrected granted rate= Number registered x expected granted rate. To conclude, when all countries are considered, the corrected granted number is the number of the registered times of the expected granted rate according to the past records. The periods 2007–2008 for USPTO and 2004–2005 for the EPO were defined as the reference years.

[154] This is not surprising, since this is the last year and the closest to 2015.

Table A.1 *Pendency Time Comparison between the Two Methods (2002–2013)*

Filing year	Method 1 USA	Method 2 USA	Method 1 Europe	Method 2 Europe
2002	100	114	100	111
2003	100	108	100	105
2004	100	103	100	102
2005	100	98	100	98
2006	100	97	93	91
2007	100	98	88	87
2008	100	101	80	82
2009	100	100	72	72
2010	90	93	60	59
2011	78	79	48	46
2012	54	51	30	24
2013	30	19	12	2

Table A.2 *Patent Data Pooling: USPTO and EPO (1996–2013)*

Obs	Earliest_filing_year	EP_UE	EP_UE_only	US_UE	US_UE_only
1	1996	0.31036	0.10200	0.89800	0.68964
2	1997	0.30547	0.10878	0.89122	0.69453
3	1998	0.30023	0.11339	0.88661	0.69977
4	1999	0.29966	0.11635	0.88365	0.70034
5	2000	0.26933	0.08627	0.91373	0.73067
6	2001	0.24037	0.08002	0.91998	0.75963
7	2002	0.23215	0.08174	0.91826	0.76785
8	2003	0.23279	0.08584	0.91416	0.76721
9	2004	0.22933	0.08650	0.91350	0.77067
10	2005	0.21673	0.08769	0.91231	0.78327
11	2006	0.21593	0.09168	0.90832	0.78407
12	2007	0.21456	0.09537	0.90463	0.78544
13	2008	0.20409	0.09796	0.90204	0.79591
14	2009	0.21819	0.10305	0.89695	0.78181
15	2010	0.21273	0.09605	0.90395	0.78727
16	2011	0.20969	0.09417	0.90583	0.79031
17	2012	0.19953	0.08884	0.91116	0.80047
18	2013	0.19166	0.08540	0.91460	0.80834

The final corrected numbers were then calculated. The percentage of granted patents in the data file (UE) during years 2004–2005 was considered as the expected granted rate labeled as: *Corrected granted number = Number registered x expected granted rate.* Thus, the expected granted rate for the overall list of patents was found to be (0.666+0.634)/2=0.65.

Table A.3 *Combined Patent Grant Rate (Percentage) (2002–2013)*

Earliest filing year	Grant rate
2002	0.73388
2003	0.69997
2004	0.66627
2005	0.63428
2006	0.62539
2007	0.62968
2008	0.64500
2009	0.63343
2010	0.58959
2011	0.50195
2012	0.32527
2013	0.11423

Table A.3 presents the combined percentage of granted patent rates obtained by applying this method to the UE data (USPTO and EPO countries together).

In practice, the correction was applied for each year and country according to its granted rate during 2004–2005. No correction was applied when the number after correction exceeded the number recorded as granted exceeded the corrected number.

A.2   GERD Data

The 1996–2013 yearly data for Purchasing Power Parity (PPP), expressed in US dollars and held constant at 2005 values,[155] were collected from the Science & Technology (S&T) database[156] of UNESCO. The PPP currency conversion largely eliminates the differences in price levels between countries and country groups.[157] Table A.4 below presents the data of all countries pooled together.

It should be recalled that the statistical analysis is based on 1996–2011 yearly ratios for each country, reflecting the number of granted patents of the next year divided by the GERD of current year. Table A.5 shows the available GERD data for each country and year; X denotes available data.

[155] United Nations Educational, Scientific and Cultural Organization (UNESCO), Glossary – 63 terms for science & technology, at: http://uis.unesco.org/en/glossary. As the UNESCO report explains, this methodology was adapted from OECD (2002), Frascati Manual, §423.
[156] United Nations Educational, Scientific and Cultural Organization (UNESCO) Science & Technology (S&T) database at: http://data.uis.unesco.org/.
[157] Ibid.

Table A.4 *Gross Domestic Expenditure on R&D (GERD) Combined (1996–2013)*

Year	N	Mean	Std Dev	Minimum	Maximum
Y1996	51	11812390.92	36094833.86	1944.00	237217325
Y1997	67	9539274.34	33502305.28	1348.00	250814421
Y1998	63	10429529.75	36237651.23	9993.00	264715164
Y1999	65	10850977.05	37798730.86	4867.00	282110915
Y2000	68	11394739.88	39549524.78	8609.00	302755131
Y2001	76	10714758.57	38254758.15	319.00	307788767
Y2002	87	9520059.64	35640207.94	946.00	302759910
Y2003	85	10137120.47	37135676.17	1515.00	311647193
Y2004	86	10594317.02	37576153.14	2046.00	315474390
Y2005	85	11140212.58	39656811.49	3970.00	328128000
Y2006	80	12739460.75	42918277.87	5118.00	342796382
Y2007	91	11716484.33	42608475.90	6980.00	359414718
Y2008	92	12531853.97	44579360.22	449.00	377452983
Y2009	91	12714229.59	45356711.71	409.000	373469274
Y2010	89	13595106.34	46844065.38	7287.00	372233631
Y2011	87	14675900.25	49708819.86	534.00	382105683
Y2012	74	17326387.59	56901633.48	6953.00	396710307
Y2013	61	14265624.72	42318754.30	6629.00	287444867

Due to missing data, imputation had to be applied. Since the required data for the analyses were only available through 2011, data for 2012 and 2013 were utilized in the imputation process. Nevertheless, some countries had to be deleted since their data were extremely sparse and imputation could not yield reliable estimates.

Though the GERD data has the structure of multiple longitudinal data, the imputation was performed for each country independently from the data of the other countries. The reason for this was that the main purpose of the analysis was to identify clusters of countries based on their longitudinal data. The use of cross-sectional data for the imputation could lead to falsely exaggerated similarities between countries. Two single imputation procedures were used, both drawn from the *imputeTS* R library,[158] which specializes in univariate time series imputation. These are the missing values imputation by *Kalman smoothing*, using *na.kalman* function, and missing value imputation by *interpolation*, using *na.interpolation*. The use of two different methods enabled the verification in the next stage of the clustering results yielded as based on imputed data.

---

[158] See Steffen Moritz, ImputeTS: Time Series Missing Value Imputation, R package version 1.5, at: https://CRAN.R-project.org/package=imputeTS (2016).

Table A.5 *Gross Domestic Expenditure on R&D (GERD) per Country (1996–2013)*

Group	1996	1997	1998	1999	2000	2001	2002	2003	2004	2005	2006	2007	2008	2009	2010	2011	2012	2013	Freq
1	X	X	X	X	X	X	X	X	X	X	X	X	X	X	X	X	X	X	29
2	X	X	X	X	X	X	X	X	X	X	X	X	X	X	X	X	X	.	4
3	X	X	X	X	X	X	X	X	X	X	X	X	X	X	X	X	.	.	2
4	X	X	X	X	X	X	X	X	X	X	.	.	.	.	.	.	.	X	1
5	X	X	X	X	X	X	X	X	.	.	.	.	.	.	.	.	.	.	1
6	X	X	X	X	X	.	X	.	X	.	X	.	X	X	X	X	X	.	1
7	X	X	X	X	X	.	X	X	X	X	X	X	X	X	X	X	X	X	1
8	X	X	X	X	X	X	X	X	X	X	X	X	X	X	X	X	X	X	1
9	X	X	X	X	X	X	X	X	X	X	X	X	X	X	X	X	X	X	1
10	X	X	X	.	.	X	X	X	X	X	X	X	X	X	X	X	.	.	1
11	X	X	X	.	X	X	X	X	X	X	X	X	X	X	X	X	X	X	1
12	X	X	X	.	X	X	X	X	X	X	X	X	X	X	X	X	X	X	1
13	X	X	X	.	.	X	X	X	X	X	X	X	X	X	X	.	.	X	1
14	X	X	.	.	.	.	.	.	X	X	X	X	X	X	.	.	.	.	1
15	X	X	.	X	X	.	X	.	X	X	X	X	X	.	.	.	.	X	1
16	X	X	X	.	.	.	.	.	X	X	X	X	X	X	X	X	X	X	1
17	X	X	X	.	.	.	.	.	.	.	.	.	X	.	.	.	.	.	1
18	X	X	.	.	.	.	.	.	.	.	X	.	.	.	.	.	.	.	1
19	X	.	.	.	.	.	.	.	.	.	.	.	.	.	.	.	X	X	7
20	.	X	X	X	X	X	X	X	X	X	X	X	X	X	X	X	X	X	2
21	.	X	X	X	X	X	X	X	X	X	X	X	X	X	X	X	.	.	1
22	.	X	X	X	X	X	X	X	X	X	X	X	X	X	X	X	X	X	1
23	.	X	X	X	X	X	X	X	.	X	X	X	X	X	.	.	X	.	1
24	.	X	.	.	.	.	.	.	.	.	.	.	.	.	.	.	.	.	1
25	.	X	X	X	.	.	.	X	X	X	X	X	X	X	X	X	X	X	1
26	.	X	.	.	.	.	.	X	X	X	X	X	X	X	X	X	X	X	2

```
                    1 1 1 2 1 1 1 1 1 2 1
      .  .  × ×  .  .  .  ×  .  × ×  .
      .  × × ×  .  × × × × × ×
      × × × ×  .  × × × × × ×
      .  × × × × × × × × × ×
      × × × ×  .  × × × × × ×
      .  × × ×  .  × × × × × ×
      × × × ×  .  × × × × × ×
      .  ×  .  × ×  .  × × × × ×
      × ×  .  × ×  .  .  × × × × ×
      .  ×  .  × ×  .  .  × × × × ×
      × ×  .  × × ×  .  × × × × ×
      .  .  .  × × ×  .  × ×  .  × × ×
      × ×  .  × × ×  .  × ×  .  ×  .  .
      .  .  .  × ×  .  .  × × ×  .  .  .
      ×  .  .  × ×  .  .  ×  .  .  .  .  .
      .  .  .  × × × ×  .  .  .  .  .  .
      × × ×  .  .  .  .  .  .  .  .  .  .
      .  .  .  .  .  .  .  .  .  .  .  .
27 28 29 30 31 32 33 34 35 36 37 38 39
```

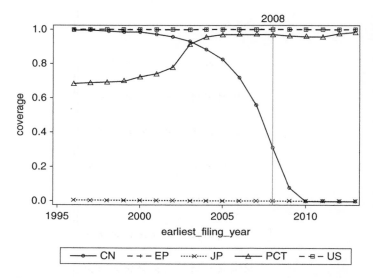

Graph A.3 Patent Filings Coverage for JPO, SIPO, EP, USPTO, and PCT (1996–2013)
Legend: CN – SIPO, EP – EPO, JP – JPO, PCT – PCT, US – USPTO

A.3   JPO and SIPO Missing Data

The value of patents connotes an association with the geographical scope of patent protection, measured by the number of jurisdictions in which patent protection is sought.[159] Yet, while focusing on the leading patent offices, the data on patent filings in the Japanese and Chinese patent offices were not included in the analysis. For each year and patent office, the record notes whether the inventors' country is known or unknown. Graph A.3 describes the yearly coverage for each of the five patent offices, namely the Japanese Patent Office (JPO), the Chinese State Intellectual Property Office (SIPO), EPO, USPTO, and PCT patent filings. The coverage for the year and authority was defined as the proportion of applications (for year – patent office combination) for which the inventors' country is known (calculated as the number of applications minus the number of applications for which the inventors' country is unknown), divided by the total number of applications for the year – patent office combination.

---

[159] See Jean O. Lanjouw, Ariel Pakes, and Jonathan Putnam, How to Count Patents and Value Intellectual Property: The Uses of Patent Renewal and Application Data, *Journal of Industrial Economics*, 46(4) 404 (1998). In correlation, large international patent families have been found to be particularly valuable. See Dietmar Harhoff, Frederic M. Scherer, and Katrin Vopel, supra note 114.

As Graph A.3 shows, both the USPTO and EPO exhibit high coverage values, and PCT has also shown high coverage since 2004. The high coverage before 2002 for the USPTO may be misleading, since there is no official information before March 2001 regarding non-granted applications.

The coverage of the JPO is practically zero, while that of SIPO is decreasing dramatically toward zero. The association between patents and various country parameters were examined. Accordingly, without knowing the inventors' country the data are useless.

A.4   Data Methodology and Interpretation

The methodology used throughout this book adheres to the conceptualization and critique put forth by two formative OECD statistical manuals: the Frascati Manual (2002) on R&D & GERD-related statistics[160] and the OECD/Eurostat Oslo Manual (2005) on innovation-related statistics.[161] In principle, both manuals emphasize the need to move beyond normative posturing by stakeholders, role players, and policy-makers and to adopt empirical observations. For the past 50 years, the OECD's Frascati Manual has constituted the gold standard for the internationally comparable measurement of R&D & GERD of OECD member states and associated observer states.[162] This standard is applied with reference to two additional manuals. The first is UNESCO's Technical Paper No. 5, titled: Measuring R&D: Challenges Faced by Developing Countries (2010).[163] This manual provides guidance relating to several methodological challenges that are relevant to developing countries and that may have not been elaborated clearly enough in the Frascati Manual. The second is the OECD's Patent Statistics Manual of 2009,[164] which provides users and producers of patent statistics with basic guidelines for compiling and analyzing such data. Both manuals confirm the Frascati Manual as the most widely accepted international standard practice for R&D & GERD-related surveys.[165]

---

[160] OECD (2002), Proposed Standard Practice for Surveys on Research and Experimental Development (Paris: OECD) (Frascati Manual).

[161] OECD and Eurostat (2005), Oslo Manual: Guidelines for Collecting and Interpreting Innovation Data (Paris: OECD) (Oslo Manual).

[162] See generally, Benoît Godin, On the Origins of Bibliometrics, *Scientometrics*, 68 (1) 109–33 (2006).

[163] United Nations Educational, Scientific and Cultural Organization (UNESCO) (2010), Technical Paper No. 5, Measuring R&D: Challenges Faced by Developing Countries, ibid.

[164] OECD, Patent Statistics Manual (2009), at: www.oecdbookshop.org/en/browse/title-detail/?ISB=9789264056442.

[165] United Nations Educational, Scientific and Cultural Organization (UNESCO) (2010), Technical Paper No. 5, supra note 163. This chapter adheres to these methodologies while entailing a series of statistical analysis using Statistical Analysis System (SAS) software.

Appendix B   Clustering Procedure: Technical Description

B.1   Imputation

The clustering procedure was applied on the imputed data. As explained above, two single imputation procedures were both drawn from the *imputeTS* R library,[166] which specializes in univariate time series imputation. These are the missing values imputation by *Kalman smoothing*, using *na.kalman* function, and the missing value imputation by *interpolation*, using *na.interpolation*. The use of two different methods enabled the verification in the next stage of the clustering results, yielded as based on imputed data.

B.2   Clustering Analysis

B.2.1   Assessing Clustering Tendency

The first step was to examine whether the datasets are clusterable. The *get_clust_tendency* function of *factoextra* R library was used to estimate the Hopkins statistics.[167] The result of 0.23 is below the critical value of 0.5, and accordingly it was concluded that the datasets are clusterable.

B.2.2   Clustering Methods

Of the large number of clustering methods available, k-means and hierarchical clustering are currently two of the most popular. Both were applied in our study. The k-means method aims to partition the points into k groups such that the sum of squares from points to the assigned cluster centers is minimized. Hierarchical Ward's clustering method, which is a minimum variance method, aims to locate compact spherical clusters. The *k-means* R function of stats R library was used to implement the k-means method and the *hclust* R function of stats R library to implement Ward's clustering method.[168]

---

[166] See Steffen Moritz, supra note 158.
[167] Alboukadel Kassambara and Fabian Mundt, Factoextra: Extract and Visualize the Results of Multivariate Data Analyses, R package version 1.0.3. www.sthda.com/english/rpkgs/factoextra (2016).
[168] See R Core Team, R: A Language and Environment for Statistical Computing, R Foundation for Statistical Computing, Vienna, Austria, at: www.R-project.org/ (2016).

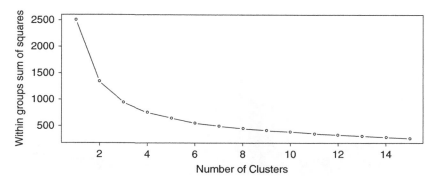

Graph B.1 Groups Sum of Squares vs. the No. of Clusters by the K-Means

B.3 Determining the Optimal Number of Clusters

None of the methods for determining the optimal number of clusters is considered optimal,[169] and none of the suggested indices are completely satisfactory.[170] By using several criteria which gave the same result for the optimal number of clusters, a unanimous answer has been reached – namely, that there is a correct number of clusters. The visual inspection of two clustering methods was undertaken. The first measures the groups' sum of squares versus the number of clusters for k-means and looks for an elbow. The second clustering method uses a hierarchical cluster analysis dendogram. Based on the results of each of the two clustering methods, namely the k-means and hierarchical clustering, the NbClust function was applied providing several indices that are used to assess the optimal numbers of clusters.[171] The results of the two methods are described below.

B.3.1 K-Means

Graph B.1 displays a visual inspection of the groups' sum of squares versus the number of clusters for k-means and looking for an elbow.

[169] See Brian S. Everitt, Sabine Landau, Morven Leese, and Daniel Stahl, *Cluster Analysis* (Arnold, 5th edn., 2011); Glenn W. Milligan and Martha C. Cooper, An Examination of Procedures for Determining the Number of Clusters in a Data Set, *Psychometrika*, 50(2), 159 (1985).

[170] Yosung Shim, Ji-won Chung and In-chan Choi, A Comparison Study of Cluster Validity Indices using a Nonhierarchical Clustering Algorithm, Proceedings – International Conference on Computational Intelligence for Modeling, Control and Automation, CIMCA 2005 and International Conference on Intelligent Agents, Web Technologies and Internet, vol. 1, 199 (2005).

[171] See Malika Charrad, Nadia Ghazzali, Véronique Boiteau, and Azam Niknafs, NbClust: An R Package for Determining the Relevant Number of Clusters in a Data Set, *Journal of Statistical Software*, vol. 61(6) (2014).

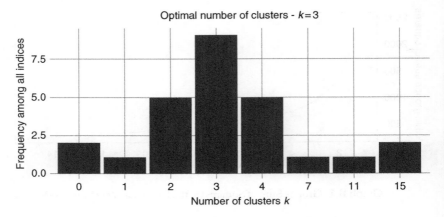

Graph B.2 Optimal Number of Clusters by K-Means

According to *NbClust*, a majority of nine indices suggested three as the best number of clusters.[172] Thus, according to majority rule, the best number of clusters is three. Graph B.2 displays these results.

B.3.2   Hierarchical Clustering

According to *NbClust*, a majority of nine indices proposed three as the best number of clusters.[173] In conclusion, according to the majority rule, the best number of clusters according to hierarchical clustering is three. Graph B.3 displays these results by this clustering method.

B.4   Internal Clustering Validation

Three clusters were identified based on each of the two clustering methods used (k-means and hierarchical cluster analysis with Ward's minimum variance method). An internal clustering validation was then applied using a silhouette analysis. This technique provides a succinct graphical

---

[172] The other proposals based on k-means were as follows: (1) 5 proposed 2 as the best number of clusters, (2) 5 proposed 4 as the best number of clusters, (3) 1 proposed 7 as the best number of clusters, (4) 1 proposed 11 as the best number of clusters, and (5) 2 proposed 15 as the best number of clusters.

[173] The other proposals based on hierarchical clustering were as follows: (1) 2 proposed 0 as the best number of clusters, (2) 1 proposed 1 as the best number of clusters, (3) 4 proposed 2 as the best number of clusters, (4) 1 proposed 4 as the best number of clusters, (5) 2 proposed 5 as the best number of clusters, (6) 1 proposed 6 as the best number of clusters, (7) 2 proposed 8 as the best number of clusters, (8) 1 proposed 13 as the best number of clusters, (9) 2 proposed 14 as the best number of clusters, and (10) 1 proposed 15 as the best number of clusters.

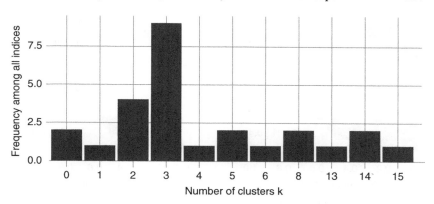

Graph B.3 Optimal No. of Clusters by Hierarchical Clustering

representation of how well each object lies within its cluster. The silhouette value obtained for each object is a measure of how similar it is to its own cluster (cohesion) compared to other clusters (separation). The silhouette values range from −1 to 1, where a high value indicates that the object is well matched to its own cluster and poorly matched to neighboring clusters.[174]

In Silhouette Plots B.1 and B.2, the left-hand side corresponds to the structure obtained by the k-means method, and the right-hand side to the hierarchical method. Next to each cluster is displayed the number of objects, i.e. countries, and the average silhouette values of its countries.

As seen in Silhouette Plot B.1, the first cluster (dark gray, upper form) includes countries with negative silhouette values in both clustering methods. Based on the silhouette values, few minor modifications were made in the clustering structure. The position of each country i that had a negative silhouette value was replaced with its neighbor cluster, namely the cluster not containing country i for which the average dissimilarity between its countries and i is minimal. Silhouette Plot B.2 depicts the clustering validation after the modifications.

After the modifications, the clusters obtained by the two clustering methods were identical, thus validating the said three clusters by patent propensity and GERD intensity.

---

[174] See Peter J. Rousseeuw, Silhouettes: A Graphical Aid to the Interpretation and Validation of Cluster Analysis, *Journal of Computational and Applied Mathematics*, 20, 53 (1987).

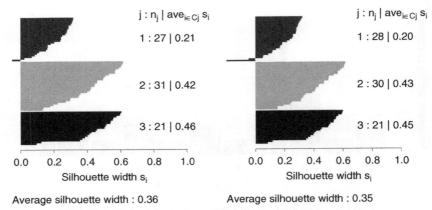

Silhouette Plot B.1 Clustering Validation (by *k*-means and hierarchical clustering)

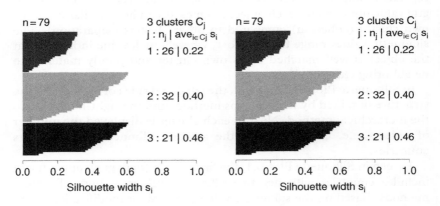

Silhouette Plot B.2 Clustering Validation (by *k*-means and hierarchical clustering) without Negative Values

B.5   Characterizing Clustering Structure: Applying CART Algorithm

In order to characterize the clustering structure more precisely, the CART algorithm was applied using *rpart* function of *rpart* R library.[175]

---

[175] Terry Therneau, Beth Atkinson, and Brian Ripley, rpart: Recursive Partitioning and Regression Trees (2015) (R package version of clustering based on the imputed data by interpolation).

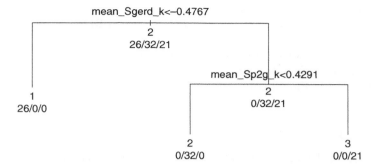

Chart B.1 Classifying Clusters by Mean Standardized Square Root (Patents to GERD) and Mean of Standardized Log (GERD)

As an input for each country, two variables were used for creating the clusters, namely the mean standardized log GERD and the mean of the standardized square root of the ratio.

As the tree obtained in Chart B.1 below clearly shows, the marginalized cluster (dark gray, no. 1) with the 26 countries is characterized by low GERD values, while between the other two clusters the discriminating variable is the ratio of patents by GERD which is low for the followers cluster (light gray, no. 2) of 32 countries and high for the leaders cluster (dark, no. 3) with 21 countries.[176]

Appendix C   Relationship between Patent Activity Intensity Indicators and Clusters

Legend

Code	Country code
AdEco	Economy category
Cont	Continent
Cluster_km_3b	Patenting cluster (according to *k*-means)

---

[176] The same stages of the clustering procedure were conducted on the imputed data by interpolation, as we did on the imputed data by the Kalman smoothing. For brevity, we shall display only part of the results, since they were essentially identical to those obtained previously, that were based on the imputed data by the Kalman smoothing. Among all indices by *k*-means according to NbClust, the majority of 11 proposed 3 as the best number of clusters. Among all indices by hierarchical clustering according to NbClust, a majority of 10 proposed 3 as the best number of clusters. Exactly the same clusters were formed both by the K means and the hierarchical method based on both versions of imputation.

Mean_Sp2g_k	Mean of the standardized square root of the ratio: number of patents divided by the GERD of previous year
Mean_SGerd_k	Mean standardized log GERD
S_p2gg_pr.country	Country name
M_p_UE_all	Patent grant rate in the UE (in %)

LEAF 1

The following is the list of the 20 countries that were correctly classified in the marginalized cluster (labeled 1) and are Other developing countries (labeled 3).

	code	register_rd	AdEco	Cont	cluster_ km_3b	mean_ Sp2g_k	mean_ SGerd_k S_p2gg_pr	country
1	AM	0.231932773	3	3	1	0.74696787	-1.4433210	ARMENIA
4	AZ	0.087438424	3	3	1	-0.89018495	-0.7678047	AZERBAIJAN
6	BF	0.000000000	3	4	1	-1.44239905	-1.4551008	BURKINA FASO
9	BO	0.291666667	3	1	1	-0.94033977	-0.9712226	BOLIVIA
16	CR	0.443243243	3	1	1	0.75471921	-0.7582162	COSTA RICA
22	EC	0.220000000	3	1	1	-0.34012915	-0.8866879	ECUADOR
29	GE	0.069078599	3	3	1	0.48537778	-1.3405683	GEORGIA
40	KG	0.000000000	3	3	1	-1.32010556	-1.5628973	KYRGYZSTAN
42	KW	0.941722973	3	3	1	0.45116926	-0.5940816	KUWAIT
44	LK	0.219219219	3	3	1	-0.24866899	-0.7187510	SRI LANKA
48	MD	0.004172235	3	3	1	-0.51556321	-1.2026794	MOLDOVA
49	MG	0.333333333	3	4	1	-0.98802348	-1.3270258	MADAGASCAR
50	MK	0.055555556	3	2	1	-1.09065542	-1.2495002	MACEDONIA
51	MN	0.118055556	3	3	1	-0.68363196	-1.4160180	MONGOLIA
54	MU	0.111111111	3	4	1	-1.31369244	-1.2722990	MAURITIUS
60	PA	0.422222222	3	1	1	-0.08207209	-0.9833690	PANAMA
71	SV	0.250000000	3	1	1	-0.19455517	-1.3966584	EL SALVADOR
73	TJ	0.000000000	3	3	1	-1.53309083	-1.9253368	TAJIKISTAN
76	TT	0.326388889	3	1	1	0.30250112	-1.5416874	TRINIDAD AND TOBAGO
79	UY	0.221951219	3	1	1	-0.25427328	-0.8250091	URUGUAY

The following is the list of the 9 countries that were wrongly classified in the marginalized cluster (labeled 1) and are Other developing countries (labeled 3).

	code	m_p_UE_all	AdEco	Cont	cluster_ km_3b	mean_ Sp2g_k	mean_ SGerd_k S_p2gg_pr	country
11	BY	0.04885535	3	3	2	-0.8168995	-0.17840076	BELARUS
15	CO	0.12382519	3	1	2	-0.5140681	-0.16100282	COLOMBIA
17	CU	0.02013557	3	1	2	-1.3076374	-0.15048139	CUBA
24	EG	0.25906433	3	4	2	-0.6963779	0.14921245	EGYPT
32	HR	0.08803908	3	1	2	-0.5412137	-0.21722531	CROATIA
43	KZ	0.02227361	3	3	2	-1.2638272	-0.34096013	KAZAKHSTAN
47	MA	0.10639666	3	4	2	-1.0416434	-0.07411728	MOROCCO
65	RS	0.05256036	3	2	2	-0.9068045	-0.39896955	SERBIA
74	TN	0.22366071	3	4	2	-1.0007306	-0.27712597	TUNISIA

LEAF 2

Leaf 2 lists 17 countries, all of which are emerging economy countries (labeled 2), that were correctly classified in the followers cluster (labeled 2). The following is the list of these countries.

	code	m_p_UE_all	AdEco	Cont	cluster_ km_3b	mean_ Sp2g_k	mean_ SGerd_k S_p2gg_pr	country
7	BG	0.20134576	2	2	2	0.25051118	-0.40236267	BULGARIA
10	BR	0.28289032	2	1	2	-0.98224745	1.25917901	BRAZIL
14	CN	0.10502239	2	3	2	-0.48031173	1.71291054	CHINA
33	HU	0.17779260	2	1	2	-0.10665264	0.15861754	HUNGARY
36	IN	0.44423206	2	3	2	-0.08011805	1.28825128	INDIA
55	MX	0.18505081	2	1	2	-0.46591631	0.64751157	MEXICO
56	MY	0.29613267	2	3	2	0.25562197	0.34759064	MALAYSIA
61	PK	0.23863636	2	3	2	-1.00082915	0.14682725	PAKISTAN
62	PL	0.07777716	2	2	2	-0.44983002	0.49294284	POLAND
64	RO	0.06993718	2	2	2	-0.36638629	-0.02800305	ROMANIA
66	RU	0.01481855	2	3	2	-0.72234888	1.16920086	RUSSIA
72	TH	0.42360905	2	3	2	-0.17709976	0.14584301	THAILAND
75	TR	0.11333315	2	2	2	-0.76283492	0.59920174	TURKEY
77	UA	0.02365113	2	3	2	-0.71073732	0.41382451	UKRAINE
80	ZA	0.15842491	2	4	2	-0.61214936	0.50788007	SOUTH AFRICA

Leaf 2 also includes two countries that are emerging economy countries (labeled 2) and which were incorrectly classified in the followers cluster (labeled 1), while they were actually in the followers cluster (labeled 2). The following is the list of these countries.

	code	m_p_UE_all	AdEco	Cont	cluster_ km_3b	mean_ Sp2g_k	mean_ SGerd_k S_p2gg_pr	country
23	EE	0.3082933	2	2	1	0.003956716	-0.6845334	ESTONIA
46	LV	0.0720624	2	2	1	-0.211877559	-0.8483230	LATVIA

LEAF 3

Leaf 3 includes nine countries, all advanced economies (labeled 1), but low in patent grant rates in percentage in the UE. Those countries were correctly classified in the followers cluster (labeled 2). The following is the list of these countries.

	code	m_p_UE_all	AdEco	Cont	cluster_ km_3b	mean_ Sp2g_k	mean_ SGerd_k S_p2gg_pr	country
19	CZ	0.2307749	1	2	2	-0.20531707	0.3961222	CZECH REPUBLIC
25	ES	0.2534840	1	1	2	-0.14153728	1.0201479	SPAIN
30	GR	0.2055182	1	1	2	-0.39235625	0.1655359	GREECE
58	NO	0.2780320	1	2	2	0.09034316	0.5197102	NORWAY
63	PT	0.1772633	1	2	2	-0.67253554	0.2963512	PORTUGAL
69	SI	0.1514822	1	2	2	-0.35861486	-0.1398156	SLOVENIA
70	SK	0.1446864	1	2	2	-0.52103835	-0.2493788	SLOVAKIA

Leaf 3 also includes Cyprus, which is an advanced economy (labeled 1) with a low patent grant rate in the UE. It was incorrectly classified in the leaders cluster (labeled 1), but actually lies in the marginalized cluster (labeled 3).

	code	m_p_UE_all	AdEco	Cont	cluster_ km_3b	mean_ Sp2g_k	mean_ SGerd_k S_p2gg_pr	country
18	CY	0.2744361	1	2	1	0.3202078	-1.125948	CYPRUS

Leaf 3 also includes South Korea, which is an advanced economy (labeled 1) with a low patent grant rate in the UE. It was incorrectly classified in the followers cluster (labeled 2), but actually lies in the leaders cluster (labeled 3).

	code	m_p_UE_all	AdEco	Cont	cluster_ km_3b	mean_ Sp2g_k	mean_ SGerd_k S_p2gg_pr	country
41	KR	0.1208257	1	3	3	2.011998	1.376326	REPUBLIC OF KOREA

LEAF 4

Leaf 4 includes 24 countries, all of which are advanced economies (labeled 1) with a high patent grant rate in the UE. Twenty of those countries were correctly classified in the leaders cluster (labeled 3). The following is the list of these countries.

	code	m_p_UE_all	AdEco	Cont	cluster_ km_3b	mean_ Sp2g_k	mean_ SGerd_k S_p2gg_pr	country
2	AT	0.3885877	1	1	3	1.0194936	0.74657069	AUSTRIA
5	BE	0.4598161	1	1	3	0.8570452	0.76956422	BELGIUM
12	CA	0.6066608	1	1	3	1.6042957	1.24434680	CANADA
13	CH	0.4534808	1	2	3	1.6658175	0.83602265	SWITZERLAND
20	DE	0.3154631	1	2	3	1.4725198	1.71296686	GERMANY
21	DK	0.3466832	1	2	3	0.6726657	0.61003559	DENMARK
26	FI	0.3572485	1	1	3	0.9901177	0.68249303	FINLAND
27	FR	0.3032434	1	1	3	0.6757233	1.50820562	FRANCE
28	GB	0.3300398	1	1	3	0.6113102	1.44014903	UNITED KINGDOM
31	HK	0.6570821	1	3	3	1.7379545	0.16397639	HONG KONG
34	IE	0.4316739	1	1	3	1.0087690	0.26698334	IRELAND
35	IL	0.5643077	1	3	3	1.4891059	0.76973806	ISRAEL
38	IT	0.4737708	1	1	3	0.8252644	1.19343163	ITALY
39	JP	0.5330784	1	3	3	1.9399828	1.96365730	JAPAN
45	LU	0.5187700	1	2	3	0.8369549	-0.27176384	LUXEMBOURG
57	NL	0.3119621	1	2	3	1.0102601	0.97803037	NETHERLANDS
59	NZ	0.2852241	1	4	3	0.6191730	0.06344327	NEW ZEALAND
67	SE	0.3212841	1	2	3	0.6026290	0.96088605	SWEDEN
68	SG	0.4521211	1	3	3	0.7745592	0.61815809	SINGAPORE
78	US	0.8464213	1	1	3	1.9769280	2.35782183	UNITED STATES

Leaf 4 also includes three countries that were incorrectly classified in the leaders cluster (labeled 3), but actually lies in the marginalized cluster (labeled 1). The following is the list of these countries:

	code	m_p_UE_all	AdEco	Cont	cluster_km_3b	mean_Sp2g_k	mean_SGerd_k S_p2gg_pr	country
37	IS	0.3571303	1	1	1	0.4856342	-0.5510799	ICELAND
52	MO	0.4166667	1	3	1	0.1928672	-1.6422122	MACAO
53	MT	0.4423913	1	2	1	0.5651882	-1.3490090	MALTA

Leaf 4 also includes Australia, which is an advanced economy (labeled 1) with a patent grant rate in the UE. It was incorrectly classified in the leaders cluster (labeled 3), though it actually lies in the followers cluster (labeled 2).

	code	m_p_UE_all	AdEco	Cont	cluster_km_3b	mean_Sp2g_k	mean_SGerd_k S_p2gg_pr	country
3	AU	0.3119423	1	4	2	0.03969361	1.02928	AUSTRALIA

The following lists help to clarify the relationship between the economy type and the clustering structure:

i)   Advanced Economies

	code	m_p_UE_all	AdEco	Cont	cluster_km_3b	mean_Sp2g_k	mean_SGerd_k S_p2gg_pr	Country
18	CY	0.274436090	1	2	1	0.320207786	-1.12594827	CYPRUS
37	IS	0.357130313	1	1	1	0.485634195	-0.55107987	ICELAND
52	MO	0.416666667	1	3	1	0.192867156	-1.64221221	MACAO
53	MT	0.442391304	1	2	1	0.565188226	-1.34900901	MALTA
3	AU	0.311942286	1	4	2	0.039693607	1.02928025	AUSTRALIA
19	CZ	0.230774880	1	2	2	-0.205317070	0.39612216	CZECH REPUBLIC
25	ES	0.253483971	1	1	2	-0.141537285	1.02014792	SPAIN
30	GR	0.205518200	1	1	2	-0.392356249	0.16553586	GREECE
58	NO	0.278032010	1	2	2	0.090343158	0.51971020	NORWAY
63	PT	0.177263287	1	2	2	-0.672535537	0.29635120	PORTUGAL
69	SI	0.151482227	1	2	2	-0.358614859	-0.13981561	SLOVENIA
70	SK	0.144686411	1	2	2	-0.521038355	-0.24937875	SLOVAKIA
2	AT	0.388587732	1	1	3	1.019493647	0.74657069	AUSTRIA
5	BE	0.459816101	1	1	3	0.857045164	0.76956422	BELGIUM
12	CA	0.606660788	1	1	3	1.604295665	1.24434680	CANADA
13	CH	0.453480826	1	2	3	1.665817498	0.83602265	SWITZERLAND
20	DE	0.315463125	1	2	3	1.472519840	1.71296686	GERMANY
21	DK	0.346683187	1	2	3	0.672665693	0.61003559	DENMARK

(cont.)

	code	m_p_UE_all	AdEco	Cont	cluster_km_3b	mean_Sp2g_k	mean_SGerd_k S_p2gg_pr	Country
26	FI	0.357248528	1	1	3	0.990117700	0.68249303	FINLAND
27	FR	0.303243390	1	1	3	0.675723269	1.50820562	FRANCE
28	GB	0.330039847	1	1	3	0.611310227	1.44014903	UNITED KINGDOM
31	HK	0.657082148	1	3	3	1.737954487	0.16397639	HONG KONG
34	IE	0.431673878	1	1	3	1.008768971	0.26698334	IRELAND
35	IL	0.564307719	1	3	3	1.489105941	0.76973806	ISRAEL
38	IT	0.473770817	1	1	3	0.825264410	1.19343163	ITALY
39	JP	0.533078417	1	3	3	1.939982834	1.96365730	JAPAN
41	KR	0.120825730	1	3	3	2.011997570	1.37632556	REPUBLIC OF KOREA
45	LU	0.518770007	1	2	3	0.836954950	-0.27176384	LUXEMBOURG
57	NL	0.311962146	1	2	3	1.010260093	0.97803037	NETHERLANDS
59	NZ	0.285224103	1	4	3	0.619173038	0.06344327	NEW ZEALAND
67	SE	0.321284118	1	2	3	0.602628971	0.96088605	SWEDEN
68	SG	0.452121107	1	3	3	0.774559174	0.61815809	SINGAPORE
78	US	0.846421297	1	1	3	1.976928004	2.35782183	UNITED STATES

ii)   *Emerging Economies*

	code	m_p_UE_all	AdEco	Cont	cluster_km_3b	mean_Sp2g_k	mean_SGerd_k S_p2gg_pr	Country
23	EE	0.308293269	2	2	1	0.003956716	-0.68453340	ESTONIA
46	LV	0.072062402	2	2	1	-0.211877559	-0.84832296	LATVIA
7	BG	0.201345756	2	2	2	0.250511185	-0.40236267	BULGARIA
10	BR	0.282890325	2	1	2	-0.982247450	1.25917901	BRAZIL
14	CN	0.105022395	2	3	2	-0.480311732	1.71291054	CHINA
33	HU	0.177792596	2	1	2	-0.106652644	0.15861754	HUNGARY
36	IN	0.444232059	2	3	2	-0.080118048	1.28825128	INDIA
55	MX	0.185050813	2	1	2	-0.465916311	0.64751157	MEXICO
56	MY	0.296132666	2	3	2	0.255621973	0.34759064	MALAYSIA
61	PK	0.238636364	2	3	2	-1.000829150	0.14682725	PAKISTAN
62	PL	0.077777161	2	2	2	-0.449830016	0.49294284	POLAND
64	RO	0.069937177	2	2	2	-0.366386287	-0.02800305	ROMANIA
66	RU	0.014818552	2	3	2	-0.722348879	1.16920086	RUSSIA
72	TH	0.423609046	2	3	2	-0.177099760	0.14584301	THAILAND
75	TR	0.113333153	2	2	2	-0.762834917	0.59920174	TURKEY
77	UA	0.023651129	2	3	2	-0.710737322	0.41382451	UKRAINE
80	ZA	0.158424908	2	4	2	-0.612149360	0.50788007	SOUTH AFRICA

iii)   *Other Developing Countries*

	code	m_p_UE_all	AdEco	Cont	cluster_ km_3b	mean_ Sp2g_k	mean_ SGerd_k S_p2gg_pr	Country
1	AM	0.231932773	3	3	1	0.746967867	-1.44332098	ARMENIA
4	AZ	0.087438424	3	3	1	-0.890184954	-0.76780467	AZERBAIJAN
6	BF	0.000000000	3	4	1	-1.442399046	-1.45510077	BURKINA FASO
9	BO	0.291666667	3	1	1	-0.940339765	-0.97122264	BOLIVIA
16	CR	0.443243243	3	1	1	0.754719213	-0.75821618	COSTA RICA
22	EC	0.220000000	3	1	1	-0.340129145	-0.88668792	ECUADOR
29	GE	0.069078599	3	3	1	0.485377779	-1.34056832	GEORGIA
40	KG	0.000000000	3	3	1	-1.320105557	-1.56289728	KYRGYZSTAN
42	KW	0.941722973	3	3	1	0.451169255	-0.59408156	KUWAIT
44	LK	0.219219219	3	3	1	-0.248668990	-0.71875103	SRI LANKA
48	MD	0.004172235	3	3	1	-0.515563210	-1.20267944	MOLDOVA
49	MG	0.333333333	3	4	1	-0.988023484	-1.32702578	MADAGASCAR
50	MK	0.055555556	3	2	1	-1.090655421	-1.24950020	MACEDONIA
51	MN	0.118055556	3	3	1	-0.683631956	-1.41601800	MONGOLIA
54	MU	0.111111111	3	4	1	-1.313692437	-1.27229896	MAURITIUS
60	PA	0.422222222	3	1	1	-0.082072088	-0.98336900	PANAMA
71	SV	0.250000000	3	1	1	-0.194555169	-1.39665839	EL SALVADOR
73	TJ	0.000000000	3	3	1	-1.533090827	-1.92533682	TAJIKISTAN
76	TT	0.326388889	3	1	1	0.302501123	-1.54168740	TRINIDAD AND TOBAGO
79	UY	0.221951219	3	1	1	-0.254273281	-0.82500908	URUGUAY
11	BY	0.048855355	3	3	2	-0.816899509	-0.17840076	BELARUS
15	CO	0.123825188	3	1	2	-0.514068105	-0.16100282	COLOMBIA
17	CU	0.020135566	3	1	2	-1.307637355	-0.15048139	CUBA
24	EG	0.259064328	3	4	2	-0.696377950	0.14921245	EGYPT
32	HR	0.088039084	3	1	2	-0.541213679	-0.21722531	CROATIA
43	KZ	0.022273613	3	3	2	-1.263827163	-0.34096013	KAZAKHSTAN
47	MA	0.106396657	3	4	2	-1.041643388	-0.07411728	MOROCCO
65	RS	0.052560363	3	2	2	-0.906804494	-0.39896955	SERBIA
74	TN	0.223660714	3	4	2	-1.000730621	-0.27712597	TUNISIA

A macro review of all the previously reported results yielded insights concerning the relationships between the clustering structure and the characteristics of the countries in each cluster.

3 Institutions, GERD Intensity, and Patent Clusters

Introduction

This chapter offers an empirical and theoretical critique of the principal impact of institutions on patent intensity across countries. To date, innovation-based economic growth theory has emphasized that R&D, and particularly internationalized R&D, should be promoted by MNCs worldwide.[1] Such R&D activity is also strongly associated with a higher yield of patenting activity, as measured by comparable national patent propensity rates. Yet across the board, current literature supporting this approach focuses on advanced or developed countries. Accordingly, it is unsurprising that there are a large number of scientific studies on this phenomenon that clearly reflect the experience of advanced economies, nor that several of these studies show the increasing internationalization of innovative activity, primarily through R&D by MNCs in such countries.[2] Across the development divide, numerous other examples reinforce the impression that internationalized R&D and the propensity to patent in emerging economies is a leading institutional choice. Examples include Motorola's first foreign-owned R&D lab in China, opened in 1993; the India-based R&D activities of General Electric in areas as diverse as aircraft engines, consumer durables, and medical equipment; and the presence of pharmaceutical companies such as Abbott, Sanofi,

---

[1] See, e.g., Frieder Meyer-Krahmer and Guido Reger, New perspectives on the innovation strategies of multinational enterprises: Lessons for technology policy in Europe, *Research Policy*. vol. 28, 751 (1999), at 758–63. See also discussion herein.

[2] OECD, Compendium of Patent Statistics, Economic Analysis and Statistics Division of the OECD Directorate for Science, Technology and Industry (2004); Daniele Archibugi and Alberto Coco, The Globalization of Technology and the European Innovation System, IEEE Working Paper DT09/2001. No (2001); Parimal Patel and Modesto Vega, Patterns of internationalization of corporate technology: location vs. home country advantages, *Research Policy*. vol. 28, No. 145–55 (1999); Alexander Gerybadze and Guido Reger, Globalization of R&D: Recent changes in the management of innovation in transnational corporations, *Research Policy*, vol. 28, No. 2–3 (special issue) 251 (1999); Parimal Patel, Localized Production of Technology for Global Markets, *Cambridge Journal of Economics*, vol. 19(1), 141 (1991).

GlaxoSmithKline, Cipla, and Pfizer, all of which run clinical research activities in India.[3]

Not surprisingly, this has also been the policy of different UN organs in recent years, albeit implicitly in most instances. Notably, this view is to be found in the *2005 United Nations Millennium* Project,[4] the view of WIPO,[5] and even the United Nations Economic Commission for Africa.[6]

Rooted in dependency theories of development that regard developing countries flatly as dependent on developed ones, the TRIPS Agreement includes an implicit pledge for a "freer trade" leading role for the business sector in fostering domestic innovation backed by patenting activity. In this respect, TRIPS corresponded and continues to correspond with the World Bank and UNCTAD's labeling of technology transfer as a reactive form of innovation-based economic growth for developing countries.[7] Thus, rather than promoting domestic innovation in developing countries through local technological capacity (with its clear limitations), innovation

---

[3] See, e.g., IBEF – India Brand Equity Foundation, Pharmaceutical Companies in India, at: www.ibef.org/industry/pharmaceutical-india/showcase. See also: *The Rise of Indian Multinationals: Perspectives on Indian Outward Foreign Direct Investment* (Karl Sauvant, Jaya Pradhan, Ayesha Chatterjee, and Brian Harley, eds.) (Palgrave Macmillan, 2010), at 15–17 (on the rise of India MNCs against the backdrop of foreign MNEs operating in India).

[4] United Nations Millennium Project, Innovation: Applying Knowledge in Development, London: Task Force on Science, Technology and Innovation, Earthscan (2005) ("thriving private sector depends fundamentally on adequate infrastructure, human capital, and research and development ... Through support for higher education and for research and development outlays, the government lays the groundwork for economic growth through technological advance"), at 123.

[5] WIPO, Economic Aspects of Intellectual Property in Countries with Economies in Transition, Ver. 1, the Division for Certain Countries in Europe and Asia, WIPO (2012) *(hereinafter WIPO, Economic Aspects) (focusing on developing countries mostly while* re-emphasizing that R&D is the most important economic indicator on how effective the innovation process is), at 22. See broadly also Recommendation no. 26 of the WIPO, 45 Adopted Recommendations under the WIPO Development Agenda (2007) (hereinafter, *WIPO Development Agenda*) ("To encourage Member States, especially developed countries, to urge their research and scientific institutions to enhance cooperation and exchange with research and development institutions in developing countries, especially LDCs").

[6] See United Nations Millennium Project, ibid. (Emphasizing the role of innovation and underlying investment needs as a basis for economic transformation).

But see for critique, e.g., Rasigan Maharajh and Erika Kraemer-Mbula, Innovation Strategies in Developing Countries, in *Innovation and the Development Agenda* (Erika Kraemer-Mbula and Watu Wamae, eds.) (OECD, 2009), at 136; Andreanne Léger and Sushmita Swaminathan, Innovation Theories: Relevance and Implications for Developing Country Innovation, German Institute for Economic Research (DIW) Discussion paper 743 (November 2007).

[7] See World Bank, Innovation Policy: A Guide for Developing Countries (2010), at 116; UNCTAD and ICTSD (International Centre for Trade and Sustainable Development), Intellectual Property Rights: Implications for Development, Intellectual Property Rights and Sustainable Development Series Policy Discussion Paper, ICTSD, Geneva (2003).

was to be received and at most was to be adapted.[8] Henceforth, the business sector was meant to foster technologically based trade.

However, a more careful look suggests that MNCs and the business sector in general have not met these high expectations in terms of their role in promoting an internationalized form of innovation in the developing world, as assumed by the UN's internationalized R&D view.

In order to examine this argument, this chapter undertakes a comparison between the three patent clusters across the development divide. The chapter analyzes statistical connections between the government and business sectors, both domestically and from abroad, with patent intensity serving as a significant proxy for domestic innovation. It follows the institutional wisdom whereby any effective innovation strategy requires coordination of multiple layers of institutional policies.[9] A focus of concern is the impact of the role these institutional actors take in promoting patenting activity on domestic innovation across countries. The empirical discussion below further analyzes GERD-related variables, namely GERD by sector of performance and GERD by source of funds. The former, as noted, relate to business enterprises, the government, private non-profit, and the higher education sectors. GERD by source of funds also include a fifth sector, namely funding from abroad.[10]

---

[8] Ibid.

[9] See Isabel Maria Bodas Freitas and Nick von Tunzelmann, Alignment of Innovation Policy Objectives: A Demand Side Perspective, DRUID Working Paper No. 13–02 (2008); Sanjaya Lall and Morris Teubal, Market-stimulating technology policies in developing countries: A framework with examples from East Asia, *World Development, Elsevier*, vol. 26 (8) 1369 (1998) (for the context of East Asia); Bengt-Åke Lundvall and Susana Borrás, The globalizing learning economy: Implications for innovation policies, Science Research Development, European Commission (1997). In developing countries, these layers of intervention need to be adjusted and coordinated to alleviate poverty. In his seminal book modeling the fastest growing markets among the billions of poor people at the bottom of the financial pyramid, C. K. Prahalad models innovation through distributive justice policies that are also profitable, while adhering to the central role of institutions and governments in particular. Coimbatore Krishnarao Prahalad, *The Fortune at the Bottom of the Pyramid: Eradicating Poverty through Profits* (Wharton School Publishing, 2005), at 81, 84. See also Rasigan Maharajh and Erika Kraemer-Mbula, supra note 6, at 142. See previous discussion in Daniel Benoliel, The Impact of Institutions on Patent Propensity Across Countries, *Boston University International Law Journal*, vol. 33(1) 129 (2015), 134–47 (accounting for differences between advanced and emerging economies).

[10] This analysis uses the United Nations Educational, Scientific and Cultural organization (UNESCO) Science and Technology (S&T) Statistical report (hereafter, *UNESCO S&T*) referring to *Table 27: Gerd by sector of performance* and *Table 28: Gerd by source of funds*. See both tables at: http://stats.uis.unesco.org/unesco/ReportFolders/ReportFolders.aspx. Table 27 does not include data on performance by entities from abroad. Thus, the summation of domestic and abroad business sectors occurs for table 28 only. See also: Adapted from OECD, Frascati Manual: Proposed Standard Practice for Surveys on Research and Experimental Development (2002) (hereinafter, OECD, Frascati Manual) (for source of funds), §423.

The 2010 UNESCO *Technical Paper No. 5* report on developing countries[11] already offered an alternative course. As the report suggests, developing countries' innovation systems and associated R&D or GERD measurement systems exhibit wide variety over countries, and particularly in the case of developing countries. The report suggests that this variance probably includes the irregular absorption of R&D performers, as well as an uneven empirical aptitude to measure R&D intensity.[12]

3.1   Innovation-Based Growth and Institutional Analysis

Over the past 20 years, OECD countries have witnessed the increasing impact of business R&D. On balance, foreign or R&D from abroad, mostly associated with multinational enterprises, has been stable, while public R&D has decreased.[13]

This pattern regarding advanced economies is less clear when developing countries are examined. In the case of developing countries, and emerging economies in particular, little result-based analysis has been undertaken to assess innovation and patent-related policy, including a retrospective examination. This section consolidates the existing analyses of all types of institutions promoting patenting activity, including the government public sector, local business sector, and the internationalized form of R&D performance and financing by MNCs.

3.1.1   Multinational Corporations

The funding of R&D has historically been conducted by two separate financing institutions – government and private businesses. In recent years, a third source of finance has assumed importance in several countries, namely overseas finance for R&D conducted in the domestic economy. By way of illustration, an OECD report indicates that between 1995 and 2004 the share of R&D expended by Western European multinationals outside their home country increased from 26 percent to 44 percent.[14] Similarly, between 1995 and 2004, the share of R&D spent

---

[11] United Nations Educational, Scientific and Cultural Organization (UNESCO) (2010), Technical Paper No. 5, (hereinafter, *UNESCO, Technical Paper 5*).

[12] Ibid.

[13] See, e.g., Dominique Guellec and Bruno Van Pottelsberghe de la Potterie, The Impact of Public R&D Expenditure on Business R&D, *Economics of Innovation and New Technology*, vol.12(3) (2003), at 228; and see previous discussion in Daniel Benoliel, The Impact of Institutions on Patent Propensity Across Countries, *Boston University International Law Journal*, vol. 33(1) 129 (2015), at 134–147.

[14] See OECD/OCDE, Background report to the Conference on internationalization of R&D, Brussels (March 2005).

outside the home country by Japanese multinationals rose from 5 percent to 11 percent, and by North American multinationals from 23 percent to 32 percent.[15] This has been followed by the growth of investments by these same multinationals in developing economies, especially Brazil, India, and China.[16] A report published by Goldman Sachs in 2010 identifies present and upcoming R&D facilities in China, India, and Brazil.[17] The MNCs mentioned in the report include financing agents such as Ford, IBM, Pfizer, Microsoft, Intel, Cisco, and Boeing.[18] A fourth source of funding is the set of "nonprofit institutions," including charitable trusts, some of which were set up by wealthy individuals following success in industry.[19]

The core theoretical idea within innovation theory attributes a central role to MNCs in fostering innovation.[20] The initiating argument concerning economic growth through innovation originated with Cambridge University economist Nicholas Kaldor as early as 1957. As Kaldor theorized, differing rates in the adoption of technology explain differences in development stages across countries.[21] The underlying idea was that investment and learning are interrelated and that the rate at which they take place determines technical progress.[22] The dominant underlying notion was that for determining the speed and orienting the direction of technological change for all countries alike, investment in R&D is required.[23]

---

[15] Ibid.

[16] See Goldman Sachs Group, The new geography of global innovation, Global Markets Institute report, September 20, 2010 (2010), at 8.

[17] Ibid.   [18] Ibid., at 6.

[19] See Christine Greenhalgh and Mark Rogers, *Innovation, Intellectual Property and Economic Growth* (Princeton University Press, 2010), at 89.

[20] For UNESCO S&T, supra note 10 (data and indicators on R&D funding from abroad, analyzed in the statistical model), at Section 3.3, referring to OECD Frascati Manual, supra note 10, §229. In the context of R&D statistics within the UNESCO dataset analyzed in the empirical model in Section 3.3, "Abroad" refers to "All institutions and individuals located outside the political borders of a country; except vehicles, ships, aircraft and space satellites operated by domestic entities and testing grounds acquired by such entities." In addition it includes: "All international organizations (except business enterprises), including facilities and operations within a country's borders." For additional discussion see the OECD, Frascati Manual, at 72–73. Such funding sources include overseas business enterprise, other national governments, private non-profit, higher education, and overseas international organizations.

[21] Nicholas Kaldor, A Model of Economic Growth, *Economic Journal*, Dec. 1957, 591 (1957). The latter analysis has been later measured using rampant patent statistics methodology. To illustrate, Stanford University Charles Jones and Paul Romer recently exemplified the usage of patent statistics over Kaldor's growth theory. See Charles I. Jones and Paul M. Romer, The New Kaldor Facts: Ideas, Institutions, Population, and Human Capital, NBER Working Paper Series (2009) (Offering cross-country patent statistics for measuring international flows of ideas alongside trade and FDI as key facets for economic growth.), at 8.

[22] Kaldor, supra note 21.   [23] Ibid.

During the 1990s, the internationalization of R&D activities increased significantly.[24] In a widely cited study of the trade-related impact of international R&D spillovers on a country's total factor productivity (TFP) published in 1995,[25] Coe and Helpman further emphasize the importance of foreign R&D capital stocks.[26] Focusing again solely on developed countries, they measured the importance of the R&D capital stock by the elasticity of a country's TFP with respect to the R&D capital stock. They present evidence suggesting that close links exist between productivity and R&D capital stocks. Not only does a country's TFP depend on its own R&D capital stock, but, as suggested by the theory, it also depends on the R&D capital stocks of its trade partners.[27] Simply put, roughly one-quarter of the total remuneration of R&D investment in a G7 country is attributed to its trade partners.[28] In a somewhat frail aside, the scholars suggest by way of analogy that foreign R&D capital stock may be at least as important as domestic R&D capital stock in the smaller developing countries.[29] In the larger

---

[24] Rajneesh Narula and Antonello Zanfei, Globalization of Innovation: The Role of Multinational Enterprises, Chapter 12 in *The Oxford Handbook of Innovation* (Jan Fagerberg and David Mowery, eds.) (Oxford University Press, 2005); Wolfgang Keller, International Technology Diffusion, *Journal of Economic Literature* 42, 3 (2004).

[25] TFP is a function of the domestic R&D capital stock and a measure of the foreign R&D capital stock, where all the measures of R&D capital were constructed from the business sectors' R&D activities. See David T. Coe and Elhanan Helpman, International R&D Spillovers, *European Economic Review*, vol. 39, 859 (1995).

[26] As they explain, foreign R&D mostly has a stronger effect on domestic productivity the larger the share of domestic imports in GDP. See Coe and Helpman, ibid. (Estimating the own rate of return to R&D as 123% for the G-7, and 85% for other 15 countries. Equally importantly, the spillover return from the G-7 of 32%, implies that roughly a quarter of the benefits from R&D in G-7 countries accrues to their trading partners), at 874.

[27] Ibid., 875.   [28] Ibid., 874.

[29] Ibid., 861. For critique of their findings, see Keller, supra note 24 (casting doubt on the trade-related of Coe and Helpman's finding in David T. Coe and Elhanan Helpman, International R&D Spillovers, *European Economic Review* 39: 859–87 (1995), concerning the effect of foreign R&D spillovers, by showing that significant foreign R&D spillovers can be obtained when the weights in the construction of the spillover are random rather than based on import shares); Frank R. Lichtenberg, and Bruno van Pottelsberghe de la Potterie, International R&D Spillovers: A Comment. *European Economic Review* 42(8): 1483 (1998) (criticizing Coe and Helpman's weighting of the foreign R&D stocks by means of the proportion of total imports originating from the foreign R&D sources, it being too sensitive to the aggregation of the data and propose. That is instead of normalizing the imports from the recipient country by the GDP of the sending country); Bruno van Pottelsberghe and Frank R. Lichtenberg, Does foreign direct investment transfer technology across borders? *Review of Economics and Statistics* 83(3), 490 (2001) (providing evidence for outward FDI as an overlooked channel of international R&D spillovers); Chihwa Kao, Min-Hsien Chiang, and Bangtian Chen, International R&D Spillovers: An Application of Estimation and Inference in Panel Cointegration, *Oxford Bulleting of Economic and Statistics* 61(S1): 691–709 (1999) (using a different empirical methodology thus finding cointegration between the TFP and R&D variables, using cointegration tests that are appropriate for panel data. When they re-estimate the Coe and Helpman

G7 countries, by contrast, the domestic R&D capital stock may be more significant.[30]

To date, innovation-based economic growth literature has emphasized that R&D, and in particular internationalized R&D, should be promoted by MNCs worldwide.[31] Yet across the board, current literature merely focuses on advanced or developed countries. It is thus not surprising that there are a large number of scientific studies on this occurrence evidently merging the experience mostly of advanced economies or that several of these studies show an increasing internationalization of innovative activity, mainly R&D by MNCs in such countries.[32] In practice, numerous examples reinforce the impression that internationalized R&D has triumphed in emerging economies. However, a more careful look suggests that the role of MNCs in promoting innovation in the developing world, and the entire UN innovation voice-over thereof, seem to have failed to meet these high expectations in three key respects. Firstly, MNCs currently invest less than expected in developing countries, and in emerging economies in particular. In the absence of more updated UN-level or other core research on the subject, the UNCTAD's seminal *2005 World Investment Report* provides a relatively recent illustration. This report shows that only China, the Republic of Korea, Taiwan Province of China, and Brazil, in that descending order, came close to or exceeded the US$5 billion in total GERD as of 2002 (the latest available year in UNCTD's report).[33] What is also disappointing about these results from the standpoint of emerging economies is that they further result in what has been hailed as a successful internationalization R&D process. As UC Berkeley economist Bronwyn Hall further describes, this internationalization R&D process has been measured during two different recent time periods: 1999 and 2005. The measurement related to some 40 large OECD and non-OECD countries.[34]

specification with a dynamic ordinary least squares (DOLS) estimator (which is not biased in small samples, unlike the ordinary estimator) they no longer obtain a significant effect for the trade-related foreign R&D spillover).

[30] See Coe and Helpman, supra note 25 above, at 861. For a critique of their findings, see Keller, supra note 24; Frank R. Lichtenberg and Bruno van Pottelsberghe de la Potterie, International R&D Spillovers: A Comment. *European Economic Review* 42: 1483 (1998); Bruno van Pottelsberghe and Frank R. Lichtenberg, Does Foreign Direct Investment Transfer Technology Across Borders? *Review of Economics and Statistics* 83, 490 (2001); Kao, Chiang and Chen, supra note 29.

[31] Frieder Meyer-Krahmer and Guido Reger, supra note 1.

[32] OECD, supra note 2; Daniele Archibugi and Alberto Coco, supra note 2; Alexander Gerybadze and Guido Reger, supra note 2; Parimal Patel, supra note 2.

[33] See UNCTAD, World Investment Report (Geneva, 2005), at 119–20, and see table III.1 therein.

[34] As Hall explains, two basic facts about the distribution of GDP and R&D performance are apparent during these periods. The first is that R&D performance is slightly more concentrated than GDP (Gini coefficients of 0.78 in 1999 and 0.75 in 2005 as opposed to

Yet even in large emerging economies, such as India, Mexico, and the Russian Federation, MNCs have invested in R&D at a rate well below the comparable figure of US$5 billion. An even starker example is provided by the relatively poorer emerging economies of South-East Europe and the former Soviet Bloc's Commonwealth of Independent States (CIS), where MNCs have invested much less.[35] This reality further explains why in the USPTO most patents assigned to entities in 25 selected developing countries in the new millennium were rarely owned by foreign affiliates.[36] Instead, they were owned by domestic enterprises, and at times by public institutions.[37]

Secondly, it is suggested that MNCs' internationalized R&D model has generally poorly met the expectation of promoting innovation in the developing world. This argument concerns the marginal number of MNCs originating from the developing world, with an emphasis on emerging economies. UNCTAD's 2005 investment report again shows that more than 80 percent of the 700 largest R&D spending firms come from just 5 advanced economies: the United States, Japan, Germany, the United Kingdom, and France, in descending order.[38] Only 1 percent of the top 700 are based in developing countries, South-East Europe, and the CIS.[39] Within the list of developing countries' MNCs, almost all the

0.69 in both years for GDP). Second, R&D has been becoming less concentrated over time, even during this brief six year period, in contrast to the GDP concentration, which has remain essentially unchanged. This change, although it appears small, reflects the internationalization of R&D that has taken place during the same period. See Bronwyn H. Hall, The Internationalization of R&D, UC Berkeley and University of Maastricht (March 2010), at 3, referring to and Figure "Concentration of R&D and GDP," at 22 and figure 1. See also: Greenhalgh and Rogers, supra note 19, at 344.

[35] UNCTAD, World Investment Report (Geneva, 2005), at xxvi.

[36] Ibid., at 134 (for date collected for the years 2001–2003), referring in table IV.11 to South Africa, Egypt, Kenya, Taiwan Province of China, Republic of Korea, China, Singapore, Hong Kong (China), India, Malaysia, Turkey, Thailand, Philippines, Saudi Arabia, Indonesia, Brazil, Mexico, Argentina, Bahamas, Bermuda, Cuba, Chile, Russian Federation, Ukraine, and Bulgaria. Only in Bulgaria and Brazil did foreign affiliates account for more than 20% of all patents assigned. In India and Cuba, public research institutions accounted for the largest shares (68% and 84% respectively) of those countries' totals. Public research institutions in Singapore, the Russian Federation, and Ukraine also receive a significant proportion of the patents assigned by the USPTO.

[37] Ibid.

[38] Ibid. See table IV.2. The IMF's Balance of Payments Manual (fifth edition, 1993) and the OECD Benchmark Definition of Foreign Direct Investment (third edition, 1995) provide agreed guidelines for compiling FDI flows. The largest TNCs remain geographically concentrated in a few home countries. The United States dominated the list with 25 entries. Five entries as well as Singapore remained the most important home economies, with ten and nine entries in the list respectively, with eight companies in the top 50. Taiwan Province of China, became the home economy with the third largest contingent of TNCs on the list. It largely owned this rank to its electronics companies. See UNCTAD, supra note 33, at 16–17.

[39] Ibid., see table IV.1. Several countries have moved up the ranks since the late 1990s. Ibid.

firms come from Asia, notably from the Republic of Korea and Taiwan Province of China.[40]

A third related finding follows. For the same 700 largest R&D spenders, most are concentrated in a relatively small number of industries, offering little innovative diversification and hence little adaptability for the plethora of innovative activities occurring in emerging economies. In 2003, more than half of them were to be found in just three industries: information technology (IT) hardware, the automotive industry, and the pharmaceuticals/biotechnology industries.[41] It has, of course, never been suggested that such industrial concentration in the entire group of emerging economies is recommended or satisfactory.

In conclusion, the role of MNCs in fostering innovation, backed by a high yield of patenting propensity rates in emerging economies, is disputable at best. As the UNCTAD 2005 Investment report itself indicates, through 2005 only a small number of developing countries and economies in transition participate in the process of R&D internationalization.[42] Furthermore, it remains questionable whether MNCs contribute a relatively high marginal growth rate to these countries' propensity to patent as a proxy for meaningful domestic innovation. The empirical analysis presented in Section 3.3 offers sobering support for the distrust shown in these countries, in the face of the possibly opposite reality in the advanced economies.

3.1.2   The Business Sector

Institutional analysis highlights a second industrial sector involved in fostering innovation, namely the business sector.[43] Regardless of the question of its specific profile in developing countries as opposed to developed ones, the business sector unquestionably remains highly

---

[40] Ibid., See table IV.2. On balance, only one MNC comes from Africa and two are from Latin America.

[41] Ibid. See table IV.3.

[42] See UNCTAD, supra note 33 (adding that the fact that some are now perceived as attractive locations for highly complex R&D indicates permit countries to develop the capabilities that are needed to connect with the global R&D systems of TNCs), at Overview at XXIV.

[43] For UNESCO S&T, supra note 10, data and indicators for Business enterprise intramural expenditure on R&D (BERD), analyzed in the statistical model in Part III, see: http://uis.unesco.org/en/glossary, referring to OECD, Frascati Manual, supra note 10, §163. R&D expenditure in the business sector, where the business sector in the context of R&D statistics includes: "All firms, organizations and institutions whose primary activity is the market production of goods or services (other than higher education) for sale to the general public at an economically significant price." In addition it includes "The private non-profit institutions mainly serving them." Ibid. For additional discussion see OECD, Frascati Manual, at 54–56.

influential in its propensity to patent by whichever country groups abridge the North–South divide. In the institutional realm of imperfect alternatives, the question remains: What is the relative role of the business sector in promoting domestic innovation-based archetypical patenting activity in the three patent clusters? That is, in comparison with both the government sector and MNCs. The answer to this inquiry relates to the impact on the patent intensity as a proxy for domestic innovation in all three clusters.

Notwithstanding its deep-rooted innovation implications, the TRIPS Agreement serves as a useful point of departure regarding the role of the business sector. Rooted in dependency theories of development, whereby developing countries were flatly perceived to be dependent on developed ones, the TRIPS Agreement has widely been accepted as a trade-related compromise.[44] Even more than internationalized R&D via MNCs, freer trade was said to impoverish countries of the "periphery."[45] Yet TRIPS' idealistic pledge of "freer trade" may also have undermined the role of the business sector in directly fostering innovative activity. The reason for this is that TRIPS corresponded, and continues to correspond, with the World Bank and UNCTAD's labeling of technology transfer as a reactive form of innovation-based economic growth for developing countries.[46] Accordingly, rather than promoting domestic innovation through the development of local technological capacity, innovation was to be received and at best adapted.[47] The business sector henceforth was meant to foster technologically based trade. The enhancement of domestic innovation based on enhancing the patenting yield of developing countries was initially contained.

[44] See Jayashree Watal, *Intellectual Property Rights in the WTO and Developing Countries* (Kluwer, 2001) (explaining how developed countries agreed to phase out their quotas under the ATC on the most sensitive items of textiles and clothing, in exchange for developing countries acceptance of the phasing-in of product patents for pharmaceuticals which is perceived as the most important patent-related good), at 20. See also: Frederick M. Abbott, The WTO TRIPS Agreement and Global Economic Development, in *Public Policy and Global Technological Integration*, 39 (Frederick M. Abbott and David J. Gerber, eds.) (Springer, 1997), at 39–40. See also: Charles S. Levy, Implementing TRIPS – A Test of Political Will, 31 *Law and Policy in International Business* 789 (2000), at 790.

[45] See, e.g., Raul Prebisch, International Trade and Payments in an Era of Coexistence: Commercial Policy in the Underdeveloped Countries, 49 *American Economic Review* 251 (1959) (offering examples of reasoning used by developing "periphery" countries fostering an aversion to increasing free trade), at 251–52. For a seminal Latin American perspective, see Fernando Henrique Cardoso and Enzo Faletto, *Dependency and Development in Latin America* (University of California Press, 1979) (depicting the tension between Latin American nationalist and populist political agendas and its impact on related international trade policies), at 149–71

[46] See World Bank, supra note 7, at 116; UNCTAD and ICTSD, supra note 7.

[47] Ibid.

Against the backdrop of the limited adherence to institutional aspects of innovation enhancement in developing countries as a policy concern, WIPO appears to have gone further. Although it plausibly has missed out on an opportunity to carefully adhere to the institutional aspects of innovation in its archetypical 2007 Development Agenda, WIPO has done so directly elsewhere, albeit in a loose manner. In its 2012 report *Economic Aspects of Intellectual Property in Countries with Economies in Transition*, WIPO itemizes the main factors that decorate the process of innovation activity in the developing world.[48] The first item in this list, reflecting an important institutional choice, is the poor involvement of the business sector in innovative activity in developing countries.[49] The report acknowledges the need for private–public partnership,[50] but does so without specifically opting for a separate direct regulatory role reserved for the government sector in performing or even financing innovative activity in developing countries. The general conclusion is that the effect of business R&D is mostly found to be larger in size than the impact of public R&D, since the latter deemed to "take a long time to materialize."[51] Notwithstanding WIPO's proclivity for business over government as an institutional choice for the developing world, a future policy challenge is evident here. Greater openness to trade and capital flows, as the TRIPS dialectics demand, should not be considered to lessen the imperative of local innovative efforts.[52] On the contrary, liberalization and the associated open market environment make it necessary for the business sector in developing countries to acquire the technological and innovative capabilities needed to become or remain competitive.[53]

A more accurate, through again partial, institutional choice can be seen in innovation-based economic growth literature. On this theoretical front, the effect of business R&D on productivity has been investigated intensively concerning developed countries. This investigation has addressed all aggregation levels, namely business unit, firm, industry, and country levels. As already noted, however, most of the empirical analysis focused rather predictably on the advanced economies, and particularly on the United States. All these studies have not only confirmed that business

---

[48] See WIPO Development Agenda, supra note 5, at 9.

[49] Ibid. (stating as its first recommendation: "1. Poor involvement of the business community in innovation policy elaboration and implementation (including funding of innovation projects."

[50] Ibid. (stating as its second recommendation: "Poor development of public-private partnerships.") Similarly, the report's fifth recommendation states: "Inadequate level of interaction between public and private research centres."

[51] See WIPO, World Intellectual Property Report 2011 ("The contribution of public R&D can take also a long time to materialize"), at 142.

[52] Cf.: UNCTAD, supra note 33, at Overview at XXV.   [53] Ibid.

R&D matters, but also that the estimated elasticity of output with respect to business R&D varies from 10 percent to an impressive 30 percent rate of return on business sector R&D.[54] The earliest panel data analysis on a principled level was undertaken by economists Luc Soete and Parimal Patel for 5 countries, confirming the impact of business sector R&D on innovation-based economic growth.[55] Columbia University economist Frank Lichtenberg later appears to have pioneered the use of large country dataset analysis by using a cross-section of 53 countries to corroborate the impact of business-related R&D on labor productivity.[56] Soon after, economists David Coe and Elhanan Helpman, alongside Walter Park, were the first to combine a large number of countries with long-term series.[57] These panel data analyses all converge toward the conclusion that the "social" return to business R&D is significant in fostering productivity.[58] In his panel data analysis of 10 OECD countries, Park compared business sector R&D with public sector R&D, concluding that public R&D loses its significant impact on productivity growth when business R&D is included among the explanatory variables.[59]

[54] See Dominique Guellec and Bruno Van Pottelsberghe de la Potterie, supra note 13 (offering estimates based on a panel dataset composed of 16 major OECD countries over the period 1980–98 suggesting that in these countries the domestic business sector, the government, and foreign R&D contribute significantly to output on multifactor productivity growth), at 229, generally referring to Ishaq Nadiri, Innovations and Technological Spillovers, NBER Working Paper Series, 4423, Cambridge, MA (1993). This large variation naturally is due to the fact that studies differ over the econometric specification, data sources, number of economic units, measurement methods for R&D, etc. Similarly, Griliches and Mairesse found that US manufacturing firms' rates of return to private R&D were around 20–40%. See Zvi Griliches and Jacques Mairesse, R&D and Productivity Growth: Comparing Japanese and US Manufacturing Firms, in *Productivity Growth in Japan and United States* (Charles R. Hulten, ed.) (University of Chicago Press, 1991) (finding rates of return in the range 30–40% also for the Japanese business sector). Hall and Mairesse found returns to French firms in the 1980s between 22% and 34%. See Bronwyn Hall and Jacques Mairesse, Exploring the relationship between R&D and productivity in French manufacturing firms. *Journal of Econometrics* 65, 263 (1995). Finally, Harhoff found a rate of return of around 20% for German firms from 1979 to 1989. See Dietmar Harhoff, R&D and Productivity in German Manufacturing Firms, *Economics of Innovation and New Technology* 6:22 (1998).

[55] Luc Soete and Parimal Patel, Recherche-Développement, Importations Technologiques et Croissance Economique, *Revue Economique*, vol. 36, pp. 975–1000 (1985).

[56] Frank R. Lichtenberg, R&D Investment and International Productivity Differences, in *Economic Growth in the World Economy* (Horst Siebert, ed.) (Mohr, 1993), at 47–68.

[57] Namely, 22 industrialized countries from 1970 to 1990 for Coe and Helpman and 10 OECD countries from 1970 to 1987 for Park.

[58] David T. Coe and Elhanan Helpman, supra note 25 (domestic R&D contributes significantly to productivity growth and that this impact is substantially higher for the G7 than for other developed countries); Walter G. Park, International R&D spillovers and OECD economic growth, *Economic Inquiry*, vol. 33, 571 (1995).

[59] See Ishaq Nadiri and Theofanis P. Mamuneas, The Effects of Public Infrastructure and R&D Capital on the Cost Structure and Performance of US Manufacturing Industries, *Review of Economics and Statistics*, vol. 76, 22–37 (1994); Walter G. Park, ibid.

At about the same time, Bronwyn Hall employed a separate market value approach to assess the returns on R&D in United States manufacturing firms over the period 1973–91.[60] For the full sample, R&D spending was strongly and positively associated with share market value. In fact, current R&D spending reported had a stronger association than the R&D stock (calculated by depreciating past R&D at 15 percent), suggesting that the share market considers current R&D a better indicator of future performance. Hall's ultimate conclusion was that the magnitude of the association suggests that the returns on R&D were two to three times those on normal investment.[61] Lastly, in 2006, Bronwyn Hall and Raffaele Oriani expanded their analysis and underlying conclusion to the business sectors in France, Germany, Italy, the United Kingdom, and the United States for the period 1989–98.[62]

Studies on the effect of business R&D on productivity have thus yielded positive overall findings. Regrettably, though, none of these findings includes a comparison between developing and developed countries, regardless of whether domestic innovation is predominantly patent-based or not in emerging economies, South-East Europe, and the Soviet Bloc's CIS countries.[63] In effect, as UNCTAD's 2005 World Investment Report tellingly indicates, the share of the latter group of developing countries in business sector R&D reached just 5.4 percent in 1996 and 7.1 percent in 2002.[64] As this chapter shows, innovation-based economic growth embodies greater reliance on government R&D in these countries by patent clusters. The late Alice Amsden offered further corroboration of this insight. As she explains in her seminal book *The Rise of "The Rest,"* based on extensive post-war national experiences in developing countries, institutions in the form of markets are largely rudimentary

---

[60] Bronwyn Hall, The stock market valuation of R&D investment during the 1980s, *American Economic Review* 83(2), 259 (1993).

[61] Ibid.; Bronwyn Hall, Industrial research during the 1980s: did the rate of return fall? *Brookings Papers on Economic Activity Microeconomics* (2):289–344 (1993) (connoting a temporal decline in returns in the computing/electronics sector due to the start of the personal computer revolution).

[62] Bronwyn Hall and Raffaele Oriani, Does the market value R&D investment by European firms? Evidence from a panel of manufacturing firms in France, Germany and Italy, *International Journal of Industrial Organization* 24, 971 (2006).

[63] Cf.: UNCTAD, supra note 33, at 106. CIS countries are also called the Russian Commonwealth and include 9 former Soviet bloc members out of its total of 15. The CIS members are Armenia, Azerbaijan, Belarus, Kazakhstan, Kyrgyzstan, Moldova, Russia, Tajikistan, and Uzbekistan. See About Commonwealth of Independent States, at: www.cisstat.com/eng/cis.htm.

[64] UNCTAD, supra note 33, at 106.

in the early stages of development.[65] Thus, the configuration of protected property rights forms part of the progress toward deeper and more ideal market structures.[66] As with the current TRIPS trade-based narrative of economic growth in developing countries, this narrative also encourages the creation of firm-specific proprietary skills that are distortionary (price exceeds marginal cost), since these gradually confer innovation-based market power.[67]

3.1.3   The Government Sector

Lastly, institutional analysis identifies a third industrial sector fostering innovation, namely the government sector.[68] In contrast to research into the impact of the business sector or MNC-based R&D in innovation, very few studies have examined the effects of alternative governmental sector R&D in fostering domestic innovation in general or patent activity in particular.[69] Only a few components of public research have been empirically analyzed. Again focusing mostly on advanced economies, and particularly the United States, James Adams finds that fundamental stocks of knowledge, proxied by accumulated academic scientific papers, significantly contribute to productivity growth in US manufacturing industries.[70] Another important study by Erik Poole and Jean-Thomas Bernard examined the subject of military innovations in Canada and

---

[65] See Alice H. Amsden, *The Rise of "The Rest": Challenges to the West from Late-Industrializing Economies* (Oxford University Press, 2001), at 286–87.

[66] Ibid.   [67] Cf. ibid.

[68] For UNESCO Science and technology data and indicators, analyzed in the statistical model in Section 3.3, see: http://uis.unesco.org/en/glossary, referring to OECD; Frascati Manual, supra note 10, at §184. Government intramural expenditure on R&D (GOVERD) or R&D expenditure in the government sector includes "all departments, offices and other bodies which furnish, but normally do not sell to the community, those common services, other than higher education, which cannot otherwise be conveniently and economically provided, as well as those that administer the state and the economic and social policy of the community. Public enterprises are included in the business enterprise sector." It further includes "the non-profit institutions (NPIs) controlled and mainly financed by government but not administered by the higher education sector."

[69] For an historical account focusing upon the United States in the twentieth century, see generally, David C. Mowery and Nathan Rosenberg, *Technology and the Pursuit of Economic Growth* (Cambridge University Press, 1989). For contributions dealing with particular sectors and industries see Richard R. Nelson, *Government and Technical Progress: A Cross-Industry Analysis*, Pergamon, New York (1982). For the post-Cold War climate affecting government support, especially in the United States, see Linda R. Cohen and Roger G. Noll, The Technology Pork Barrel, The Brookings Institution Press, Washington, DC (1997).

[70] James Adams, Fundamental Stocks of Knowledge and Productivity Growth, *Journal of Political Economy*, vol. 98, 673 (1990).

present evidence showing that a defense-related stock of innovation had a significantly negative effect on the multifactor productivity growth of four industries over the period 1961–85.[71] Ishaq Nadiri and Theofanis Mamuneas formally include the stock of public R&D, along with the stock of public infrastructure, as a determinant of the cost structure of US manufacturing activities.[72] Their results confirm that public R&D capital has important effects on industrial activity and is associated with a considerable "social" rate of return. On the other hand, Walter Park notes that public R&D loses its significant impact on productivity growth when business R&D is included in the explanatory variables. Park's findings are based on a panel data analysis of ten OECD countries. This important finding does not transcend the boundaries of advanced economies, as noted. Similarly, earlier findings on the negative productivity growth payoff from government expenditures for industrial R&D emerged from econometric studies, identified particularly with the seminal work of Harvard University economist Zvi Griliches,[73] as well as with Eric Bartelsman,[74] Frank Lichtenberg, and Donald Siegel.[75] Numerous other scholars have also found close to zero and statistically insignificant coefficients for federally funded R&D.[76] It must again be emphasized that none of these studies have met the challenge of comparing their evidence with that for developing countries in general, and emerging economies in particular. As Amsden explains in the broader context of development in the post-war era, the mutual control apparatus of countries at early stages of development thus transformed the incompetence associated with government interference into communal good, "just as the 'invisible hand' market-driven control mechanism transformed the chaos and selfishness of market forces into general well-being."[77]

---

[71] Erik Poole and Jean-Thomas Bernard, Defense Innovation Stock and Total Factor Productivity Growth, *Canadian Journal of Economics*, vol. 25, 438 (1992).

[72] Ishaq Nadiri and Theofanis P. Mamuneas, supra note 59.

[73] See Zvi Griliches, R&D and productivity: Econometric results and measurement issues, in (Paul Stoneman, ed.), *The Handbook of the Economics of Innovation and Technological Change* (Blackwell, 1995); Zvi Griliches and Frank R. Lichtenberg, R&D and productivity growth at the industry level: is there still a relationship? in R&D, *Patents and Productivity* (Zvi Griliches, ed.) (University of Chicago Press, 1984).

[74] Eric J. Bartelsman, Federally Sponsored R&D and Productivity Growth, Federal Reserve Economics Discussion Paper No. 121, Federal Reserve Board of Governors, Washington, DC (1990).

[75] Frank Lichtenberg and Donald Siegel, The impact of R&D investment on productivity – new evidence using linked R&D-LRD data, *Economic Inquiry* 29, 203 (1991).

[76] See Paul A. David, Bronwyn H. Hall, and Andrew A. Toole, Is Public R&D a Complement or Substitute for Private R&D? A Review of the Econometric Evidence, *Research Policy*, vol. 29 (4–5) 497 (2000), at 498 (adding additional sources).

[77] See Alice H. Amsden, *The Rise of "The Rest": Challenges to the West from Late-Industrializing Economies* (Oxford University Press, 2001), at 8.

During this early historical period, government's role was largely reactive and oriented toward securing the best terms for a post-factum form of technology transfer, while at the same time slowly increasing investments in R&D and formal education.[78] Analyzing the transition in Brazil, Argentina, India, and Mexico, and historically also in South Korea, Amsden explains that what began to differ sharply was the shift to proactive innovation policy in the post-war era.[79] The history of industrialization identifies numerous emerging economies that began to develop the new technology that was perceived as a precondition for sustainable national enterprise.[80]

According to Gerschenkron's theory of economic backwardness and the process of catching up, any country experiencing industrialization will have a diverse practice depending on its "degree of economics backwardness" when industrialization begins.[81] Accordingly, the later a country industrializes, the greater the economic interventions of its government.[82] Government interventions increase because production methods allegedly become more capital-intensive. Bigger absolute capital requirements over time bring forth new institutional arrangements that entail a larger role for government intervention in economic growth.[83]

Gerschenkron's concept of catch-up flexibility brings to mind a series of case studies. At one extreme we find what political scientist Eswaran Sridharan has defined as the state-promoted electronics industry case of Brazil. In this country, virtually all R&D efforts initially came from state enterprises and national firms. It was only much later that the Brazilian electronics industry witnessed MNCs-led R&D in innovation as the result of considerable policy pressure to secure such investments.[84] Yet, as Gerschenkron perceived, economic backwardness differs widely across developing countries, as does governmental intervention in their economic promotion. Malaysia, by way of example, has not achieved

---

[78] Ibid., 239.   [79] Ibid., 240–45.   [80] Ibid.

[81] Economics backwardness is not clearly defined in Gerschenkron, but he relates it to: income per capita, amount of social overhead capital, literacy, savings rates, and level of technology. Since many of these are positively correlated, it is often proxied by income per capita. See Alexander Gerschenkron, *Economic Backwardness in Historical Perspective* (Harvard University Press, 1962) (His analysis came as a reaction to uniform development stages theories like Walt Whitman Rostow's *The Stages of Economic Growth: A Non-Communist Manifesto* (Cambridge University Press, 1960).

[82] See Gerschenkron, *Economic Backwardness in Historical Perspective* (1962).   [83] Ibid.

[84] See Eswaran Sridharan, The Political Economy of Industrial Promotion: Indian, Brazilian, and Korean Electronics in Comparative Perspective 1969–1994 (analyzing the political economy and the role of the state in the electronics industry in India, Brazil, and Korea), at 89.

economic growth despite more than two decades of government-led and innovation-based protection. The Malaysian International Trade and Industry Minister recently acknowledged that public efforts to expand the local automotive industry, with an emphasis on the "National Car," have failed to yield the desired results.[85] A third central example is the Indian space program. Since the 1950s, this expensive program has been heavily subsidized and has yet to yield any commercial success.[86] As Gerschenkron predicted, the distance from the world technological frontier and the degree of government intervention do not necessarily act in a uniform manner in a latecomer developing country.[87] Be that as it may, it clearly remains to be seen whether newer attempts to industrialize innovation-based economies will continue to entail a categorical role for government intervention based on patenting activity or other forms of proprietary protection of domestic innovation.[88]

3.2   The Empirical Analysis

3.2.1   Methodology

The empirical analysis in this chapter assesses possible statistical connections between the institutions or sectors financing and performing GERD with patenting intensity rates (assessed surely as the propensity to patent by GERD intensity) for all three patent clusters.

Our analysis will examine two R&D-related indicators: GERD by sector of performance and GERD by sector of finance.[89] These two UNESCO S&T datasets incorporate yearly time series concerning sectoral GERD expenditure in one of the following five performance sectors:

---

[85] Tilman Altenburg, Building Inclusive Innovation in Developing Countries: Challenges for IS research, in *Handbook of Innovation Systems and Developing Countries* (Bengt-Åke Lundvall et al., eds.) (Elgar, 2009), at 38.

[86] Ibid., referring to Angathevar Baskaran, From Science to Commerce: The Evolution of Space Development Policy and Technology Accumulation in India, *Technology in Society*, 27 (2), 155–79 (2005).

[87] See Alice H. Amsden, *The Rise of "The Rest": Challenges to the West from Late-Industrializing Economies* (Oxford University Press, 2001), at 286. Amsden adds that instead what probably does increase is the role of the foreign firm in relation to the relative decline in the role of the state in fostering economic growth. This important concern thus remains outside the scope of this chapter on emphasis on innovation-based economic growth, as explained. She further offers a completing aphorism whereby the later a country industrializes in chronological history, the greater the probability that its major manufacturing firms will be foreign-owned.

[88] Cf.: Amsden, *The Rise of "The Rest": Challenges to the West from Late-Industrializing Economies*, at 285.

[89] This analysis uses the UNESCO S&T, supra note 10.

Business enterprise;[90] government;[91] higher education;[92] private non-profit[93] and abroad;[94] and not specified.[95]

Prior to the analysis, imputation was performed for each of the analyzed series for each country separately, as outlined in Appendix A.[96]

Regression models were used to study the time and cluster effects on various GERD components. In order to meet the required assumption of normality for the dependent variables, the suitability of Box-Cox transformation was examined for each cluster and year within each of the sectors. The results indicated the square root as the optimal transformation, and this was therefore applied. Time was modeled as a categorical

---

[90] In the context of GERD statistics, the business enterprise sector includes "All firms, organizations and institutions whose primary activity is the market production of goods or services (other than higher education) for sale to the general public at an economically significant price." In addition, it includes "The private non-profit institutions mainly serving them." Lastly, "The core of this sector is made up of private enterprises. This also includes public enterprises." See UNESCO S&T glossary, at: http://data.uis.unesco.org/. See also: OECD Frascati Manual: Proposed Standard Practice for Surveys on Research and Experimental Development (2002), §163.

[91] The government sector is defined as "All departments, offices and other bodies which furnish, but normally do not sell to the community, those common services, other than higher education, which cannot otherwise be conveniently and economically provided, as well as those that administer the state and the economic and social policy of the community. Public enterprises are included in the business enterprise sector." In addition it includes: "The non-profit institutions (NPIs) controlled and mainly financed by government, but not administered by the higher education sector." See UNESCO S&T glossary, ibid. See also: OECD, Frascati Manual, ibid. at §184.

[92] Higher education is defined as "All universities, colleges of technology and other institutions of post-secondary education, whatever their source of finance or legal status." It further includes "all research institutes, experimental stations and clinics operating under the direct control of or administered by or associated with higher education institutions." See UNESCO S&T glossary, ibid. See also: OECD, Frascati Manual, ibid. at §206.

[93] In the context of GERD statistics, the private non-profit sector includes "Non-market, private non-profit institutions serving households (i.e. the general public)," as well as "Private individuals or households." See UNESCO S&T glossary, supra note 90. See also: OECD, Frascati Manual, supra note 90, §194.

[94] In the context of GERD statistics, abroad refers to "All institutions and individuals located outside the political borders of a country; except vehicles, ships, aircraft and space satellites operated by domestic entities and testing grounds acquired by such entities." It further incorporates "All international organizations (except business enterprises), including facilities and operations within a country's borders." See UNESCO S&T glossary, ibid. See also: OECD, Frascati Manual, ibid., at §229.

[95] The percent of GERD included in the last category is negligible, therefore the following analyses relates to the first four.

[96] A Box-Cox function of MASS R library was used to obtain the optimal e Box-Cox power transformation See William N. Venables and Brian D. Ripley, *Modern Applied Statistics with S* (4th edn.) (Springer, 2002). The function indicates a 95% confidence interval about the power (lambda) with the highest value of the log-likelihood. For each GERD performance data, the best Box-Cox transformation for each cluster at each year was analyzed separately. A lambda=0.5 (i.e. square root transformation) was included in nearly all confidence intervals attained for all years and clusters combinations for all GERD parts. Hence, a square root transformation was used for analyzing the data.

variable, and an interaction term was included to allow for a different time effect between clusters, or equivalently, different cluster effects between times.[97]

3.2.2   Findings

3.2.2.1   Patent Clusters by Performance of GERD by Sector

Figure 3.1 displays the average GERD performed for each country by the government sector by the average GERD performed by the business sector, over the years 2000–2009. As seen, in the majority of the countries in the leaders cluster the government's share in GERD performance is relatively small, while the major contribution comes from the business sector. On the other hand, in the two other clusters a variation among their countries was observed. For example, within the followers cluster India has a large value in the government sector, while Malaysia shows a large value of GERD performance in the business sector.

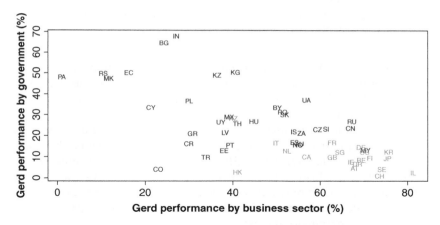

Figure 3.1 Mean of GERD Performance by Government vs. Business Sector

[97] The procedure MIXED of SAS was used. This procedure enables modeling different covariance structures. The CSH (Heterogeneous Compound Symmetry) option was used to allow both for different variances in different times and for constant correlation between observations from the same country (constant correlation between observations of a country is equivalent to including a random effect of a country). Lastly, the GROUP option has been used to allow for differences between the three patent clusters in the covariance parameters.

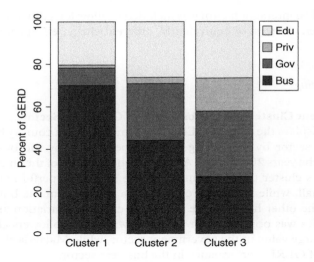

Figure 3.2 Mean of GERD by Performance Sector and Type of
Cluster (%)
Legend: *Edu* – Higher education; *Priv* – Non-government; *Gov* –
Government; *Bus* – Business sector

Figure 3.2 concludes with bar charts of the distribution of the mean
GERD values by sectors for each cluster, labeled 1–3, for leaders, followers, and marginalized respectively. A regression equation is presented
in Appendix D below.

The regression model included in Appendix D below yields numerous
conclusions. Firstly, for business sector GERD performance there are
significant differences regarding GERD performed by the business sector
between all three clusters. In the government sector, however, significant
differences are found only between the leaders cluster and the other two
clusters, while there is no statistically significant difference between the
followers and marginalized clusters.

These statistical significances support the conclusion that GERD performed by the business sector is significantly prevalent in the three clusters, in descending order. The statistical significance of government
sector GERD performance supports the conclusion that there is a substantive difference between the lesser role of the government in countries
belonging to the leaders cluster and the two other clusters, where the
government's role in performing GERD and innovation is substantially
greater.

3.2.2.2 Patent Clusters by GERD Financing by Sector

The analysis presented in Figure 3.3 concerns the relationship between the clusters and GERD by financing sector. These comprise the above-mentioned sectors, including funding from abroad. The bar chart describes the distribution of the mean GERD values by sectors for each cluster 1–3, in reference to the leaders, followers, and marginalized patent clusters, respectively.

The regression results in Appendix D also support numerous conclusions. Firstly, for the GERD financing by the business sector there are significant differences between all three clusters. This implies that the presence of a business sector is well aligned with the type of patent cluster, further corroborating the imminent role of the private sector in fostering domestic innovation. Conversely, significant differences can be seen in terms of government finance of GERD between the leaders cluster and the other two clusters. However, there is no statistically significant difference between the followers and marginalized clusters. Lastly, the regression model indicates that there are no significant differences between all three clusters over GERD financing from overseas. These results can be explained in a combined discussion.

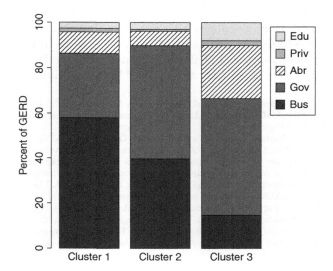

Figure 3.3 Mean of GERD by Financing Sector and Type of Cluster (%)
Legend: *Edu* – Higher education; *Priv* – Non-government; *Abr* – *Abroad;* *Gov* – Government; *Bus* – Business sector

3.2.2.3 Test Results Explanation

In accounting for decreasing patent intensity rates across the three clusters over GERD performing and financing as described, these findings correspond with earlier evidence by Kahn, Blankley, and Molotja,[98] as well as with UNESCO Technical Paper No. 5, *Measuring R&D: Challenges Faced by Developing Countries.*[99] According to these sources, the business sector in middle-income or emerging economies both finances and performs much less GERD-related innovative activity by comparison with public sector institutions.[100] Such findings may further correspond with UNESCO's 2010 Technical Paper No. 5 upholding that GERD-related innovative activity in the business sector within emerging economies is commissioned ad hoc to deal with production issues making it infrequent, informal and difficult to capture.

Lastly, these findings implicitly correspond with WIPO's *World Intellectual Property Report 2011*, identifying governments, rather than universities, as the main R&D actors in many low- and middle-income economies, in a reality in which industry often contributes little to scientific research.[101] As the WIPO report shows, government funding on average is responsible for about 53 percent of total R&D in the middle-income countries for which data are available.[102] As the level of a country's revenue diminishes, governmental endowment approaches 100 percent, in particular for R&D in the agricultural and health sectors. In Argentina, Bolivia, Brazil, India, Peru, and Romania, the share of public sector R&D often surpasses 70 percent of total R&D.[103] For instance, the public sector funded 100 percent of R&D in Burkina Faso in the last year for which data are available.[104] Econometric studies at the

---

[98] Michael Kahn, William Blankley, and Neo Molotja, Measuring R&D in South Africa and in Selected SADC Countries: Issues in Implementing Frascati Manual Based Surveys, Working Paper prepared for the UIS, Montreal (2008).

[99] United Nations Educational, Scientific and Cultural Organization (UNESCO) (2010), Technical Paper No. 5, Measuring R&D: Challenges Faced by Developing Countries ("Business enterprises that cater mainly to the local market may experience reduced competitive pressure, making systematic R&D the exception rather than the rule"), at 7.

[100] Michael Kahn, William Blankley, and Neo Molotja, Measuring R&D in South Africa and in Selected SADC Countries: Issues in Implementing Frascati Manual Based Surveys, Working Paper prepared for the UIS, Montreal (2008); United Nations Educational, Scientific and Cultural Organization (UNESCO) (2010), Technical Paper No. 5, Measuring R&D: Challenges Faced by Developing Countries.

[101] See World Intellectual Property Report 2011, supra note 51, at 140–41 and figure 4.1: Basic research is mainly conducted by the public sector.

[102] Ibid.

[103] Exceptions are Malaysia, China, the Philippines, and Thailand where, for both R&D funding and performance, the business sector has the largest share. See World Intellectual Property Report 2011, at 140–41.

[104] Ibid. More particularly, the WIPO report shows that in low- and middle-income countries for which data are available, public research is also responsible for the majority

firm and industry levels provide fewer irrefutable results as to the constructive impact of public R&D.[105] Public R&D, specifically, is not deemed to donate directly to economic growth, but has a circuitous impact through the motivation of increased private sector R&D. In other words, "crowding in" of private R&D takes place as public R&D raises the returns on private R&D.[106] However, these reports ignore the existence of significantly different rates of institutional financing and performance between the three patent clusters. In that sense, sub-optimal policy ramifications repeatedly assume an uncorroborated North–South divide, with significant institutional ramifications.

What explains these highly controversial aspects of the noted presence of the business sector in fostering innovation in developing countries? The work of the late Alice Amsden, including her thorough historical account of late-industrializing economies in *The Rise of "The Rest,"* may cast some light on this phenomenon. Amsden labels "late-industrializing economies" as newcomers from the South-East Asian Tiger economies and numerous emerging economies, or archetypically just as the rest. As she explains, governments in "the rest" have intervened in markets in a deliberate and profound way – substantively more so than the business sector. The main reason for this is that their economies lacked knowledge-based assets, particularly intellectual property assets typically associated with the business sector. Lack of knowledge-based assets, Amsden explains, impairs the ability of the archetypical "rest" to compete in world market prices, even in modern labor-intensive industries.[107]

Amsden thus follows the intellectual trail that has increasingly focused on the institutional causes of uncertainty and diversity in innovation economics. This course became part of a wider shift in economics toward understanding the role of non-market institutions in economic growth. This shift had its roots in Simon's work on organizations, and in the Behavioral School at Carnegie Mellon, as well as in the seminal work of economists Cyert and March[108] and March and Simon,[109] and the

of basic R&D. See World Intellectual Property Report 2011, at 141 (offering the examples of close to 100% in China, close to 90% in Mexico, about 80% in Chile and the Russian Federation, and about 75% in South Africa).

[105] See World Intellectual Property Report 2011, at 142 ("The contribution of public R&D can take also a long time to materialize.").

[106] For an overview of the literature, see Paul A. David and Bronwyn H. Hall, Property and the Pursuit of Knowledge: IPR Issues Affecting Scientific Research, *Research Policy*, 35, 767 (2006), 767–71. In turn, some public R&D may crowd out private R&D if it is not focused on basic (pre-commercial) R&D. Ibid.

[107] See Amsden, *The Rise of "The Rest,"* at 284.

[108] Richard M. Cyert and James G. March, *A Behavioral Theory of the Firm* (Prentice-Hall, 1963).

[109] James G. March and Herbert A. Simon, *Organizations* (Wiley, 1958).

contribution of Levinthal and March in the early 1980s regarding the function of non-market institutions in fostering economic growth.[110]

The lead role of governments in financing and performing innovation in developing countries is a second-best outcome, given the lack of better intellectual property incentives against the backdrop of knowledge-based assets, as explained. In such countries, government political intervention indeed dominates in the setting of frequent macroeconomics static efficiency flaws, motivated primarily by government rent seeking in innovative industries. Consequently, much innovation is fostered in the shadow of the IPR. In developing countries, a system of economic incentives dominated by IPR translates into temporary economic market concentrations funneled by state monopolies.

The role of government political will in fostering economic growth resonates with earlier development economics work by Pierre Schlag,[111] in particular, followed by Curtis Milhaupt and others.[112] In developing countries, government-backed economic clogs markedly offer a higher return on investment than IPRs. This alternative incentive mechanism coexists with ongoing bilateral and multilateral intellectual property agreements. Such growth also witnessed the substitution of private law, contract law, and property law as alternative "first best" solutions for the requisition of growth by such governments. Consistent with Schlag's explanations, each time such governments are short-sighted they may fall for regulatory expropriation of innovation activity; otherwise, governments may decide not to expropriate every time they would view such course of action as politically untimely, as they often

---

[110] Daniel Levinthal and James G. March, A Model of Adaptive Organizational Search, *Journal of Economic Behavior and Organization*, 2, 307–33 (1981). From the very outset, institutional theorists were committed to capitalist developed countries mostly. While building on Ronald Coase's 1937 earlier idea of transaction costs to explore the nature of institutions in The Nature of the Firm, in *Economica NS* 4, 386–405 (1937), institutional theorists thus made much headway in explaining the role of non-market institutions in economic growth in developed countries. Noticeably, these were Douglas North, *Institutional Change and Economic Performance* (Cambridge University Press, 1990), and the late Oliver Williamson in his 1975 book *Markets and Hierarchies, Analysis and Anti-Trust Implications: A Study in the Economics of Industrial Organizations*, New York: Free Press. A later expansion of work on the role of institutions in fostering economic growth is his 1985 book *The Economic Institutions of Capitalism*, New York: Free Press (1985).

[111] Pierre Schlag, An Appreciative Comment on Coase's The Problem of Social Cost: A View from the Left, *Wisconsin Law Review* 919 (1986).

[112] Gilson Ronald J., Curtis J. Milhaupt, Economically Benevolent Dictators: Lessons for Developing Democracies, *American Journal of Comparative Law* (2011), vol. 59, Issue 01, pp. 227–88; Curtis Milhaupt and Katharina Pistor, *Law and Capitalism – What Corporate Crises Reveal about Legal Systems and Economic Development around the World* (Chicago University Press, 2008).

do with the goal of becoming international exporters of innovation-based goods.[113]

The phenomena of political determination and political pull have indisputably been witnessed primarily in developed countries. Electronics, and particularly the fields of semiconductors and computers, formed a key example of this during the first two decades of the post-war era. Military and space government-led programmers operated during this period as an influential mechanism geared toward defined technological targets, as governments provided ongoing financial assistance to R&D and assured public procurement.[114] An earlier example is the appearance of synthetic chemistry in Germany, motivated by the political desire to ensure the self-reliance of the German financial system in the post-Bismarck era.[115] In some high-tech sectors, such as aerospace or pharmaceuticals, experimental data are gradually being accumulated showing that ongoing high levels of government political pull lead to state regulation, as witnessed in numerous instances in both the developed and developing countries.[116]

The underrated theorization by "political pull" of innovation activity within developing countries, where political causality is often king, may indeed be related to contemporary bureaucratic and slow-changing indications by national institutions and governments. Following Christopher Freeman, national institutions that mitigate innovation have gained a reputation as very slow movers. Freeman calculated that government innovation policies often persist for a century, despite changes in macroeconomic conditions and government policy.[117]

---

[113] Schlag, supra note 111.

[114] See Giovanni Dosi, Institutional Factors and Market Mechanisms in the Innovative Process, SERC, University of Sussex, mimeo (1979); Giovanni Dosi, Institutions and Markets in a Dynamic World, The Manchester School 56(2), 119–46 (1988); Giovanni Dosi, The Nature of The Innovation Process, Chapter 10 in *Technical Change and Economic Theory* (Giovanni Dosi, Christopher Freeman, Richard Nelson, Gerarld Silverberg, and Luc Soete, eds.) (LEM Book Series, 1988).

[115] See generally, Christopher Freeman, Technology Policy and Economic Performance, London: Pinter (1987); Vivien M. Walsh, J.F. Townsend, B.G. Achilladelis and C. Freeman, Trends in Invention and Innovation in the Chemical Industry, Report to SSRC, SPRU, University of Sussex, mimeo (1979).

[116] Thomas Lacy Glenn, Implicit Industrial Policy: The Triumph of Britain and the Failure of France in Global Pharmaceuticals, *Industrial and Corporate Change*, 1(2), 451 (1994).

[117] Christopher Freeman, *Technology Policy and Economic Performance* (Pinter, 1987). For later economic growth literature adaptations, see also: Michael Porter, *The Competitive Advantage of Nations* (Free Press, 1990); Richard R. Nelson, *National Innovation Systems: A Comparative Analysis* (Oxford University Press, 1993); Richard R. Nelson, The Co-evolution of Technology, Industrial Structure and Supporting Institutions, *Industrial and Corporate Change*, 3(1) 47–63 (1994). For the context of the expanding National Innovation Systems theory, see particularly Bengt-Åke Lundvall, Product

To conclude, countries associated with the followers and marginalized patent clusters powerfully illustrate that the business sector is sub-optimally related to the increase in patent propensity and GERD intensity as a proxy for domestic innovation. These findings are in line with the rather preliminary findings already collected by UNCTAD and WIPO, as described. Moreover, the evidence supports the diagnosis of excessive government political pulling of innovative activity in such countries. Within the confinements of the positive theory reasoning adopted here, the present supposition differs from critical studies on neopatrimonialism and rent-seeking literature in the context developing countries. The often normative evaluation of the latter approach argues that the state mostly follow its monetary and political interests by demonstrating predatory behavior.[118] In short, the present findings are not about how law ought to incentivize innovation, but simply seek to indicate what effect intervention has.

Another question remains: What may explain the relationship between government sector GERD financing and performance and patent intensity? Though this question clearly requires additional empirical evidence, WIPO's partial findings focusing on advanced economies offer some telling leads. WIPO's 2011 indications show that in high-income economies, the public sector is responsible for anywhere between 20 and 45 percent of annual total R&D expenditure. More particularly, governments usually provide the majority of the funds for patent-low intensity forms of basic research in these countries. Since basic research upholds less patenting activity, by comparison to experimental or applied research, governments in advanced economies may decrease the average rates of patent intensity in these countries. To illustrate, in 2009 on average the public sector performed more than three-quarters of all basic research in advanced economies.[119]

Innovation and User-Producer Interaction, *Serie om industriel odvikling*, 31 (1985); Bengt-Åke Lundvall, Innovation as an Interactive Process: From User-Producer Interaction to the National System of Innovation, chapter 18 in G. Dosi et al. (1988); Bengt-Åke Lundvall (ed.), *National Innovation Systems: Toward a Theory of Innovation and Interactive Learning* (Pinter, 1992).

[118] See, e.g., Tilman Altenburg, Building Inclusive Innovation in Developing Countries: Challenges for IS research, in *Handbook of Innovation Systems and Developing Countries*, at 33, referring to earlier work by Shmuel N. Eisenstadt, *Traditional Patrimonialism and Modern Neo-Patrimonialism*, London, Sage (1973). See also, Markus Loewe, Jonas Blume, Verena Schönleber, Stella Seibert, Johanna Speer, Christian Voss, The Impact of Favoritism on the Business Climate: A Study of Wasta in Jordan, DIE studies 30, Bonn (German Development Institute) (2007).

See also Curtis Milhaupt and Katharina Pistor, *Law and Capitalism – What Corporate Crises Reveal about Legal Systems and Economic Development Around the World* (Chicago University Press, 2008).

[119] See OECD, Research & Development Statistics. Depending on the country in question, it accounts for about 40% (Republic of Korea) to close to 100% (Slovakia) of all basic

The 2005 UNCTAD investment report focusing on R&D measurements worldwide is also significant in this context. Although it largely ignores the middle-income and emerging economies, the report offers two highly instructive findings concerning developing countries in general. These relate to the relatively low intensity of R&D activity measured by industry, as well as the low quality of R&D in such countries by comparison with advanced economies, as follows.

The UNCTAD report provides an important account of the relatively low intensity R&D in developing countries in comparison with advanced ones, thereby possibly explaining the former countries' lesser patent propensity rates. As the report shows, most developing economies begin modern manufacturing with the simplest technologies directed by low intensity R&D.[120] These technologies include textiles, clothing, food processing, and wood products. Some of these technologies indeed move up the scale into heavy process industries, such as metals, petroleum refining, and metal products.[121] However, hardly any such additional technologies turn into competent users of "medium-high" technologies, making added advanced intermediary and capital goods (such as chemicals, automobiles, and industrial machinery).[122] On average, only a few such industries develop competitive capabilities in high-technology industries backed by an extended patent propensity and patenting per se. Similar to advanced economies, these industries in developing countries may include aerospace, micro-electronics, and pharmaceuticals.[123]

The UNCTAD investment report offers a second highly instructive finding concerning developing countries at large. This relates to the relatively low quality of R&D in this broad spectrum of developing countries, by comparison to the advanced economies. In the former, such as Latin America and the Caribbean, MNCs have so far allocated only limited R&D funds. The report goes on to indicate that FDI in such countries is rarely in R&D-intensive activities. In cases where FDI in such countries is R&D-intense, it is mainly confined to adaptation of technology or products for local markets. In the case of the UNCTAD's Latin American example, the process is also known as "tropicalization." The Latin American example further illustrates how foreign affiliates play a relatively large role in business enterprise R&D in Brazil and Mexico, a

---

research performed. See World Intellectual Property Report 2011, at 140 and figure 4.1: Basic research is mainly conducted by the public sector.

[120] See UNCTAD, supra note 33, at 108–9. See also, table III.3: Classification of Manufacturing industries by R&D Capacity Index. Ibid., at 102.

[121] Ibid., at 102.

[122] Ibid., at 108–9. See also, Table III.3: Classification of Manufacturing industries by R&D Capacity Index. Ibid., at 102.

[123] Ibid.

moderate role in Argentina, and a low one in Chile.[124] Equally importantly, such low quality R&D in developing countries may justifiably be deemed to explain lower patent intensity rates in countries pertaining to the followers and marginalized cluster in comparison with countries in the leaders cluster.

3.3   Theoretical Ramifications

The core findings above imply important theoretical ramifications. Three such central ramifications come to mind. Firstly, there remains broad concern whether spillovers or externalities deriving from R&D activity funneled by patenting activity affect economic growth across country groups. Diffusion of GERD-related knowledge across the North–South divide introduces the idea of absorptive capacity as a conditioning factor. So far, however, empirical studies have merely investigated the question of whether R&D spillovers are internationally present between advanced economies. Little or no findings establish the scope and pattern of R&D spillovers across the development divide. Coe and Helpman most noticeably analyze 21 OECD economics between the years of 1970 and 1990.[125] Their rather limited findings uphold that R&D spillovers occur between advanced countries the greater the trade that openness prevails. Other studies have extended this work to data sets with two[126] or more countries and looked at other factors affecting R&D spillovers, such as education levels in OECD countries[127] or public sector R&D among advanced economies, as discussed earlier.[128]

---

[124] See UNCTAD, supra note 33, at 143, referring to Mario Cimoli, Networks, market structures and economic shocks: the structural changes of innovation systems in Latin America, Paper presented at the seminar on "Redes productivas e institucionales en America Latina," Buenos Aires, 9–12 April (2001).

[125] David Coe and Elhanan Helpman, International R&D Spillovers, *European Economic Review* 39: 859–87 (1995).

[126] See Rachel Griffith, Elena Huergo, Jacques Mairesse, Bettina Peters, Innovation and productivity across four European countries, NBER Working Paper 12722 (2006) (upholding substantial R&D spillovers from US manufacturing to UK firms whereby the latter undertaking R&D in the United States appear to benefit the most).

[127] See Hans-Jürgen Engelbrecht, International R&D Spillovers, Human Capital and Productivity in OECD economies: An Empirical Investigation, *European Economic Review* 41(8): 1479–88 (1997).

[128] See Dominique Guellec and Bruno Van Pottelsberghe de la Potterie, From R&D to Productivity Growth: Do the Institutional Settings and the Source of Funds of R&D Matter?, *Oxford Bulletin of Economics and Statistics, Department of Economics, University of Oxford*, vol. 66(3), 353–78, 07 (2004).

Economic studies have further examined the particular impact of academic research on business-related R&D, again solely within the context of advanced economies. Regarding further research on R&D spillovers within the context of advanced economies, see Zvi. Griliches, R&D and the Productivity Slowdown, The American *Economic*

A second theoretical ramification concerns the relationship between public and private R&D in developing countries. The advanced economies-based research pioneered from 1957 by Blank and Stigler examines a selection of data to ascertain whether the connection between public and private R&D investments is generally characterized by complementarity or by substitution. Following in this tradition, numerous contemporary econometric studies document positive, statistically significant spillover effects for the stimulation of private R&D investment by publicly funded additions to the stock of scientific knowledge.[129]

The same comment could be made regarding a significantly larger body of case studies illustrating the pressure of government-sponsored research programs and ventures on commercial technological innovation.[130] However, merely including the numbers of findings for and against that have accumulated on the matter of public–private R&D complementarity since the mid-1960s is unlikely to be particularly informative.[131]

This is even more so when developing countries are concerned. As the latter studies mostly focus on United States federally funded research performed in academic institutions or quasi-academic public institutes, they have little immediate bearing on questions concerning the impacts of publicly sponsored R&D in developing countries and on the comparison with advanced countries.[132]

A third theoretical concern is that development economics has yet to account for the particularities over the boundary between R&D and other technological innovation activities that can be found in pre-production development activities. It is true that this distinction is one that is considered difficult to apply in advanced economies.[133] The

---

Review, 70(2), 343–48 (1980); James D. Adams, supra note 70 (finding that basic research has a significant effect on increasing industry productivity, although the effect may be delayed for 20 years), at 673–702; Edwin Mansfield, Academic Research and Industrial Innovation: An Update of Empirical Findings. *Research Policy*, 26(7–8), 773–76 (1998) (surveying R&D executives from 76 randomly selected firms, estimating that 10% of industrial innovation was dependent on the academic research conducted within the 15 years prior); Mosahid Khan and Kul B. Luintel, Basic, Applied and Experimental Knowledge and Productivity: Further Evidence, *Economics Letters*, 111(1), 71–74 (2011).

[129] David, Hall, and Toole, supra note 76, at 499 and sources therein.

[130] Ibid., referring to Albert N. Link and John T. Scott, *Public Accountability: Evaluating Technology-Based Institutions* (Kluwer Academic Publishers,1998); National Research Council, Funding a Revolution: Government Support for Computing Research, Report of the NRC Computer Science and Telecommunications Board Committee on Innovations in Computing: Lessons from History. National Academy Press, Washington, DC (1999).

[131] David, Hall, and Toole, supra note 76, at 500.   [132] Ibid., at 499.

[133] See Alice Amsden and Ted Tschang, A new approach to assess the technological complexity of different categories of R&D (with examples from Singapore), *Research Policy*, 32(4), 553 (2003); Francisco Moris, R&D investments by US TNCs in emerging

distinction between "research" and "development" is particularly tricky in technology-intensive industries, since much of the R&D conducted involves close interaction between researchers in both the private and public sectors, often also including close collaboration with customers and suppliers.[134] The analogous challenge for developing countries thus remains regrettably unmet, though it may bear implications for these countries' propensity to patent at large.

Conclusion

Innovation and patent-related policy are known to have deep-rooted institutional implications. In the case of developing countries, however, UN-level organs are only loosely concerned with the role institutions take in promoting patenting activity. To date, innovation-based economic growth theory has emphasized how R&D, and particularly internationalized R&D, should be promoted by MNCs worldwide. Such R&D activity is also strongly connoted with a higher yield of patenting activity as measured by comparable national patent propensity rates. Across the board, however, contemporary literature focuses on advanced or developed countries. In practice, numerous examples have created the impression that internationalized R&D and the propensity to patent in emerging economies have triumphed. Not surprisingly, this has also been the general policy of different UN-level organs in recent years, albeit implicitly in most cases. Noticeably, this view is to be found in the 2005 UN's Millennium Project, the view of the WIPO, and even the United Nations Economic Commission for Africa. Rooted in dependency theories of development whereby developing countries were flatly perceived to be dependent on developed ones, the TRIPS Agreement implicitly pledge for a "freer trade" leading role for the business sector in directly fostering domestic innovative activity directed by a higher yield of patenting activity. This is due to the fact that TRIPS corresponded, and largely continues to do so, with the World Bank and UNCTAD's labeling of technology transfer as a reactive form of innovation-based economic growth for developing countries. Thus, rather than promoting domestic innovation by the promotion of local technological capacity,

---

and developing markets in the 1990s, Background paper prepared for UNCTAD (Arlington, VA: US National Science Foundation), mimeo (2005). See also UNCTAD, supra note 33, at 106.

[134] Ibid.

innovation was to be received and at best adapted. Thereafter, the business sector was meant to foster technologically based trade.

However, a careful look suggests that MNCs and the business sector in general have not met these high expectations – that is, in terms of their role in promoting an internationalized form of innovation in the developing world, as assumed by the United Nations' internationalized R&D view.

Against this backdrop, the comparison between the three patent clusters approximating the North–South divide was made. The analysis shows possible statistical connections between the government and the business sector (domestic and foreign) based on patent intensity measurement. In so doing the analysis offers two R&D-related indicators: the financing and the performance of GERD by three types of innovating sectors, namely government, the business sector, and private investment by multinational enterprises.

In critique of the present business sector pattern seen in developing and developed countries alike, the analysis offers two central findings. Firstly, in accounting for relatively lower patent intensity rates in the lower patent clusters in comparison with the leaders cluster, it is shown that the business sector in the former finances and performs relatively much less GERD-related innovative activity by comparison to public sector institutions. This hypothesis may substantiate UNCTAD's 2005 World Investment Report's key findings. Accordingly, the share of middle-income and emerging economies in global business R&D spending (with an emphasis on advanced economies) is lower than in total R&D spending. Moreover, these findings implicitly correspond with WIPO's 2011 report on innovation. As the WIPO report shows, governments, rather than universities, are often the main R&D actors in low- and middle-income economies, while industry often contributes little to scientific research.

To conclude, the relatively lower patent intensity witnessed in the two lower clusters seemingly relate to both a sub-optimal process of thus "second best" government political pulling of innovation activity. The latter is directed in tandem by a deficient form of IPR regulatory framework promoted by the WTO apparatus and TRIPS in particular. As a whole, the lower patent clusters illustrate how the business sector is sub-optimally related to the increase in patent intensity rates as proxy for domestic innovation.

The analysis, moreover, estimates in relative terms the central role of the government public sector in financing and performing GERD-related innovative activity in the lower patent clusters in comparison with the leaders patent cluster. Governments are repeatedly and unreservedly assumed to be benign institutions that are merely, or mostly, driven by their desire to exploit social welfare (even if their limited executing competence is frequently recognized). This supposition plainly differs from research on neopatrimonialism and from rent seeking that emphasizes the function of the state – particularly in developing countries – as entities that follow their individual monetary and political interest and may demonstrate predatory behavior.

Appendix D   GERD by Sector of Performance and Financing and Patent Clusters

D.1   GERD Performance by Sectors and Patent Clusters: Descriptive Statistics

The following plots describe for each sector and cluster the distributions along time:

The legend below explains the data presented in Figure D.1.

Legend

Code	Country code
r_cluster_km	Patenting cluster (according to k-means)
Country	Country
Business	GERD performance by Business sector (%)
Government	GERD performance by Government (%)
Private	GERD performance by Non-government organization sector (%)

	code	Cluster	Country	Business	Government
1	AT	1	AUSTRIA	68.113207	5.454131
2	AU	2	AUSTRALIA	55.286273	16.145006
3	BE	1	BELGIUM	69.590000	7.650000
4	BG	2	BULGARIA	24.330000	64.710000
5	BY	2	BELARUS	50.290000	33.560000
6	CA	1	CANADA	56.910000	9.980000
7	CH	1	SWITZERLAND	73.594909	1.054193
8	CN	2	CHINA	66.900000	23.770000
9	CO	2	COLOMBIA	22.890000	4.330000
10	CR	3	COSTA RICA	30.040468	16.255507
11	CY	3	CYPRUS	21.400000	33.470000
12	CZ	2	CZECH REPUBLIC	59.750000	22.990000
13	DE	1	GERMANY	69.500000	13.880000
14	DK	1	DENMARK	68.627578	6.725000

(cont.)

	code	Cluster	Country	Business	Government
15	EC	3	ECUADOR	16.394038	49.989023
16	EE	3	ESTONIA	38.430000	13.920000
17	ES	2	SPAIN	54.120000	16.810000
18	FI	1	FINLAND	71.260000	9.490000
19	FR	1	FRANCE	62.730000	16.700000
20	GB	1	UNITED KINGDOM	62.960000	10.110000
21	GR	2	GREECE	30.991698	21.043766
22	HK	1	HONG KONG	40.880000	2.540000
23	HU	2	HUNGARY	44.930000	26.680000
24	IE	1	IRELAND	67.420000	7.280000
25	IL	1	ISRAEL	81.380000	2.510000
26	IN	2	INDIA	27.120000	67.900000
27	IS	3	ICELAND	54.014989	21.978340
28	IT	1	ITALY	50.060000	16.500000
29	JP	1	JAPAN	75.510000	8.960000
30	KG	3	KYRGYZSTAN	40.700000	50.260000
31	KR	1	REPUBLIC OF KOREA	75.800000	12.410000
32	KZ	2	KAZAKHSTAN	36.538376	48.964439
33	LV	3	LATVIA	38.150000	21.570000
34	MK	3	MACEDONIA	11.890000	48.070000
35	MX	2	MEXICO	39.430000	28.620000
36	MY	2	MALAYSIA	70.488414	13.102155
37	NL	1	NETHERLANDS	52.460000	12.640000
38	NO	2	NORWAY	55.089140	15.499863
39	NZ	1	NEW ZEALAND	40.197971	28.187924
40	PA	3	PANAMA	1.107796	48.170000
41	PL	2	POLAND	30.140000	36.710000
42	PT	2	PORTUGAL	39.480000	14.600000
43	RO	2	ROMANIA	51.480000	31.260000
44	RS	2	SERBIA	10.580000	49.160000
45	RU	2	RUSSIA	67.260000	26.640000
46	SE	1	SWEDEN	74.198304	3.919348
47	SG	1	SINGAPORE	64.340000	11.520000
48	SI	2	SLOVENIA	61.270000	23.110000
49	SK	2	SLOVAKIA	51.820000	30.170000
50	TH	2	THAILAND	41.026827	26.000000
51	TR	2	TURKEY	33.950000	9.740000
52	UA	2	UKRAINE	56.990000	37.180000
53	US	1	UNITED STATES	70.290000	12.030000
54	UY	3	URUGUAY	37.561254	26.813545
55	ZA	2	SOUTH AFRICA	55.751387	21.419888

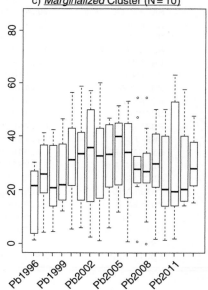

D.1 GERD Performance by Business Sector (%) (2000–2009)

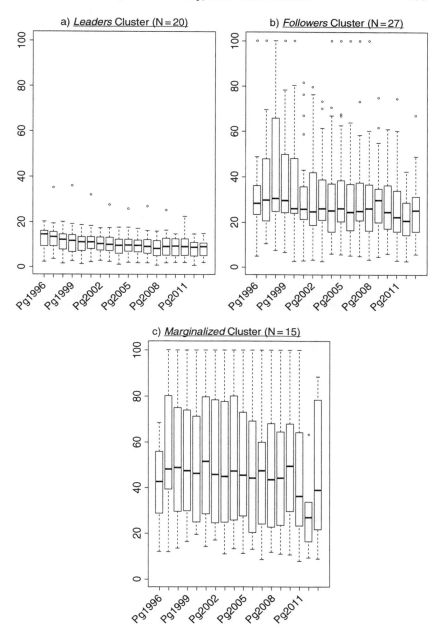

D.2 GERD Performance by Government Sector (%) (2000–2009)

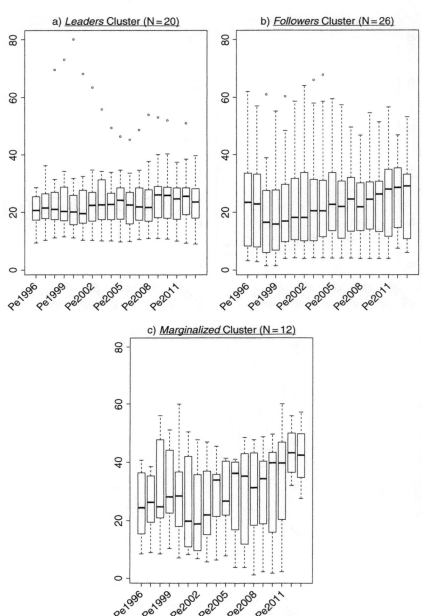

D.3 GERD Performance by Higher Education Sector (%)
(2000–2009)

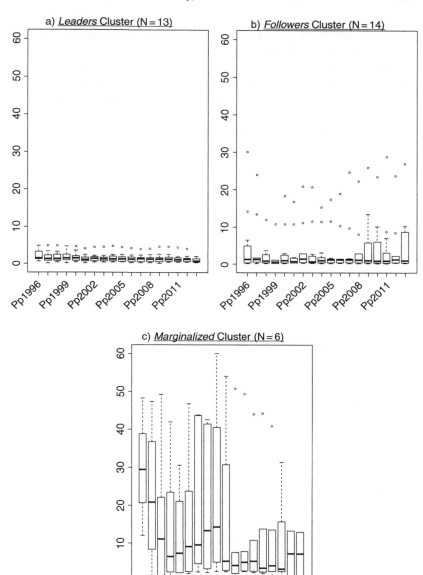

D.4 GERD Performance by Non-Profit Sector (%) (2000–2009)

The average GERD values by sector for the government, business, and private sectors is displayed for the years 2000–2009 in the following record. As seen, in the majority of the countries the non-government sector has made a relatively low contribution. The exceptions are countries in the marginalized cluster such as Panama, Ecuador, or Colombia, where there is a relatively large contribution from that sector of GERD performance.

	Code	Business	Government	Private	cluster
1	AT	68.113207	5.454131	0.3721859	1
2	AU	55.286273	16.145006	2.7901028	2
3	BE	69.590000	7.650000	1.2400000	1
4	BG	24.330000	64.710000	0.6143926	2
5	CA	56.910000	9.980000	0.4200000	1
6	CH	73.594909	1.054193	1.9710857	1
7	CO	22.890000	4.330000	20.0800000	2
8	CR	30.040468	16.255507	13.1770776	3
9	CY	21.400000	33.470000	8.7600000	3
10	CZ	59.750000	22.990000	0.4400000	2
11	DK	68.627578	6.725000	0.6051027	1
12	EC	16.394038	49.989023	21.0488334	3
13	EE	38.430000	13.920000	2.4300000	3
14	ES	54.120000	16.810000	0.3100000	2
15	FI	71.260000	9.490000	0.6000000	1
16	FR	62.730000	16.700000	1.2900000	1
17	GB	62.960000	10.110000	2.1000000	1
18	GR	30.991698	21.043766	0.9439609	2
19	IL	81.380000	2.510000	1.1000000	1
20	IS	54.014989	21.978340	2.4144479	3
21	JP	75.510000	8.960000	2.1700000	1
22	KR	75.800000	12.410000	1.3200000	1
23	KZ	36.538376	48.964439	2.5392866	2
24	MX	39.430000	28.620000	1.3600000	2
25	PA	1.107796	48.170000	44.2800000	3
26	PL	30.140000	36.710000	0.2400000	2
27	PT	39.480000	14.600000	10.4300000	2
28	RU	67.260000	26.640000	0.2300000	2
29	SE	74.198304	3.919348	0.2225000	1
30	SI	61.270000	23.110000	0.5900000	2
31	TH	41.026827	26.000000	1.9117397	2
32	US	70.290000	12.030000	4.2100000	1
33	ZA	55.751387	21.419888	1.2957242	2

The following figures describe the basic descriptive statistics of the GERD values by cluster.

Leaders *cluster (N=13)*

	mean	Sd	median	min	max	range
Business	70.07	6.49	70.29	56.91	81.38	24.47
Government	8.23	4.33	8.96	1.05	16.70	15.65
Private	1.36	1.08	1.24	0.22	4.21	3.99
Education	20.34	6.17	21.54	10.51	32.71	22.20

Followers *cluster (N=14)*

	mean	sd	median	min	max	range
Business	44.16	14.56	40.25	22.89	67.26	44.37
Government	26.58	15.16	23.05	4.33	64.71	60.38
Private	3.13	5.54	1.12	0.23	20.08	19.85
Education	26.17	13.65	27.28	5.87	52.69	46.82

Marginalized *cluster (N=6)*

	mean	sd	median	min	max	range
Business	26.90	18.35	25.72	1.11	54.01	52.91
Government	30.63	15.82	27.72	13.92	49.99	36.07
Private	15.35	15.82	10.97	2.41	44.28	41.87
Educational	26.43	16.56	28.92	7.11	45.25	38.14

D.2 GERD Finance by Sectors and Patent Clusters: Descriptive Statistics

The following analysis concerns the relationship between the clusters and GERD by financing sector. In addition to the above-mentioned sectors, these also include funding from abroad. The following plots describe the distributions along time for each source and cluster.

The following plots show the relationships of the GERD average values by cluster between the sources of government, business, and abroad. As seen, countries in the leaders cluster receive most of their GERD from business, and only a small proportion from government. Extremely low contributions from business are observed for Panama and El Salvador. The largest financing sector in Panama is from abroad. Most of Cyprus's GERD is financed by the government, as is also true of some of the countries in the followers cluster, such as Bulgaria and Russia. Similarly, we display the average GERD values by cluster of the government, business, and abroad sources, over the years 2000–2009. As we see, in the majority of the countries the abroad section makes a rather low

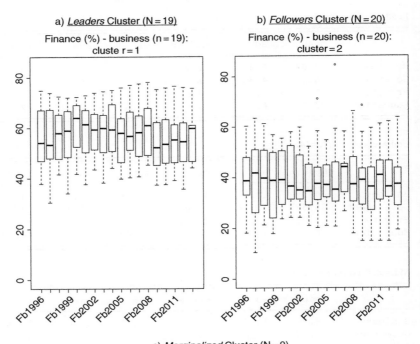

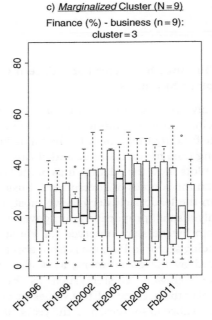

D.5 GERD Financing by Business Sector (%) (2000–2009)

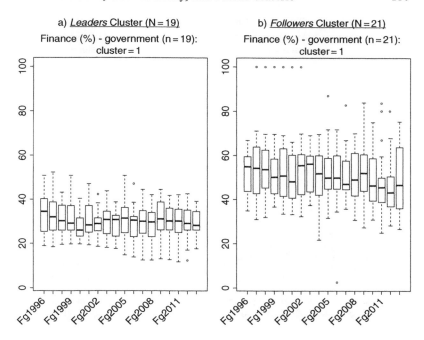

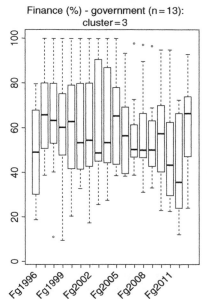

D.6 GERD Financing by Government (%) (2000–2009)

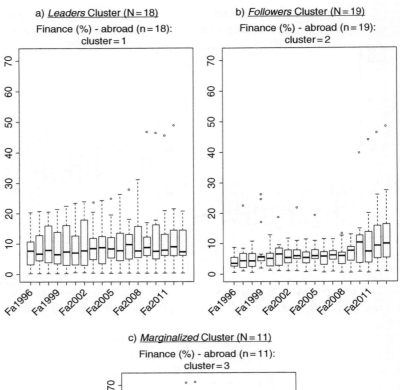

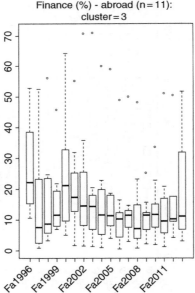

D.7 GERD Financing by Abroad Sector (%) (2000–2009)

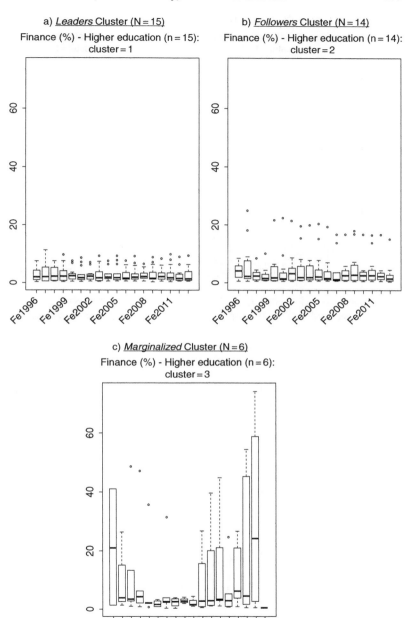

D.8 GERD Financing by Higher Education Sector (%) (2000–2009)

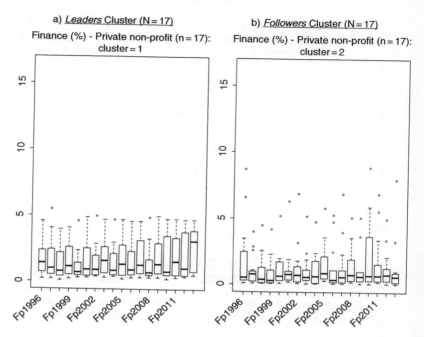

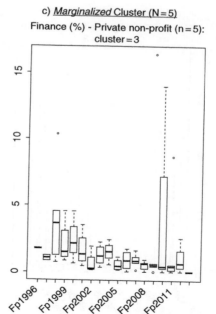

D.9 GERD Financing by Non-profit Organizations Sector (%)
(2000–2009)

contribution. The exception to this is Panama. The following record displays the data that are described.

	code	Business	Government	abroad	cluster
1	AT	45.640000	34.93000	18.790000	1
2	BE	60.740000	23.38000	12.690000	1
3	BG	28.470000	64.21000	6.420000	2
4	BY	26.620000	64.35000	7.910000	2
5	CA	49.490000	31.60000	9.600000	1
6	CH	69.073565	22.93930	5.317374	1
7	CU	31.930000	61.47000	6.600000	2
8	CY	17.160000	64.90000	12.680000	3
9	CZ	49.090000	44.13000	5.400000	2
10	DE	66.750000	29.77000	3.110000	1
11	DK	60.678919	27.20545	9.089349	1
12	EE	35.210000	49.04000	13.750000	3
13	ES	46.950000	42.06000	6.110000	2
14	FI	68.990000	25.05000	4.770000	1
15	FR	51.990000	38.44000	7.660000	1
16	GB	44.680000	31.14000	18.250000	1
17	GR	30.390158	50.65500	15.796079	2
18	HU	39.140000	49.38000	10.460000	2
19	IE	57.600000	29.72000	10.830000	1
20	L	52.350000	16.42000	27.350000	1
21	IS	47.132951	39.08460	13.116609	3
22	IT	41.617768	49.70213	8.129944	1
23	JP	75.340000	17.50000	0.360000	1
24	KG	40.593571	56.21084	5.681577	3
25	KR	73.420000	24.50000	0.360000	1
26	KZ	43.560492	43.01114	2.316720	2
27	LV	33.620000	43.79000	21.650000	3
28	MX	37.600000	54.12000	1.430000	2
29	MY	66.617215	23.41443	3.309217	2
30	NL	47.485484	39.14000	10.845743	1
31	NO	48.052941	42.68500	7.763742	2
32	NZ	39.110788	44.49500	5.744139	1
33	PA	1.980000	38.42000	55.130000	3
34	PL	30.960000	61.16000	4.910000	2
35	PT	37.430000	54.15000	4.750000	2
36	RO	38.020000	53.85000	5.730000	2
37	RU	30.530000	60.74000	8.180000	2
38	SE	64.814116	24.61943	7.385246	1
39	SG	55.850000	37.87000	5.410000	1
40	SI	57.190000	35.44000	6.860000	2
41	SK	42.450000	50.53000	6.430000	2
42	SV	4.083166	52.22667	11.803195	3
43	TH	39.286271	38.66500	2.237193	2

(cont.)

	code	Business	Government	abroad	cluster
44	TR	42.920000	47.46000	0.950000	2
45	UY	40.821053	28.38739	2.871947	32

The following tables describe the basic descriptive statistics of the GERD values by cluster.

Leaders *Cluster (N=13)*

	mean	sd	median	min	max	range
business	58.08	11.62	57.60	39.11	75.34	36.23
government	28.38	8.43	25.05	16.42	44.50	28.08
abroad	9.32	7.29	7.66	0.36	27.35	26.99
private	1.66	1.29	0.97	0.37	4.77	4.40
education	2.61	2.71	1.40	0.21	8.60	8.39

Followers *Cluster (N=12)*

	mean	sd	median	min	max	range
business	39.50	9.05	37.81	28.47	57.19	28.72
government	50.16	9.45	52.25	35.44	64.21	28.77
abroad	6.30	3.59	5.92	1.43	15.80	14.37
private	0.66	0.59	0.48	0.13	2.26	2.13
education	3.07	4.16	1.69	0.43	15.04	14.61

Marginalized *Clusters (N=4)*

	mean	sd	median	min	max	range
business	14.61	15.29	10.62	1.98	35.21	33.23
government	51.15	10.90	50.63	38.42	64.90	26.48
abroad	23.34	21.21	13.21	11.80	55.13	43.33
private	2.04	1.08	2.31	0.55	2.97	2.42
education	7.91	11.48	2.54	1.46	25.10	23.64

D.3   Regression Models

Detailed results are presented for square root of GERD performed by the business sector, in order to clarify the method for inference based on our regression model. However, since the main interest is only in the differences among clusters, only those differences will be presented for the other GERD components.

D.3.1   GERD Performed by Business Sector

Type 3 Tests of Fixed Effects

Effect	Num DF	Den DF	F Value	Pr > F
YEAR	15	780	2.78	0.0003
Cluster	2	52	21.12	<.0001
Interaction	30	780	0.74	0.8450

There was no significant interaction – namely, there was a same time effect for all three clusters, and same cluster effects in all years. Least squares means (*predicted population margins*) of both cluster and year effects were calculated and are presented in the following table:

Least Squares Means

Effect	r_cluster_km	Year	Estimate	Standard Error	DF	t Value	Pr > |t|
r_cluster_km	1		7.9654	0.1739	52	45.81	<.0001
r_cluster_km	2		6.5858	0.2487	52	26.48	<.0001
r_cluster_km	3		4.8564	0.5539	52	8.77	<.0001
year		1996	6.1655	0.2228	780	27.67	<.0001
year		1997	6.1468	0.2297	780	26.76	<.0001
year		1998	6.1954	0.2199	780	28.18	<.0001
year		1999	6.2834	0.2143	780	29.32	<.0001
year		2000	6.3775	0.2464	780	25.88	<.0001
year		2001	6.5247	0.2322	780	28.10	<.0001
year		2002	6.4487	0.2597	780	24.83	<.0001
year		2003	6.4631	0.2634	780	24.54	<.0001
year		2004	6.5500	0.2274	780	28.80	<.0001
year		2005	6.6124	0.2105	780	31.41	<.0001
year		2006	6.6777	0.2260	780	29.54	<.0001
year		2007	6.6575	0.2085	780	31.92	<.0001
year		2008	6.5829	0.2301	780	28.61	<.0001
year		2009	6.6308	0.2070	780	32.04	<.0001
year		2010	6.5648	0.2360	780	27.81	<.0001
year		2011	6.6261	0.2967	780	22.33	<.0001

Least squares means were compared using the Bonferroni adjustment for multiple comparisons. Accordingly, significant differences were observed between all three clusters. Only very few significant time changes were found. The only ones are between years 1996 and 1997 vs. 2006, 2007. The later years were found to be statistically significantly larger.

Differences of Least Squares Means

Effect	r_cluster_km	year	_r_cluster_km	_year	Estimate	Adj P
r_cluster_km	1		2		**1.3796**	<.0001
r_cluster_km	1		3		**3.1091**	<.0001
r_cluster_km	2		3		**1.7294**	0.0189
year		1996		1997	0.01876	1.0000
year		1996		1998	-0.02988	1.0000
year		1996		1999	-0.1179	1.0000
year		1996		2000	-0.2120	1.0000
year		1996		2001	-0.3592	1.0000
year		1996		2002	-0.2831	1.0000
year		1996		2003	-0.2976	1.0000
year		1996		2004	-0.3845	0.9337
year		1996		2005	-0.4469	0.1407
year		1996		2006	-0.5122	0.0423
year		1996		2007	-0.4920	0.0332
year		1996		2008	-0.4173	0.5024
year		1996		2009	-0.4653	0.0812
year		1996		2010	-0.3993	0.8838
year		1996		2011	-0.4606	1.0000
year		1997		1998	-0.04864	1.0000
year		1997		1999	-0.1367	1.0000
year		1997		2000	-0.2307	1.0000
year		1997		2001	-0.3779	1.0000
year		1997		2002	-0.3019	1.0000
year		1997		2003	-0.3164	1.0000
year		1997		2004	-0.4032	0.7936
year		1997		2005	-0.4656	0.1254
year		1997		2006	-0.5310	0.0372
year		1997		2007	-0.5107	0.0284
year		1997		2008	-0.4361	0.4362
year		1997		2009	-0.4841	0.0762
year		1997		2010	-0.4181	0.7548
year		1997		2011	-0.4794	1.0000
year		1998		1999	-0.08803	1.0000
year		1998		2000	-0.1821	1.0000
year		1998		2001	-0.3293	1.0000
year		1998		2002	-0.2533	1.0000
year		1998		2003	-0.2677	1.0000
year		1998		2004	-0.3546	1.0000

(*cont.*)

Effect	r_cluster_km year	_r_cluster_km _year	Estimate	Adj P
year	1998	2005	−0.4170	0.2389
year	1998	2006	−0.4823	0.0720
year	1998	2007	−0.4621	0.0656
year	1998	2008	−0.3875	0.7775
year	1998	2009	−0.4354	0.1346
year	1998	2010	−0.3694	1.0000
year	1998	2011	−0.4307	1.0000
year	1999	2000	−0.09405	1.0000
year	1999	2001	−0.2412	1.0000
year	1999	2002	−0.1652	1.0000
year	1999	2003	−0.1797	1.0000
year	1999	2004	−0.2666	1.0000
year	1999	2005	−0.3289	1.0000
year	1999	2006	−0.3943	0.5611
year	1999	2007	−0.3741	0.5607
year	1999	2008	−0.2994	1.0000
year	1999	2009	−0.3474	0.9772
year	1999	2010	−0.2814	1.0000
year	1999	2011	−0.3427	1.0000
year	2000	2001	−0.1472	1.0000
year	2000	2002	−0.07117	1.0000
year	2000	2003	−0.08565	1.0000
year	2000	2004	−0.1725	1.0000
year	2000	2005	−0.2349	1.0000
year	2000	2006	−0.3002	1.0000
year	2000	2007	−0.2800	1.0000
year	2000	2008	−0.2054	1.0000
year	2000	2009	−0.2533	1.0000
year	2000	2010	−0.1874	1.0000
year	2000	2011	−0.2486	1.0000
year	2001	2002	0.07602	1.0000
year	2001	2003	0.06154	1.0000
year	2001	2004	−0.02532	1.0000
year	2001	2005	−0.08769	1.0000
year	2001	2006	−0.1530	1.0000
year	2001	2007	−0.1328	1.0000
year	2001	2008	−0.05818	1.0000
year	2001	2009	−0.1061	1.0000
year	2001	2010	−0.04017	1.0000
year	2001	2011	−0.1014	1.0000
year	2002	2003	−0.01448	1.0000
year	2002	2004	−0.1013	1.0000
year	2002	2005	−0.1637	1.0000
year	2002	2006	−0.2291	1.0000
year	2002	2007	−0.2088	1.0000
year	2002	2008	−0.1342	1.0000
year	2002	2009	−0.1822	1.0000

(*cont.*)

Effect	r_cluster_km year	_r_cluster_km _year	Estimate	Adj P
year	2002	2010	−0.1162	1.0000
year	2002	2011	−0.1775	1.0000
year	2003	2004	−0.08687	1.0000
year	2003	2005	−0.1492	1.0000
year	2003	2006	−0.2146	1.0000
year	2003	2007	−0.1944	1.0000
year	2003	2008	−0.1197	1.0000
year	2003	2009	−0.1677	1.0000
year	2003	2010	−0.1017	1.0000
year	2003	2011	−0.1630	1.0000
year	2004	2005	−0.06237	1.0000
year	2004	2006	−0.1277	1.0000
year	2004	2007	−0.1075	1.0000
year	2004	2008	−0.03286	1.0000
year	2004	2009	−0.08082	1.0000
year	2004	2010	−0.01484	1.0000
year	2004	2011	−0.07611	1.0000
year	2005	2006	−0.06535	1.0000
year	2005	2007	−0.04512	1.0000
year	2005	2008	0.02951	1.0000
year	2005	2009	−0.01844	1.0000
year	2005	2010	0.04753	1.0000
year	2005	2011	−0.01374	1.0000
year	2006	2007	0.02023	1.0000
year	2006	2008	0.09486	1.0000
year	2006	2009	0.04691	1.0000
year	2006	2010	0.1129	1.0000
year	2006	2011	0.05161	1.0000
year	2007	2008	0.07463	1.0000
year	2007	2009	0.02668	1.0000
year	2007	2010	0.09265	1.0000
year	2007	2011	0.03138	1.0000
year	2008	2009	−0.04795	1.0000
year	2008	2010	0.01802	1.0000
year	2008	2011	−0.04325	1.0000
year	2009	2010	0.06597	1.0000
year	2009	2011	0.0047011	.0000
year	2010	2011	−0.06127	1.0000

For the sake of brevity, the three estimated covariance matrices for time are not presented here. Only small differences were found between clusters with regard to the correlation between times (0.9, 0.8, 0.8, respectively), with some differences in the variances. In the leaders cluster the variances were smaller than in the other two clusters. These differences were easily observed in the graphical displays of the data, (the side by side box plots).

Type 3 Tests of Fixed Effects

Effect	Num DF	Den DF	F Value	Pr > F
year	15	780	2.78	0.0003
r_cluster_km	2	52	21.12	<.0001
*r_cluster_km*year*	30	780	0.74	0.8450

Least Squares Means

Effect	r_cluster_km	Estimate	Standard Error	DF	t Value	Pr > |t|
r_cluster_km	1	7.9654	0.1739	52	45.81	<.0001
r_cluster_km	2	6.5858	0.2487	52	26.48	<.0001
r_cluster_km	3	4.8564	0.5539	52	8.77	<.0001

Differences of Least Squares Means

Effect	r_cluster_km	_r_cluster_km	Estimate	Standard Error	DF	t Value	Pr > |t|
r_cluster_km	1	2	1.3796	0.3034	52	4.55	<.0001
r_cluster_km	1	3	3.1091	0.5806	52	5.35	<.0001
r_cluster_km	2	3	1.7294	0.6072	52	2.85	0.0063

Differences of Least Squares Means

Effect	r_cluster_km	_r_cluster_km	Adjustment	Adj P
r_cluster_km	1	2	Bonferroni	<.0001
r_cluster_km	1	3	Bonferroni	<.0001
r_cluster_km	2	3	Bonferroni	0.0189

In conclusion, there are significant differences for GERD performed by the business sector between all three clusters.

D.3.2   GERD Performed by the Government

Type 3 Tests of Fixed Effects

Effect	Num DF	Den DF	F Value	Pr > F
Year	15	885	7.39	<.0001
r_cluster_km	2	59	33.72	<.0001
r_cluster_km*year	30	885	0.56	0.9741

Least Squares Means

Effect	r_cluster_ km	Estimate	Standard Error	DF	t Value	Pr > |t|
r_cluster_km	1	3.0607	0.2227	59	13.75	<.0001
r_cluster_km	2	5.4571	0.3226	59	16.92	<.0001
r_cluster_km	3	6.8187	0.5127	59	13.30	<.0001

Differences of Least Squares Means

Effect	r_cluster_ km	_r_cluster_ km	Estimate	Standard Error	DF	t Value	Pr > |t|
r_cluster_km	1	2	-2.3964	0.3920	59	-6.11	<.0001
r_cluster_km	1	3	-3.7580	0.5589	59	-6.72	<.0001
r_cluster_km	2	3	-1.3617	0.6057	59	-2.25	0.0283

Differences of Least Squares Means

Effect	r_cluster_km	_r_cluster_ km	Adjustment	Adj P
r_cluster_km	1	2	Bonferroni	<.0001
r_cluster_km	1	3	Bonferroni	<.0001
r_cluster_km	2	3	Bonferroni	0.0850

In conclusion, in the government sector there are significant differences between the leaders cluster and the other two clusters. However, there is no statistically significant difference between the followers and marginalized clusters.

D.3.3 GERD Financed by the Business Sector

Type 3 Tests of Fixed Effects

Effect	Num DF	Den DF	F Value	Pr > F
Year	15	675	0.89	0.5807
r_cluster_km	2	45	19.94	<.0001
r_cluster_km*year	30	675	0.46	0.9944

Least Squares Means

Effect	r_cluster_ km	Estimate	Standard Error	DF	t Value	Pr > |t|
r_cluster_km	1	7.4991	0.1690	45	44.37	<.0001
r_cluster_km	2	6.2108	0.1801	45	34.48	<.0001
r_cluster_km	3	4.4427	0.6781	45	6.55	<.0001

Differences of Least Squares Means

Effect	r_cluster_ km	_r_cluster_ km	Estimate	Standard Error	DF	t Value	Pr > |t|
r_cluster_km	1	2	1.2883	0.2470	45	5.22	<.0001
r_cluster_km	1	3	3.0564	0.6989	45	4.37	<.0001
r_cluster_km	2	3	1.7681	0.7017	45	2.52	0.0154

Differences of Least Squares Means

Effect	r_cluster_km	_r_cluster_km	Adjustment	Adj P
r_cluster_km	1	2	Bonferroni	<.0001
r_cluster_km	1	3	Bonferroni	0.0002
r_cluster_km	2	3	Bonferroni	0.0461

In conclusion, there are significant differences between all three clusters for the GERD financing by the business source.

D.3.4   GERD Financed by the Government

Type 3 Tests of Fixed Effects

Effect	Num DF	Den DF	F Value	Pr > F
Year	15	750	3.58	<.0001
r_cluster_km	2	50	23.89	<.0001
r_cluster_km*year	30	750	0.60	0.9544

Least Squares Means

Effect	r_cluster_km	Estimate	Standard Error	DF	t Value	Pr > |t|
r_cluster_km	1	5.5160	0.1738	50	31.73	<.0001
r_cluster_km	2	7.0893	0.2000	50	35.44	<.0001
r_cluster_km	3	7.6250	0.3825	50	19.94	<.0001

Differences of Least Squares Means

Effect	r_cluster_km	_r_cluster_km	Estimate	Standard Error	DF	t Value	Pr > |t|
r_cluster_km	1	2	-1.5733	0.2650	50	-5.94	<.0001
r_cluster_km	1	3	-2.1089	0.4201	50	-5.02	<.0001
r_cluster_km	2	3	-0.5356	0.4316	50	-1.24	0.2204

Differences of Least Squares Means

Effect	r_cluster_km	_r_cluster_km	Adjustment	Adj P
r_cluster_km	1	2	Bonferroni	<.0001
r_cluster_km	1	3	Bonferroni	<.0001
r_cluster_km	2	3	Bonferroni	0.6612

In conclusion, there are significant differences for government finance of GERD between leaders cluster and the other two clusters. Again, however, there is no statistically significant difference between the followers and marginalized clusters.

D.3.5   GERD Financed from Abroad

Type 3 Tests of Fixed Effects

Effect	Num DF	Den DF	F Value	Pr > F
Year	15	675	1.41	0.1365
r_cluster_km	2	45	2.57	0.0881
r_cluster_km*year	30	675	1.34	0.1061

Least Squares Means

Effect	r_cluster_ km	Estimate	Standard Error	DF	t Value	Pr > |t|
r_cluster_km	1	2.7565	0.2772	45	9.95	<.0001
r_cluster_km	2	2.3053	0.1738	45	13.27	<.0001
r_cluster_km	3	3.3139	0.4678	45	7.08	<.0001

Differences of Least Squares Means

Effect	r_clus- ter_km	_r_clus- ter_km	Estimate	Standard Error	DF	t Value	Pr > |t|
r_cluster_km	1	2	0.4512	0.3271	45	1.38	0.1747
r_cluster_km	1	3	-0.5574	0.5437	45	-1.03	0.3108
r_cluster_km	2	3	-1.0085	0.4990	45	-2.02	0.0492

Differences of Least Squares Means

Effect	r_cluster_km	_r_cluster_km	Adjustment	Adj P
r_cluster_km	1	2	Bonferroni	0.5240
r_cluster_km	1	3	Bonferroni	0.9324
r_cluster_km	2	3	Bonferroni	0.1477

In conclusion, there are no significant differences between all three clusters in terms of GERD financing from overseas.

4   GERD by Type, Patenting, and Innovation

Introduction

As countries search for patent policies that can take them nearer to the technological frontier, incentivizing different types of R&D dictates different policy considerations. While there is extensive literature on the research-policy links in OECD countries from disciplines as varied as economics, political science, sociology, and international relations,[1] there has yet again been much less emphasis on research-policy links in developing countries.[2] More specifically, there is still no systematic understanding of how basic research feeds into development policies.[3] Against this backdrop, economic theory offers two major contributions to the overall policy debate concerning the management of types of R&D. These are the contribution of basic research to economic growth and the varying impacts of types of R&D on the strength of

[1] See, e.g., Maja de Vibe, Ingeborg Hovland and John Young, Bridging Research and Policy: An Annotated Bibliography, ODI Working Paper No 174, ODI, London (2002) (covering an annotated bibliography summarizing 100 interdisciplinary studies on the research-industry gap); Emma Crewe and John Young, *Bridging Research and Policy: Context, Links and Evidence* (2002); Mercedes Botto, *Research and International Trade Policy Negotiations: Knowledge and Power in Latin America* (Routledge/IDRC, 2009); Fred Carden, *Knowledge to Policy: Making the Most of Development Research* (Sage Publications, 2009). See also UNCTAD, Capacity building for academia in trade for development: A study on contributions to the development of human resources and to policy support for developing countries (2010), at 8–9.

   Numerous research centers worldwide also explore this topic. Most noticeably is the work of the International Institute for Environment and Development (IIED), or the Overseas Development Institute (ODI) that has been researching research-policy linkages since 1999. ODI set up the Research and Policy in Development (RAPID) program focusing specifically on the uptake of research into policy. One should mention the contribution of the Global Development Network (GDN), the International Food Policy Research Institute (IFPRI), as well as the UK Department for International Development (DFID), or the Canadian International Development Research Centre (IDRC). These institutions noticeably have put effort on the matter from the prism of developing countries.

[2] See Julius Court and John Young, Bridging research and policy in international development, in *Global Knowledge Networks and International Development* (Diane Stone with Simon Maxwell, eds.) (Routledge, 2005), chapter 5, at 1.

[3] See Court and Young, ibid.

international spillovers. Both contributions may plausibly have significant patent-policy implications.

4.1 Scientific Research, Economic Growth, and Patent Policy

4.1.1 The Role of Patents in Safeguarding Scientific Research

To begin with, basic research is said to be understood and justified largely by
its contribution to economic growth.[4] There is extensive evidence incorporating basic research into R&D-driven economic growth models.[5]
Contemporary theory still bears the footprints of the groundbreaking work
of Cornell University economist Karl Shell, starting from 1967,[6] regarding
technical knowledge, which he treats as a public good corresponding closely
to the present-day notion of basic research. Shell's work was followed in the
1990s by economists Hiroshi Osano,[7] Jose Bailén,[8] and others,[9] jointly
modeling different types of research within economic growth theory.

---

[4] The term "basic research" used herein is referred to in the literature by various terms,
including science, basic science, academic research, university research, and public
research. See Hans Gersbach, Ulrich Schetter, and Maik Schneider, How Much
Science? The 5 Ws (and 1 H) of Investing in Basic Research, CEPR Discussion Papers
10482, CEPR Discussion Papers (2015), at 2; Amnon J. Salter and Ben R. Martin, The
Economic Benefits of Publicly Funded Basic Research: A Critical Review, *Research Policy*
30(3), 509 (2001) (for the use of the term *basic science*), at 526.

[5] See, e.g., Arnold Lutz G., *Basic and applied research*, Finanzarchiv, vol. 54, 169 (1997);
Guido Cozzi and Silvia Galli, Privatization of knowledge: Did the US get it right? MPRA
Paper 29710 (2011); Guido Cozzi and Silvia Galli, Science-based R&D in Schumpeterian
Growth, *Scottish Journal of Political Economy*, 56, 474 (2009) (hereinafter, Cozzi and Galli,
Science-based R&D); Guido Cozzi and Silvia Galli, Upstream Innovation Protection:
Common Law Evolution and the Dynamics of Wage Inequality, MPRA Paper 31902
(2011); Hans Gersbach, Gerhard Sorger and Christian Amon, Hierarchical Growth:
Basic and Applied Research, CER-ETH Working Papers 118, CER-ETH – Center of
Economic Research at ETH Zürich (2009); Amnon J. Salter and Ben R. Martin, supra
note 4 (the authors review and assess the literature on the economic benefits associated
with basic research), at 509.

[6] See Karl Shell, Toward a Theory of Inventive Activity and Capital Accumulation,
American Economic Review, 56(1/2), 62 (1966); Karl Shell, A Model of Inventive
Activity and Capital Accumulation, in *Essays on the Theory of Optimal Economic Growth*
(Karl Shell, ed.) (MIT Press, 1967) (focusing solely on basic research and still ignoring the
relation to applied research by private firms).

[7] Hiroshi Osano, Basic Research and Applied R&D in a Model of Endogenous Economic
Growth, *Osaka Economic Papers*, 42(1–2) 144 (1992) (composing basic and applied
research while assessing their effect on growth). Osano's model is similarly incomplete
as it ignores the public basic-research sector.

[8] Jose Maria Bailén, Basic research, Product Innovation, and Growth, Economics Working
Papers 88, Department of Economics and Business, Universitat Pompeu Fabra (1994)
(incorporating to Osano's model also publicly financed basic research).

[9] See Alessandra Pelloni, Public Financing of Education and Research in a Model of
Endogenous Growth, *Labour*, 11(3) 517 (1997) (analyzing the role of the government

Yet from the point of view of developing countries, and especially emerging economies and their understated patent intensity, this body of literature is limited in three respects. Firstly, virtually all studies use developed countries as their case studies, while refraining from providing a through and methodical comparison between the different country groups across the development divide. Moreover, the relative importance of the different forms of economic benefit distinguished in the literature varies with scientific fields, technology, and industrial sectors. As a result, given great heterogeneity in the relationship between basic research and economic growth, no simple model of the nature of the economic benefits from basic research has been established, and such a model may not necessarily be possible.[10] Lastly, most of these contributions focus on the optimal level of basic research in closed economies at the firm or country level.[11] There is simply no correspondence with country-group clustering analysis, beyond a limited usage of singled-out developing countries when broadly compared with advanced ones. As a result, there is still no clear understanding as to which R&D type derives a higher innovation-based growth based on countries' patent intensity rates.

Notwithstanding profound measurement constraints, it is generally believed that in aggregate, basic research exerts a positive and significant impact on economic growth.[12] Empirical studies thus point to two

in creating conditions for growth by financing higher education and scientific research, both being perfect substitutes); Walter G. Park, A Theoretical Model of Government Research and Growth, *Journal of Economic Behavior & Organization*, 34(1), 69 (1998) (analyzing cross-country spillovers of basic research in a growth model with two identical countries, determining the efficient size of a country's basic research sector in relation to the economy's openness). In the 2000s, more complex models followed. See, e.g., Maria Morales, Research Policy and Endogenous Growth, *Spanish Economic Review*, 6 (3): 179 (2004) (combining the approaches of Osano and Bailén in an endogenous growth model of vertical innovation incorporating both basic and applied research performed by both private firms and the government); Maria Rosaria Carillo and Erasmo Papagni, Social Rewards in Science and Economic Growth, MPRA Paper 2776, University Library of Munich, Germany (2007) (considering basic research as a global public good limited to identical levels of technology across countries).

[10] Amnon J. Salter and Ben R. Martin, supra note 4, at 526.

[11] See Hans Gersbach and Maik T. Schneider, On the Global Supply of Basic Research, CER-ETH Economics working paper series 13/175, CER-ETH – Center of Economic Research (CER-ETH) at ETH Zürich (2013), at 4.

[12] See Andrew A. Toole, The Impact of Public Basic Research on Industrial Innovation: Evidence from the Pharmaceutical Industry, *Research Policy*, 41(1) 1 (2012) (considering the impact of publicly funded basic research on the pharmaceutical industry. His analyses suggest that public basic research significantly spurs innovation, with the rate of return to these public investments being as high as 43%); see Amnon J. Salter and Ben R. Martin, supra note 4 (providing a summary table from estimated rates of return to publicly funded R&D, mostly in the agricultural sector. Most of these estimated rates of return were around 30% or even higher). Cf. also Bronwyn Hall, Jacques Mairesse, and

important growth-related findings: high social rates of return on investments in basic research,[13] and, more specifically, the positive effects on productivity and GDP-based growth.[14] In these two contexts, it is well established that basic R&D generates economic growth, while underlying numerous patent-policy considerations.

Patent law, particularly in the United States and Europe, has incorporated numerous policies safeguarding scientific research against undue patent proprietarization. In both patent law systems, the use of patented inventions in research is not per se exempted from infringement liability. That said, three main policies manifest the mentioned research safeguarding rationale more clearly perceived in developed countries.[15]

Pierre Mohnen, Measuring the Returns to R&D, Working Paper 15622, National Bureau of Economic Research (2009) (for a survey of the literature measuring the returns to R&D in general); Richard C. Levin, Alvin K. Klevorick, Richard Nelson, and Sidney Winter, Appropriating the Returns from Industrial Research and Development, *Brookings Papers on Economic Activity*, 3:1987 at 783 (survey of 650 R&D directors in large US firms systematic analyses of the benefits of basic research).

[13] See Amnon J. Salter and Ben R. Martin, supra note 4 (providing a summary table from estimated rates of return to publicly funded R&D, mostly in the agricultural sector. Most of these estimated rates of return were around 30% or even higher. Cf.: Bronwyn Hall, Jacques Mairesse, and Pierre Mohnen, supra note 12, for a survey of the literature measuring the returns to R&D in general.

[14] Cf. Edwin Mansfield, Basic Research and Productivity Increase in Manufacturing, *American Economic Review*, 70(5) 863 (1980); Albert N. Link, Basic Research and Productivity Increase in Manufacturing: Additional Evidence, *American Economic Review*, 71(5) 1111 (1981); Zvi Griliches, Productivity, R&D, and basic research at the firm level in the 1970's, *American Economic Review*, 76(1) 141 (1986); James D. Adams, Fundamental Stocks of Knowledge and Productivity Growth, *Journal of Political Economy*, 98(4) 673 (1990); Fernand Martin, The Economic Impact of Canadian University R&D, *Research Policy*, 27 (7) 677 (1998); Dominique Guellec and Bruno Van Pottelsberghe de la Potterie, From R&D to Productivity Growth: Do the Institutional Settings and the Source of Funds of R&D Matter? *Oxford Bulletin of Economics and Statistics*, 66(3) 353 (2004); Kul B. Luintel and Mosahid Khan, Basic, Applied and Experimental Knowledge and Productivity: Further Evidence, *Economics Letters*, 111(1) 71 (2011); Dirk Czarnitzki and Susanne Thorwarth, Productivity Effects of Basic Research in Low-Tech and High-Tech Industries, *Research Policy*, 41(9) 1555 (2012).

There are, however, numerous critiques concerning measurement of R&D inputs and the resulting outputs, the reliance on simplified production functions, and the possibility of reversed causality. See, e.g., Zvi Griliches, R&D and Productivity: The Econometric Evidence (University of Chicago Press, 1998); See Amnon J. Salter and Ben R. Martin, supra note 4; Iain M. Cockburn, Rebecca M. Henderson, Publicly Funded Science and the Productivity of the Pharmaceutical Industry, in, *Innovation Policy and the Economy*, vol. 1, 1 (Adam B. Jaffe, Josh Lerner, and Scott Stern, eds.) (MIT Press, 2001); Bronwyn Hall, Jacques Mairesse, and Pierre Mohnen, supra note 12.

[15] A fifth safeguard mechanism against patenting core scientific research largely practiced until recently has been the American *mental steps doctrine*, deeming unpatentable claims based on calculations, formulas, or other types of scientific-related algorithms. See Diamond v. Diehr, 450 US 175 (1981), at 195–96. See also Miriam Bitton, Patenting Abstractions, *North Carolina Journal of Law and Technology*, vol. 15(2) 153 (2014) (emphasizing the recent doctrine's demise due to vagueness and little use by courts), at 169 and discussion at 168–71, ibid.

Both the United States Federal Supreme Court[16] and the European Patent Convention (EPC) and Boards of Appeal[17] consistently reiterate that basic scientific ideas, including laws of nature, natural phenomena, and abstract ideas, do not constitute patentable subject matter. These exclusions reflect the fundamental concept that patents are issued only for new means to achieve useful results, rather than for something akin to the discovery of a new scientific principle.[18]

A second policy restraining undue patenting of scientific research concerns the use of the so-called exemption for research use. This exemption is used by universities in the United States,[19] in most EU countries,[20] and in numerous OECD countries,[21] including such emerging economies as Mexico[22] and Turkey.[23] The exception is further consistent with the TRIPS Agreement concerning first-best research exemption policies, either formally or informally.[24] Universities have

---

[16] See Gottschalk v. Benson, 409 US 63 (1972), at 67; Funk Bros. Seed Co. v. Kalo Inoculant Co., 333 US 127 (1948), at 130.

[17] For the European context, see EPC, Article 52(2) (Patentable Inventions) concerning scientific research in sub-section 52(2)(a): "(2)The following in particular shall not be regarded as inventions within the meaning of paragraph 1: (a) Discoveries, scientific theories and mathematical methods."

See also EPO Boards of Appeal decision T 388/04, OJ 1/2007, 16 (The Board confirmed that under Article 52(2) and (3) to the EPC, unpatentable subject-matters or activities remain so "even where they imply the possibility of making use of unspecified technical means"), at headnote II and reasons 3.

[18] Gottschalk v. Benson, supra note 16.

[19] Case law suggests that experimental use defense may be available only for pure research with no commercial implications. See Roche Products v. Bolar Pharmaceutical Company, 733 F.2d 858 (Fed. Cir.). cert. denied, 469 US 856 (1984) (Court rejected the arguments of a generic drug manufacturer that the experimental use defense should apply to its use of a patented drug to conduct clinical trials during the patent term). Congress has partially abrogated the decision of the Federal Circuit in Roche v. Bolar in the specific context of clinical trials of patented drugs by an amendment to the patent statute. As amended, the statute explicitly permits using patented inventions for developing and submitting information under laws regulating the manufacture, use, or sale of drugs. The amendment did not address the broader question of when the experimental use defense would be available outside of that very narrow setting. See Drug Price Competition and Patent Term Restoration Act of 1984, Public Law 98–417, codified in pertinent part at 35 USC § 271(e). See also Rebecca Eisenberg, Intellectual Property Rights and the Dissemination of Research Tools in Molecular Biology: Summary of a Workshop Held at the National Academy of Sciences, February 15–16, 1996.

[20] Most EU countries have statutory exemptions that implement Article 27(b) of the Community Patent Convention (CPC). See OECD, Research Use of Patented Knowledge: A Review by Chris Dent, Paul Jensen, Sophie Waller and Beth Webster, STI Working Paper 2006/2, at 17–18.

[21] OECD, supra note 25 (depicting the wide variation amongst OECD countries over research purposes exemptions), at 17–22.

[22] The Industrial Property Law (Mexico), Art 22.

[23] The Patents Decree Law (Turkey), Section 75.

[24] TRIPS, Article 30 ("Exceptions to Rights Conferred") (requiring that any exemption to patent rights satisfies certain requirements).

therefore traditionally been exempted from paying fees for patented inventions they use in their own research.[25]

The rationale for this exemption defines universities as public good agents. An enormous amount of basic research is hence produced every year in the United States and other advanced countries without the benefit of patentability. In the United States, and to a significant extent also in emerging economies, the government funds most basic research. By 2000, half of all basic research in the United States was funded by the federal government, and of the balance, 29 percent was financed by universities and other non-profit research establishments out of their own funds.[26] However, the extent and status of the exemption for research use differs across countries and is often subject of policy debate and litigation.[27]

A third patent policy restraining the blocking of downstream research of basic scientific inventions relates to US patent law prevention of patenting research tools, namely inventions used in a research or laboratory setting.[28] To be sure, research tools are not absolutely excluded from patent protection, except insofar as they lack US patentable utility.[29] Nevertheless, the requirement of utility can be understood to have numerous economic purposes, one of which is to exclude patents

---

[25] OECD, Patents and Innovation Trends and Policy Challenges: Trends and Policy Challenges (2004) (hereinafter, OECD, Patents and Innovation), at 21.

[26] See William M. Landes and Richard A. Posner, *The Economic Structure of Intellectual Property Law* (Harvard University Press, 2003), at 306.

[27] OECD, supra note 25, at 21. In the United States the Madey *v.* Duke case narrowed the scope of research exemptions. See Madey *v.* Duke, 307 F 3d 1351 (Fed. Cir. 2002). Oppositely, the courts in Germany have taken a "very liberal" approach to the experimental use exemption. Such were the two decisions of the German Supreme Court in Klinische Versuche I and Klinische Versuche II (Clinical Trials I and II). See ACIP (2004) Issues Paper, 4.

[28] See Rebecca Eisenberg, supra note 19.

[29] In US patent law, substantial utility under the "useful invention" requirement of 35 USC. 101 exempts scientific inventions carrying out further research to identify or reasonably confirm a "real world" context of use. The Manual of Patent Examining Procedure (MPEP) 2107 Guidelines for Examination of Applications for Compliance with the Utility Requirement [R-11.2013], excludes from "substantial utility: "(A) Basic research such as studying the properties of the claimed product itself or the mechanisms in which the material is involved." Substantial utilities are defined when "[A]n application must show that an invention is useful to the public as disclosed in its current form, not that it may prove useful at some future date after further research." in *re Fisher*, 421 F.3d 1365, 76 USPQ2d 1225 (Fed. Cir. 2005), at 1371.

EU patent law doesn't consider utility per se as a patentability criterion, and patentability may be excluded even with utility. See T 388/04 of March 22, 2006 of the Boards of Appeal of the EPC. Specifically, article 57 of the EPC (titled industrial application) has not specifically excluded research tools, while adhering only to more specified rules of the industrial applicability patentability requirement. See, e.g. Directive 98/44/EC of the European Parliament and of the Council of 6 July 1998 on the legal protection of biotechnological inventions (stating that in the context of biotechnological inventions a mere nucleic acid sequence having no function indication is unpatentable), at rec. 23.

on basic research.[30] The rationale is that granting a patent could inhibit diffusion by increasing the usage costs of these tools in applied research. In response, the US National Institutes of Health (NIH) have espoused a policy that discourages unnecessary patenting and encourages non-exclusive licensing. Such guidelines are now being emulated by funding agencies and research institutions in other countries. The licensing of research tools is regarded as a complicated and therefore costly process, for reasons similar to the expensive and complex nature of the putative patent licensing of basic research.[31]

A reward system that involves prestigious academic appointments, lecture fees, or the prospect of prizes incentivizes basic research, while patent law incentivizes applied research.[32] Universities themselves, moreover, engage in lobbying, which is yet another form of rent seeking for government grants.[33] If patent protection extended to basic research, government would have little incentive to fund research, taxes equally would be lowered, and the allocative distortions caused by taxes would be smaller. A great deal of this research, however, has no short-term commercial application and so could not be financed by patenting.[34]

4.1.2   The Challenge of Basic Research Funding

Economic theory has witnessed a second large body of literature on types of R&D, again outside the scope of development economics. This focuses on their impact on the innovation process and on the strength of international spillovers worldwide.[35] The main finding concerns the sub-optimal effect of business sector subsidy for basic research.[36] In 1959, Richard

[30] See William M. Landes and Richard A. Posner, supra note 26, at 302.
[31] Ibid., at 316, referring to Report of the National Institute of Health (NIH) Working Group on Research Tools, June 4 1998, at: www.mmrrc.org/about/NIH_research_tools_policy/; Michael A. Heller and Rebecca S. Eisenberg, Can Patents Deter Innovation? The Anticommons in Biomedical Research, 280 *Science* 698 (1998). But see the contrary arguments, reflecting the industry view, in Richard A. Epstein, *Steady the Course: Property Rights in Genetic Material* (University of Chicago Law School, John M. Olin Law and Economics Working Paper No. 152 [2d set.], June 2002).
[32] See William M. Landes and Richard A. Posner, supra note 26, at 306. But see Pamela Samuelson, Benson Revisited: The Case Against Patent Protection for Algorithms and Other Computer Program-Related Inventions, 39 *Emory Law Journal* 1025 (1990) (criticizing the perceived need for an incentive to encourage the inventions of new mental processes, against the backdrop of patent enforcement constraints in mental processes at large), at 1145.
[33] See William M. Landes and Richard A. Posner, supra note 26, at 306.   [34] Ibid.
[35] See Hans Gersbach and Maik T. Schneider, supra note 11, at 4.
[36] Keith Pavitt, What Makes Basic Research Economically Useful?, *Research Policy*, 20 (1991), 109, 110.

Nelson published his pioneering paper *The Simple Economics of Basic Research*, in which he established the notion that a competitive market, left to itself, will invest sub-optimally in basic research, hence creating the need for public subsidy.[37] Nelson's insights have been developed and modified over the past 30 years, notably by the Nobel Prize for Economics laureate Kenneth Arrow,[38] as well as Harvey Averch[39] and others.[40] The case for public funding of basic research, to conclude, is in principle well substantiated.[41]

From a policy standpoint, as scientific research exhibits key characteristics of a public good, two finance-related concerns arise. Firstly, when should governments invest in basic research over the course of economic development? Secondly, where should governments target these investments?[42] Governments worldwide continuously ponder how much to invest in basic research?[43] In their decisive 2015 study *How Much Science?*, ETH Zürich University researchers Hans Gersbach et al. conclude that, considering the dynamic participation of governments and the high value at stake, the present lack of direction regarding government participation in basic R&D is stunning.[44] As the authors suggest, this may indeed be attributable to the fact that while the benefits

---

[37] Prof. Nelson explains that this is because a profit-seeking firm cannot not be sure of capturing all the benefits of the basic science that it sponsors, given significant uncertainties about the advantages for the sponsoring firm and the difficulties it faces in extracting compensation from subsequent imitators. See Richard R. Nelson, The Link between Science and Invention: The Case of the Transistor, in *The Rate and Direction of Inventive Activity* (Richard R. Nelson, ed.) (Princeton University Press, 1962). Nelson adds that if profit-seeking firms are risk-averse, or have short-term horizons in their decisions to allocate resources, private expenditures on basic research will be even more sub-optimal.

[38] Kenneth Arrow, Economic Welfare and the Allocation of Resources for Invention, in *The Rate and Direction of Inventive Activity* (Richard R. Nelson, ed.) (Princeton University Press, 1962).

[39] Harvey Allen Averch, *A Strategic Analysis of Science and Technology Policy* (John Hopkins Press, 1985).

[40] See John Kay and Chris Llewellyn Smith, Science Policy and Public Spending, *Fiscal Studies* 6 (1985) 14–23; Partha Dasgupta, The Economic Theory of Technology Policy: An Introduction, in Paul Stoneman and Partha Dasgupta, *Economic Policy and Technological Performance* (Cambridge University Press, 1987); Paul Stoneman, *The Economic Analysis of Technology Policy* (Oxford University Press, 1987); Paul Stoneman and John Vickers, The Assessment: The Economics of Technology Policy, *Oxford Review of Economic Policy* 4 (1988), at i–xvi.

[41] See, e.g., Keith Pavitt, supra note 36, at 110.

[42] Hans Gersbach, Ulrich Schetter, and Maik Schneider, supra note 4, at 2–3. Gersbach and others discuss also the two following questions of when to invest in basic research (section 6) and where if at all (section 7). Ibid. The latter two considerations remain outside the scope of this chapter.

[43] Ibid.

[44] Ibid., at 3. See also Amnon J. Salter and Ben R. Martin, supra note 4 ("Currently, we do not have the robust and reliable methodological tools needed to state with any certainty what the benefits of additional public support for science might be, other than suggesting

of basic research on innovation and patent intensity are diverse, they are often time-lagged and indirect.[45]

Against the backdrop of such measurement constraints, past experience shows that funding fluctuations in scientific research have largely been predisposed to exogenous political demand-side economics. The 1950s and 1960s Cold War years, for example, were periods of considerable growth in funding for scientific research. During this time, there was a belief that investments in science would generate innovation.[46] This dynamic was fueled by the neoclassical economics argument that, due to the unique characteristics of basic research, the private sector would always under invest in it, by way of a representation of market failure.[47]

Toward the end of the 1960s, the belief in the inevitably benevolent potential of science began to be challenged,[48] as some of the negative consequences of scientific development on the environment and wider society began to become apparent. During the late 1960s and 1970s, many countries experienced a reduction in government expenditures on science.[49] A major change in government's attitudes toward basic research occurred during the 1980s in many countries. Following the end of the Cold War, military incentives for funding research were no longer so pressing, leading to a sharper focus on the promotion of technology and economic competitiveness. At the same time, academic studies of innovation began to question the simple linear model of the connection between science and technology.[50] The changes that began in the 1980s continue to this day.

By the 1990s, most European countries had entered an era of "steady state" science,[51] whereby funding did not keep up with rapidly rising

that some support is necessary to ensure that there is a 'critical mass' of research activities"), at 529.

[45] Hans Gersbach, Ulrich Schetter, and Maik Schneider, supra note 4, at 3.

[46] Philip Gummett, The Evolution of Science and Technology Policy: A UK Perspective, Science and Public Policy 1 (1991), at 18; Jane Calvert and Ben R. Martin, Changing Conceptions of Basic Research?, Background Document for the Workshop on Policy Relevance and Measurement of Basic Research Oslo 29–30 October 2001, SPRU – Science and Technology Policy Research (September 2001), at 2.

[47] Jane Calvert and Ben R. Martin, ibid., at 2.

[48] Jean-Jacques Salomon, Science Policy Studies and the Development of Science Policy, in Science, Technology and Society: A Cross Disciplinary Perspective (Ina Spiegel-Rösing and Derek de Solla Price, eds.) (Sage, 1977).

[49] Aant Elzinga, Research, Bureaucracy and the Drift of Epistemic Criteria, in The University Research System: The Public Policies of the Home of Scientists (Björn Wittrock and Aant Elzinga, eds.) (Almqvist and Wiksell International, 1985), at 192.

[50] David Mowery and Nathan Rosenberg, Technology and the Pursuit of Economic Growth (Cambridge University Press, 1989); Christopher Freeman, The Economics of Industrial Innovation (Harmondsworth, 1974)

[51] John Ziman, Prometheus Bound (Cambridge University Press, 1994).

research costs. In the United States, basic research is mainly conducted by the federal government and by universities and colleges, and about 80 percent of it is publicly funded.[52]

To summarize, the history of the funding of basic research from the 1950s to the 2000s has shown a move away from the idea that scientists should be supported as they engage in independent pure science, and toward the idea that they should orient their work more clearly toward social and economic objectives.

After much discussion, the classification of R&D types was adopted as noted by the OECD's Frascati Manual and the National Science Foundation (NSF)'s 1996 Science and Engineering Indicators, providing official definitions for basic research, applied research, and experimental development.[53] This classification was also integrated in commercial US law. For example, a somewhat debatable definition of basic research was adopted in the United States Department of Justice's Antitrust Guidelines of 1980, which formed the foundation for the National Cooperative Research Act of 1984 (referring explicitly to basic research activities as part of its justification for encouraging collaborative ventures). In addition, a basic research credit was included as part of the Tax Reform Act of 1986, with the goal of encouraging industrial support for university- and non-profit-based basic research and experimentation.[54]

4.2   Patent Intensity by Type of R&D: Policy Considerations

Different types of R&D are reflected in differing levels of patent intensity. As a general rule, the more scientific and less commercial the type of research is, the less patent intensity applies. This core understanding relates to all three types of R&D defined above: basic research, applied research, and experimental development. The modeling of these three innovation pillars has seen much transition since the early 1960s. The NSF has been the most advanced regulatory institution to classify these types of research.[55] As of the NSF 1955–56 survey, the definition of

---

[52] Guido Cozzi and Silvia Galli, Science-based R&D, supra note 5.

[53] Professor Albert Link adds that the NSF's classifications were further accepted by the industry. See Albert N. Link, On the classification of industrial R&D, *Research Policy*, Volume 25(3) 397 (May 1996) (presenting the results of a survey of 101 R&D directors/managers showing that 89 of the respondents agree that the NSF classification appropriately describes the scope of their R&D activities), at 399.

[54] Ibid., at 398.

[55] Ibid. For an early study marking the take-off for massive financial support for basic research in the United States after World War II, see Vannevar Bush, Science: The Endless Frontier, A Report to the President by Vannevar Bush, Director of the Office of Scientific Research and Development, July 1945 (United States Government Printing

basic research is:[56] "Basic research – research projects which represent original investigation for the advancement of scientific knowledge and which do not have specific commercial objectives, although they may be in fields of present or potential interest to the reporting company."[57] The NSF definitions were an important resource when the OECD formulated its own definitions as early as 1963, in what is now OECD's 2002 renowned Frascati Manual.[58]

The Frascati Manual also divides R&D activities into three main categories: basic research, applied research, and experimental development.[59] It defines basic research as "experimental or theoretical work undertaken primarily to acquire new knowledge of the underlying foundation of phenomena and observable facts, without any particular application or use in view." According to this definition, basic research in general does not provide (potentially commercial) patentable technological solutions for specific practical problems, but rather provides the knowledge base needed to tackle core scientific problems.[60]

Basic research is typically identified by six characteristics distinguishing it from other types of R&D (conveniently labeled "non-basic research").[61] Firstly, basic research outcomes are embryonic, in the sense that they are early-stage findings with little or no imminent commercial use, as opposed to non-basic research.[62] Secondly, basic research builds on and further

---

Office, Washington: 1945) (describing basic research as mainly "curiosity-driven" and "performed without thought of practical ends"), chapter 1.

[56] National Science Foundation (NSF), 1959, Science and Engineering in American Industry: Final Report on a 1956 Survey, National Science Foundation report (Washington, DC). For a previous equivalent definition, see National Science Foundation (NSF), 1956, Science and Engineering in American Industry: Final Report on a 1953–1954 Survey, National Science Foundation report (Washington, DC). The first definition of basic research was developed by Dearborn and others in 1953, based on industry interviews. See DeWitt C. Dearborn, Rose W. Kneznek, and Robert N. Anthony, Spending for industrial research, 1951–52, Harvard University Graduate School of Business report (Cambridge, MA) (1953).

[57] Science Foundation (NSF), 1959, Science and engineering in American industry: final report on a 1956 survey, National Science Foundation report (Washington, DC), ibid.

[58] The OECD's standard definition similarly emphasizes the absence of practical considerations, stating that "basic research is experimental or theoretical work . . . without any particular application or use in view." See OECD, The Measurement of Scientific and Technological Activities: Standard Practice for Surveys of Research and Experimental Development (1994) (Frascati Manual 1993: OECD Publications), at 13.

[59] Ibid., at 30.

[60] Hans Gersbach, Ulrich Schetter and Maik Schneider, supra note 4, at 3. The Frascati Manual's definition of basic research differs from the so-called Pasteur's Quadrant. See Donald Stokes, Pasteur's Quadrant: Basic Science and Technological Innovation, The Brookings Institution, Washington, DC (1997). The term was introduced by Princeton University professor Donald Stokes. Stokes characterizes a class of scientific research methods as having immediate commercial use, bridging the gap between basic and applied research. Ibid.

[61] Hans Gersbach, Ulrich Schetter and Maik Schneider, supra note 4, at 4.   [62] Ibid.

develops the insights provided by prior (basic) research, and is thus regarded as cumulative.[63] Thirdly, the generation of new knowledge within basic research and its reflection in new and refined products or processes involve major time lags until non-basic research occurs.[64] Fourthly, basic research outcomes are highly uncertain, as opposed to more certain non-basic research.[65] Fifthly, basic and applied research (alongside experimental development) typically have a hierarchical order, with the former preparing the ground for the latter.[66] Lastly, non-basic research not only benefits from basic research but also stimulates basic research in return, in a two-way spillover.[67] From a patent-policy perspective comparing patent intensity as proxy for domestic innovation across countries, two of these six characteristics are pivotal, namely the commercial outcomes of the research (characteristic 1) and the hierarchical order between basic and non-basic research (characteristic 5). These two features also determine the differing rates of patent propensity and patent intensity at large across the three patent clusters, as follows.

4.2.1 Commerciality of R&D and Patent Intensity

The output of basic research is nascent in nature and lacks immediate commercial use.[68] It typically requires refinement through applied research before commercialization.[69] Such are perceived to be the basic research discoveries of X-rays in the area of physics and life sciences, the invention of the method of Ribonucleic acid (RNA) interference for genetics, or the invention of the method of nuclear fission for nuclear physics.[70] All these discoveries were subsequently further developed and commercialized, eventually exerting remarkable impact in particular industries.[71]

Applied research, unlike basic research, thus sets out to "provide ... complete answers" to important practical problems,[72] thus directly contributing to the improvement and development of specific production technologies or products.[73] Applied research was traditionally said to be carried out by private firms that commercialized the

---

[63] Ibid.   [64] Ibid.   [65] Ibid.   [66] Ibid.   [67] Ibid.
[68] Hans Gersbach, Gerhard Sorger, and Christian Amon, supra note 5, at 5.   [69] Ibid.
[70] Ibid., at 5 and table 3 in the appendix for further examples.
[71] Ibid., at 6. For particular industries see, e.g., Jason Owen-Smith, Massimo Riccaboni, Fabio Pammolli, and Walter W. Powell, A Comparison of US and European University-Industry Relations in the Life Sciences, *Management Science*, 48(1), 24–43 (2002) (for the biotechnology industries); Ernst Heinrich Hirschel, Horst Prem, and Gero Madelung: *Aeronautical Research in Germany – from Lilienthal until Today*, Springer-Verlag (2004) (for the aeronautics industries).
[72] See Vannevar Bush, supra note 55, at 13.   [73] Ibid.

output of basic research by transforming it into blueprints for new products.[74] Basic research as defined above thus functioned as a central category in science policy, though it bears little relation to patent intensity.[75]

Against this backdrop, the rise in university patenting has occurred as part of a policy agenda for closer relations between applied research and the industry. The goal of this is to increase social and private returns from publicly funded basic research. Examples of this include the general strengthening of patent law enforcement and technology transfer regulation,[76] starting with the United States 1980s Bayh–Dole Act, granting recipients of federal R&D funds the right to patent inventions and license them to firms,[77] and the European "professor's privilege" to own patents equivalence.[78]

---

[74] Hans Gersbach, Gerhard Sorger and Christian Amon, supra note 5, at 1.

[75] See Amnon J. Salter and Ben R. Martin, supra note 4 (the authors review and assess the literature on the economic benefits associated with publicly funded basic research), at 510; Bureau of the Census, Evaluation of proposed changes to the survey of industrial research and development, National Science Foundation report, June (Washington, DC) (1993); Benoît Godin, Measuring Science: Is There "Basic Research" without Statistics? Project on the History and Sociology of S&T Indicators, Paper No. 3, Montreal: Observatoire des Sciences et des Technologies INRS/CIRST; Philip Gummett, supra note 46, at 18; Eileen L. Collins, Estimating Basic and Applied Research and Development in Industry: A Preliminary Review of Survey Procedures, National Science Foundation report (Washington, DC) (1990); Applied Management Sciences, Inc., NSF workshop on Industrial S&T data needs for the 1990s: final report, National Science Foundation report 1 December (Washington, DC) (1989); Howard K. Nason, George E. Manners, and Joseph A. Steger, Support of Basic Research in Industry, National Science Foundation report (Washington, DC) (1978).

　　Much of present-day literature uses other terms such as "science," "academic research," or just "research," categories that are not identical with "basic research" although they overlap considerably. See Amnon J. Salter and Ben R. Martin, supra note 4 (the authors review and assess the literature on the economic benefits associated with publicly funded basic research), at 510.

[76] There is a large diversity in the structure and organization of TTOs within and across countries, such as TTOs' on or off-campus offices, arm's length intermediaries, industry sector-based TTOs and regional TTOs. See generally, OECD, Turning Science into Business: Patenting and Licensing at Public Research Organizations, OECD (2003), at 29–37.

[77] See Bayh–Dole Act or Patent and Trademark Law Amendments Act (Pub. L. 96–517, December 12, 1980). See also United States Congressional Research Service, Patent Ownership and Federal Research and Development (R&D): A Discussion on the Bayh–Dole Act and the Stevenson-Wydler Act (December 11, 2000). For a critical assessment of the Act on science policy, see generally, David C. Mowery, Richard R. Nelson, Bhaven N. Sampat, and Arvids A. Ziedonis, *Ivory Tower and Industrial Innovation University-Industry Technology Transfer Before and After the Bayh–Dole Act* (Stanford University Press, 2004).

[78] See generally, Hans K. Hvide and Benjamin F. Jones, University Innovation and the Professor's Privilege, NBER Working Paper No. 22057 (March 2016).

4.2.2   Patents and the Hierarchy of R&D

As already explained, the innovation processes in both advanced and emerging economies are characterized by a systematic hierarchy of basic and applied research.[79] A further seminal property of basic research concerns the duration between its origin in the scientific community and its impact on industrial productivity.

In this regard, economist James Adams finds that the expansion of academic knowledge exerts a positive but lagged impact on technological change and productivity growth.[80] By applying various measures of science within a growth-accounting framework, his findings suggest that the impact of new academic knowledge on industrial productivity does not take place instantaneously, but is rather associated with time-lags of about 20 years stemming from the time necessary to search for and adopt useful scientific knowledge in industry.[81] These findings suggest a rather strict hierarchy between basic and applied research.[82]

However, other views exist regarding the hierarchy between the two R&D types. Notably, Richard Nelson and Nathan Rosenberg argue that the traditional distinction between basic and applied science is anachronistic.[83] Nelson's seminal paper on the history of the transistor serves as a case in point.[84] Rosenberg explains that the distinction between basic research and applied research is highly artificial and arbitrary. The distinction is usually made in order to highlight the motives or goals of the person performing the research. In most cases, however, this is not a very useful or illuminating distinction.[85] Fundamental breakthroughs may sometimes occur while dealing with very applied or practical problems. Moreover, attempting to

[79] Hans Gersbach, Gerhard Sorger and Christian Amon, supra note 5 (This paper develops an endogenous growth model that captures the mentioned hierarchy and relations within the innovation process. The paper analyzes a closed economy with a final and an intermediate goods sector, a basic and an applied research sector, a continuum of infinitely lived households, and a government), at 1.

[80] James D. Adams, supra note 14.

[81] Hans Gersbach, Gerhard Sorger and Christian Amon, supra note 5, at 7.   [82] Ibid.

[83] Richard R. Nelson, Reflections on "The Simple Economics of Basic Scientific Research": Looking Back and Looking Forward, *Industrial and Corporate Change*, Volume 15(6) 903 (2006), at 907; Richard R. Nelson, supra note 37 (explaining that in the transistor project the results included both an advance in fundamental physical knowledge and the invention and improvement of practical devices), at 581. Analyzing the example of the transistor Nelson was noticeably first to uphold that the distinction between basic research and applied research is fuzzy. Ibid. Nathan Rosenberg and Richard R. Nelson, The Roles of Universities in the Advance of Industrial Technology, in *Engines of Innovation* (Richard S. Rosenbloom and William J. Spencer, eds.) (Harvard Business School Press, 1996), at 91.

[84] Richard R. Nelson, supra note 37.

[85] Nathan Rosenberg, Why Do Firms Do Basic Research (With Their Own Money)?, *Research Policy* 19, 165 (1990), at 169.

draw that line on the basis of the motives of the person performing the research – whether there is a concern with acquiring useful knowledge (applied) as opposed to a purely disinterested search for new basic knowledge – is, Rosenberg argues, a "hopeless quest."[86]

More generally, science and technology modeled as applied R&D are considered much more similar than previously thought.[87] Nelson and Rosenberg explain that there is a widespread belief that modern fields of technology are in effect applied science – that is, in the sense that practice is drawn directly from and illuminated by scientific understanding. Advancing technology, the authors add, is mostly a regular task of applying new scientific consideration in order to achieve better products and processes. This view, however, respectfully overstates the illumination science provides to most technologies while suggesting that the connection between the advancement of science and the advancement of technology is far simpler than is actually the case.[88] Several other studies point out the increasing complexity in applied R&D activity. Increasingly difficult applied research also reveals an increasing interconnectedness between applied and basic research.[89] To produce new medical treatments, it

---

[86] Ibid.

[87] Richard R. Nelson, supra note 83, at 907; Nathan Rosenberg and Richard R. Nelson, supra note 83, at 91;

[88] Ibid. Reflecting on evolutionary economics, Nelson explains: "In fact, much of practice in most fields remains only partially understood scientifically ... Much of engineering design practice involves solutions to problems that professional engineers have learned 'work' without any particularly deep understanding of why," at 907; "Technological practice and understanding tend to co-evolve, with sometimes advance of understanding leading to effective efforts to improve practice and sometimes advance in practice leading to effective efforts to advance understanding," Ibid. See also Richard R. Nelson and Nathan Rosenberg, Technical Innovation and National Systems, in Richard R. Nelson (ed.), *National Innovation Systems: A Comparative Analysis* (Oxford University Press, 1993) (on the complexity of the interconnections of science and technology), at 1–21.

[89] See, e.g., Amnon J. Salter and Ben R. Martin, supra note 4 (the authors review and assess the literature on the economic benefits associated with publicly funded basic research, while arguing for an overlap between basic and applied research rather than substantial differences), at 510; Guido Cozzi and Silvia Galli, Science-Based R&D in Schumpeterian Growth, *Scottish Journal of Political Economy* vol. 56(4) 474 (2009), at 475; Jane Calvert and Ben R. Martin, supra note 46 (background document analyzing the concept of basic research, drawing on interviews with nearly 50 scientists and policymakers on the definition of basic research concluding that basic R&D is most efficient when its interactive with advanced R&D); Nathan Rosenberg, supra note 85, at 170.

But see Keith Pavitt, The Social Shaping of the National Science Base, *Research Policy* 27 793 (1998) ("despite examples of spectacularly close links between basic research and technology e.g., biotechnology, basic research builds mainly on basic research scientific papers cite other scientific papers much more frequently than patents; and technology builds mainly on technology e.g., patents cite other patents much more frequently than scientific papers"), at 795.

is crucial to have an improvement in the basic understanding of the pathologies.[90]

Nevertheless, what remains in the core of patent propensity measurement is that basic research per se is perceived as substandard for patenting, and thus ill-equipped to foster patent intensity in general. Simply put, patent intensity is recurrently correlated to the propensity to patent of applied research as opposed to basic research.

Marie Thursby and Richard Jensen illustrate the prevailing received wisdom at least in the context of advanced economies.[91] They find that more than 75 percent of university inventions, perceived as basic research, were merely a non-commercial proof of concept. Thursby and Jensen emphasize the embryonic state of most university-based technologies licensed and identify a subsequent need for inventor cooperation in commercialization.[92] They found that, at the time of patent licensing, only 12 percent of all university inventions reviewed were ready for commercialization, and manufacturing feasibility was only known for 8 percent of the inventions.[93]

In view of development with the lower patent clusters and developing countries broadly, support of research-related patent policy may demand a broader base than that provided by neoclassical economics, with its strong focus on the advanced economies.[94] Endogenous economic growth reasoning should ultimately be included, especially where development growth dialectics demands archetypal endogenous relativism. That said, even at its core present wisdom, research-related patent policy must still acknowledge possible empirical country-group discrepancies.

---

[90] Amnon J. Salter and Ben R. Martin, supra note 4; Guido Cozzi and Silvia Galli, supra note 89.

[91] Marie Thursby and Richard Jensen, Proofs and Prototypes for Sale: The Licensing of University Inventions, *American Economic Review*, American Economic Association, vol. 91(1), 240 (2001).

[92] Ibid.

[93] Ibid. See also Yixin Dai, David Popp and Stuart Bretschneider, *Journal of Policy Analysis and Management*, vol. 24(3) 579, (2005) ("As with basic research, many academics performing applied research still use publications as a main output due to academic inertia. Even if the researcher decides to patent, it is unlikely that this decision was made at the very beginning of the research process"), at 584.

[94] See Hans Gersbach, Gerhard Sorger, and Christian Amon, supra note 5 (for an endogenous growth model relating to basic and applied research); c.f. Karl Max Einhaäupl, What does "Basic Research" mean in Today's Research Environment?, Keynote address to the OECD workshop on "Basic Research: Policy Relevant Definitions and Measurement," 28–30 October, Oslo, Norway, at 5.

4.3   The Empirical Analysis

4.3.1   *Methodology*

The methodology adheres to the economic literature regarding basic research in subsuming experimental development under applied research.[95] Given that substantive data are missing for the early years of the 1996–2013 dataset, a different method of imputation was further applied. This meant that there was at least one non-missing yearly dataset in the early years of 1996–2003, and that there was at least one non-missing yearly dataset for the concluding years of 2011–2013. When characterizing the patent clusters, the yearly dataset used is for the years 2003–2011. Otherwise, when characterizing economy types with reference to advanced and emerging economies, the dataset used is between the years 1996 and 2011.

4.3.2   *Findings*

4.3.2.1 Type of GERD by Patent Activity Intensity
R&D expenditures are higher in advanced economies than in middle-income or emerging economies. This core finding is clearly shown in Figures 4.1 and 4.2 for both basic and non-basic R&D expenditure

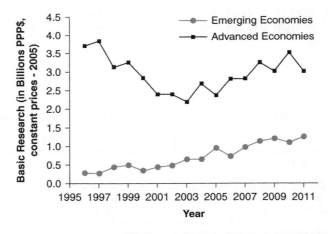

Figure 4.1 Basic Research (in Billions) PPP\$ Constant Prices – 2005 (Vertical Axis) vs. Annual Changes (horizontal axis) (1996–2011)

[95] Hans Gersbach, Ulrich Schetter and Maik Schneider, supra note 4, at 4.

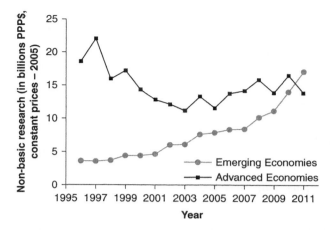

Figure 4.2 Non-basic Research (in Billions) PPP$ Constant Prices – 2005 (Vertical Axis) vs. Annual Changes (Horizontal Axis) (1996–2011)

rates. The comparison between the two economy types becomes more complex when patent activity intensity factors are accounted for, in reference to percentage of GDP and GERD. From a patent-policy standpoint, these measurements may provide much of the explanation as to why different countries and country groups differ over their patent intensity as measured by a plethora of types of R&D rates.

The discrepancies over R&D by type have rarely been measured to date. Hans Gersbach's seminal work on the macroeconomics of basic research encapsulated received wisdom concerning the geography of basic research across countries. Gersbach's findings are twofold, albeit mostly limited in empirical scope to OECD countries. Firstly, basic research is said to be undertaken mainly by industrialized countries near the technological frontier. Some emerging economies, such as Korea or Singapore, have considerably stepped up their basic research efforts.[96] Second, large industrial countries spend less on average on basic research than small economies. Within the OECD, the United States and France spend about 0.5 percent of their GDP on basic research. By contrast, Switzerland invests a substantially higher share of about 0.8 percent.[97] Figures 4.1 and 4.2 similarly depict the discrepancies over basic research fluctuations between advanced and emerging economies relating to basic and non-basic research, respectively. An analogous observation is

[96] See Hans Gersbach and Maik T. Schneider, supra note 11, at 3.   [97] Ibid.

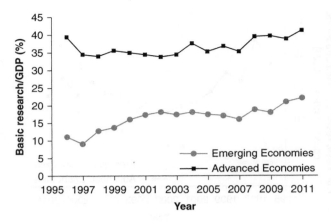

Figure 4.3 Basic Research as a Percentage of Gross Domestic Product (GDP) (Vertical Axis) vs. Annual Changes (Horizontal Axis) (1996–2011)

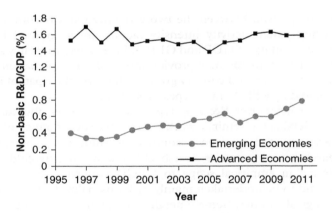

Figure 4.4 Non-basic Research as a Percentage of Gross Domestic Product (GDP) (Vertical Axis) vs. Annual Changes (Horizontal Axis) (1996–2011)

witnessed in Figures 4.3 and 4.4, representing basic and non-basic research expenditures as a percentage of GDP in both advanced and emerging economies, respectively. In all four categories of R&D by research type, advanced economies are significantly advantageous in comparison with emerging economies.

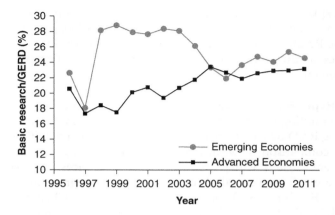

Figure 4.5 Basic Research as a Percentage of Gross Domestic Expenditure on R&D (GERD) (Vertical Axis) vs. Annual Changes (Horizontal Axis) (1996–2011)

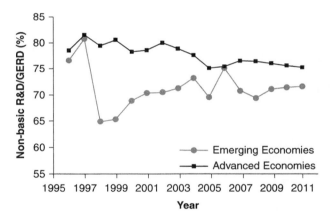

Figure 4.6 Non-basic Research as a Percentage of Gross Domestic Expenditure on R&D (GERD) (Vertical Axis) vs. Annual Changes (Horizontal Axis) (1996–2011)

However, different findings emerge when accounting for R&D fluctuations as a percentage of GERD within each economy type. In emerging economies, GERD expenditures are more significantly based on basic R&D expenditures rates than in advanced economies, with the exception of 2005–2006, as indicated in Figure 4.5. The opposite finding emerges for non-basic R&D, as Figure 4.6 details. These findings may ultimately serve to explain why emerging economies show lower propensity to patent rates than advanced economies. Since basic R&D is less patentable than non-basic R&D, the latter representing more commercial types of R&D, the figures are therefore more indicative of lower rates of patent propensity in general.

4.3.2.2 Experimental Development Advantage by Patent Clusters
When accounting for the three patent clusters, the findings arising from comparing R&D by type rates turn more narrow in scope. Different than with the comparison between advanced and emerging economies, the three patent clusters witness no significant differences over basic and applied research between all three clusters, as the regression models in Appendix E: Gerd by Type of R&D and Patent Clusters foretell. Figures 4.7 and 4.8 and Tables 4.1 and 4.2 show that the only statistically

Table 4.1 *Patent Cluster Plot Type of GERD Data*

	Cluster 1	Cluster 2	Cluster 3
Applied research	29.90401	38.51996	44.60513
Basic research	20.89564	23.63227	27.07568
Experimental development	49.20035	37.84776	28.31919

Table 4.2 *List of Countries by Patent Cluster Distributed by Type of GERD (%)*

	code	Country	Cluster	m_RA	m_RB	m_RE
1	AT	AUSTRIA	1	35.19	17.54	45.37
2	BG	BULGARIA	2	50.77	28.62	20.61
3	CH	SWITZERLAND	1	33.00	28.08	39.65
4	CN	CHINA	2	16.20	5.19	78.60
5	CU	CUBA	2	50.00	10.00	40.00
6	CY	CYPRUS	3	58.30	19.80	21.90
7	CZ	CZECH REPUBLIC	2	28.27	30.45	41.28
8	DK	DENMARK	1	26.04	16.11	56.16
9	EC	ECUADOR	3	61.49	20.77	11.89
10	EE	ESTONIA	3	24.41	29.37	46.22
11	FR	FRANCE	1	38.35	24.57	37.07
12	HR	CROATIA	2	36.65	35.27	28.08
13	HU	HUNGARY	2	33.37	25.89	37.65
14	IE	IRELAND	1	36.07	22.48	41.45
15	IL	ISRAEL	1	4.85	15.44	79.71
16	IS	ICELAND	3	38.89	17.83	43.17
17	IT	ITALY	1	44.87	27.04	28.26
18	JP	JAPAN	1	21.34	11.95	61.75

Table 4.2 (*cont.*)

	code	Country	Cluster	m_RA	m_RB	m_RE
19	KR	REPUBLIC OF KOREA	1	20.32	15.73	63.95
20	LV	LATVIA	3	43.61	26.08	30.31
21	NZ	NEW ZEALAND	1	36.39	30.57	33.61
22	PA	PANAMA	3	32.05	36.43	31.54
23	PT	PORTUGAL	2	36.77	22.39	40.83
24	SG	SINGAPORE	1	30.40	18.86	50.74
25	SI	SLOVENIA	2	65.30	12.50	22.21
26	SK	SLOVAKIA	2	30.95	44.70	24.36
27	SV	EL SALVADOR	3	47.24	35.45	9.24
28	ZA	SOUTH AFRICA	2	35.72	20.59	43.69

Legend:

Code – Country code

Cluster – 1-3 per *Leaders, Followers* and *Marginalized*, respectively

M_RA – Mean of Applied Research (%)

M_RB – Mean of Basic Research (%)

M_RE – Mean of Experimental Development (%)

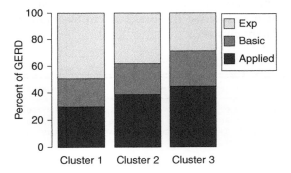

Figure 4.7 GERD by Type of R&D and Patent Cluster (%)
(2003–2001)

significant difference found is over experimental development as a percentage of GERD, it being the more patentable and commercial type of
non-basic R&D. Even here, the only significant different is between the
leaders cluster (labeled cluster 1) and marginalized cluster (labeled cluster 3). This finding, limited as it may be, nevertheless supports the finding
whereby advanced economies or more narrowly countries belonging
to the leaders patent cluster are more significantly characterized by
added commercial and thus patentable types of R&D activity. The latter

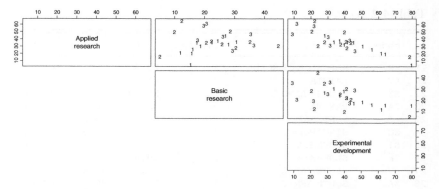

Figure 4.8 Scattered Plots of Type of GERD by Patent Clusters (%)
Legend: 1 – Leaders cluster; 2 – Followers cluster; 3 – Marginalized cluster

finding may ultimately further explain why advanced countries closely associated with the leaders patent cluster show an altogether higher patent intensity rate.

Figure 4.8 and the records presented in Table 4.2 describe the scattered plots of all combinations of R&D activity by patent clusters.

The following figures describe the basic descriptive statistics of the GERD values by patent cluster.

Leaders *Cluster:*

	vars	n	mean	sd	median	trimmed	mad	min	max	range	skew	kurtosis
m_RA	1	11	29.71	11.05	33.00	30.79	7.94	85	44.87	40.01	-0.81	-0.21
m_RB	2	11	20.76	6.11	18.86	20.65	5.36	11.95	30.57	18.62	0.22	-1.57
m_RE	3	11	48.88	15.33	45.37	47.75	15.99	28.26	79.71	51.45	0.51	-0.94
	se											
m_RA	3.33											
m_RB	1.84											
m_RE	4.62											

Followers *Cluster:*

	vars	n	mean	sd	median	trimmed	mad	min	max	range	skew	kurtosis
m_RA	1	10	38.40	13.74	36.19	37.81	9.75	16.20	65.3	49.09	0.40	-0.70
m_RB	2	10	23.56	12.09	24.14	23.21	12.93	5.19	44.7	39.50	0.08	-1.20
m_RE	3	10	37.73	16.80	38.83	34.76	11.57	20.61	78.6	58.00	1.19	0.74
	se											
m_RA	4.35											
m_RB	3.82											
m_RE	5.31											

Marginalized Cluster:

	vars	n	mean	sd	median	trimmed	mad	min	max	range	skew	Kurtosis
m_RA	1	7	43.71	13.37	43.61	43.71	17.13	24.41	61.49	37.08	-0.01	-1.65
m_RB	2	7	26.53	7.53	26.08	26.53	9.31	17.83	36.43	18.59	0.18	-1.89
m_RE	3	7	27.75	14.31	30.31	27.75	19.07	9.24	46.22	36.98	-0.02	-1.79
	se											
m_RA	5.05											
m_RB	2.84											
m_RE	5.41											

4.3.2.3 Test Results Explanation

In emerging economies, basic R&D expenditures as a percentage of GERD are significantly higher in comparison with advanced ones, as Figure 4.5 shows. In absolute terms and as a percentage of GDP, as Figures 4.3–4.4 show, the opposite is predictably true, with advanced economies trumping all type of GERD indicators. In moderate support, these differences, however, were significantly witnessed in Figures 4.7–4.8 and Tables 4.1–4.2 concerning experimental development non-basic R&D as a percentage of GERD, dividing the leaders and marginalized patent clusters. The regression models in Appendix E: Gerd by Type of R&D and Patent Clusters, further support this finding. As basic research is substantively less patentable than applied research and experimental development, the overall propensity to patent in emerging economies is consequently lowered than in advanced economies. A question remains: Why do emerging economies allocate relatively higher GERD rates to basic R&D in the first place? A completing question follows – namely, what explains the sustainability of this significance gap across the time series?

These related concerns are yet to be adequately studied. Yet one core explanation already offers considerable insight. The main practical benefits of basic research, all told, are difficult to transmit across countries. Such type of knowledge being transmitted is tacit or non-codifiable knowledge and often times is passed through personal mobility and informal human interaction.[98] The benefits deriving from basic research thus knowingly gravitate toward geographical, cultural-centrist, and linguistic locales.[99] This important observation has recently been confirmed in a long plethora of empirical studies by economists, sociologists, and bibliometricians following the seminal work of Adam Jaffe,[100] Francis

---

[98] Keith Pavitt, supra note 89, at 797.   [99] Ibid.
[100] See Adam Jaffe, The Real Effects of Academic Research, *American Economic Review*, 79 (5) 957 (1989); Adam Jaffe, Manuel Trajtenberg, and Rebecca Henderson, Geographic Localization of Knowledge Spillovers as Evidence by Patent Citations, *Quarterly Journal of Economics* 108, 577 (1993).

Narin,[101] and others, albeit focused on advanced economies.[102] More conceptually, although academic research has some attributes of a public good, it is arguably not a free good as it is costly to transfer across countries, and noticeably so to developing ones.[103] As a result, countries and firms operating outside Northern innovation centers benefit from basic research mostly if they belong to the international professional networks that exchange knowledge.[104] This requires at the very least high-quality foreign research training, and at the most a strong world presence in basic research.[105]

Economic theory surely has witnessed a large body of literature on the strength of international spillovers also across countries.[106] Yet the main policy consideration consequently concerns the importance of public subsidy for basic research for Northern, developed countries constituting these same innovation centers at the first place.[107] Given subsidy costs of basic R&D aimed at compensating for the mentioned transaction costs associated with its international transfer across countries, perhaps the most interesting trend among emerging economies is the building up of local technology-dependent industries. Emerging economies noticeably do so through the use of preferential procurement policies and their industrial policy mechanisms.[108] Against the backdrop of basic R&D transaction costs and the role preferential procurement policies play instead, it seems that for emerging economies patent systems (measured by patent intensity) are hence less pivotal in comparison with advanced ones.[109]

---

[101] Francis Narin, Kimberly S. Hamilton, and Dominic Olivastro, The Increasing linkage between US Technology and Public Science, *Research Policy* 26:317 (1997).

[102] E.g., J. Sylvan Katz, Geographical Proximity and Scientific Collaboration, *Scientometrics*, 31 (1), 31 (1994);

[103] Keith Pavitt, supra note 36, at 111, referring to David. Mowery, Economic Theory and Government Technology Policy, *Policy Sciences* 16, 27 (1983).

[104] Diana Hicks, Published papers, Tacit Competencies and Corporate Management of the Public/Private Character of Knowledge, *Industrial and Corporate Change*, 4: 401 (1995). Cf. Keith Pavitt, supra note 89 (indicating that OECD countries which spend the highest percentage of their GDP on academic research are in fact small countries, suggesting that the costs of joining international networks are relatively high), at 798 and table 3 (comparing OECDs while signaling out small countries for higher expenditures on basic R&D); Keith Pavitt, supra note 36 (adding that the constraints of distance and language have meant that nation-based transfers between science and technology have been the rule rather than the exception), at 116.

[105] See Keith Pavitt, Academic Research, Technical Change and Government Policy, in *Companion Encyclopaedia of Science in the Twentieth Century* (John Krige and Dominique Pestre, eds.) (Routledge, 2013) 143, at 153.

[106] See Hans Gersbach and Maik T. Schneider, supra note 11, at 4.

[107] Keith Pavitt, supra note 36, at 110.

[108] See Frederick. Abbott, Carlos Correa and Peter Drahos, eds., Emerging Markets and the World Patent Order: The Forces of Change, in *Emerging Markets and the World Patent Order*, at 32;

See also Tetyana Payosova, Russian Trip to the TRIPS: Patent Protection, Innovation Promotion and Public Health, in *Emerging Markets and the World Patent Order*, at 249 (for the case of Russia with drug development).

[109] See Abbott, Correa, and Drahos, supra note 108, at 32; Payosova, supra note 108.

Numerous related initiatives, to illustrate, have taken root in the international and regional contexts in emerging economies over the past two decades. These demonstrate the emphasis on collaboration of emerging economies with other developing countries over basic research, also outside Northern innovation centers, and are mostly threefold. First is the fourth BRICS Summit in New Delhi in 2012, where greater cooperation among developing countries in the areas of science, technology, and innovation has been established. This summit is most importantly known for its Delhi Declaration highlighting the need to promote science and technology and related knowledge exchange in the South.[110] Recognizing the broader relevance of knowledge sharing, paragraph 40 of the Delhi Declaration states that there is a pool of "knowledge, knowhow, capacities and best practices available in our countries that we can share and on which we can build meaningful cooperation for the benefit of our peoples." The specific sectors for cooperation set out in paragraph 43 of the Declaration include the priority areas of food, pharmaceuticals, health, and energy, as well as basic research in emerging interdisciplinary fields such as nanotechnology, biotechnology, and advanced materials science.

A second initiative has been India, Brazil, and South Africa's (IBSA) Initiative, which is a trilateral initiative between the three countries. The three emerging economies recently have forged the New Delhi Agenda for Cooperation and Plan Action, which aims to enhance trilateral trade and cooperation between the three countries over science and technology on plentiful fields.[111] Surely, South–South coalition building has been engaging the interest of developing countries and especially emerging economies on a variety of platforms. These include climate-change negotiations, the Convention on Biological Diversity (CBD), the Nagoya Protocol on Access to Genetic Resources and the Fair and Equitable Sharing of Benefits Arising from their Utilization (ABS) to the CBD, and, of course, the TRIPS Council or industrial products liberalization.[112] There is much reason to believe that this

---

[110] See Declaration of the Fourth BRICS Summit: BRICS Partnership for Stability, Security and Prosperity, March 29, 2012, at: www.itamaraty.gov.br/en/press-releases/9428-fourth-brics-summit-new-delhi-29-march-2012-brics-partnership-for-global-stability-security-and-prosperity-delhi-declaration.

[111] See India-Brazil-South Africa (IBSA) Dialogue Forum: New Delhi Agenda for Cooperation, March 4–5, 2004 (Article 20 reads: "Recalling that the Brasilia Declaration had identified trilateral cooperation among the three countries as an important tool for achieving the promotion of social and economic development, the Ministers agreed that the three countries, with rich untapped natural resources and emerging infra-structural requirements, could in a spirit of South-South cooperation, share expertise in several areas. With this view, working level discussions for enhancing trilateral cooperation in the spheres of S&T, Information Technology, Health").

[112] See Rajeev Kher, India in the World Patent Order, in *Emerging Markets and the World Patent Order*, at 218–19.

approach can and should be extended to all types of research and development, but especially basic R&D.[113]

A third noticeable initiative that stretches the need for lowered transaction costs over basic research with much relevancy to Southern economies is loosely associated with an inspirational draft agreement proposal presented by John Barton and Keith Maskus. The two have proposed an Agreement for Access to Basic Science and Technology, as a knowledge generation and diffusion platform, where both the input (scientific and technological capabilities) and output (new scientific insights and basic technologies) should be shared by and be broadly accessible to the international community.[114]

Such inclination by emerging economies to preferential procurement policies against the backdrop of the demise of patent propensity rates as proxy for domestic innovation has seen much concrete manifestations in leading emerging economies. Brazil sets a fine case in point. Upon its joining to the WTO agreement, Brazil has adopted three initiatives concerning governmental preference to basic research against the backdrop of relatively low patent propensity rates. First, in December 2010 the Brazilian government adopted the so-called Buy Brazil Act, which gives substantial preferences in government procurement to products produced in Brazil and that reflect investments in Brazil.[115] This act specifically refers to R&D in that preferential context.[116]

Second, resulting from a long-standing, successful research program funded mostly by the Brazilian government,[117] and especially oriented to regional development, Plant Variety Protection (PVP) fillings are much more locally oriented.[118] Third, Brazil has been successful in stimulating

---

[113] Ibid.

[114] See John H. Barton and Keith E. Maskus, Economic Perspective on a Multilateral Agreement on Open Access to Basic Science and Technology 349, in *Economic Development and Multilateral Trade Cooperation* (Bernard Hoekman and Simon J. Evenett, eds.) (2006) (justifying the treaty primarily in light of the significant difficulties in the diffusion of information to developing countries and the prospects for building a capacity in science and technology in these countries), at 363.

[115] Noticeably, Brazil is not party to the otherwise limiting WTO Government Procurement Agreement. On this backdrop, on July 19, 2010, the Brazilian government amended Law No 8666 with a provisional measure ("Medida Provisória" MP 495) giving preference of up to 25% to Brazilian-owned firms under specific conditions to achieve technological innovation by investing in research and development in Brazil, among other things. The final "buy Brazil act" 12.349/10 of 15 December 2010.

[116] Ibid.

[117] Randall D. Schnepf, Erik Dohlman, and Christine Bolling, Agriculture in Brazil and Argentina: Developments and Prospects for Major Fields Crops, Agriculture and Trade Report No. WRS013, Economic Research Service, USDA 85 (2001), at 61.

[118] See Denis Borges Barbosa, Patents and the emerging markets of Latin America – Brazil, in *Emerging Markets and the World Patent Order*, at 135 ("[M]ost PVP granted for cotton, beans (except soy) and wheat were filled by public or private entities in Brazil"), at 141,

local technological innovation without relying on the patent system in Brazil's bio-fuel program. Launched in the 1970s, this important substitution program has benefited from governmental supported research and subsidies.[119] These initiatives reflect a trend among emerging economies not to over-rely on stimulating domestic R&D and innovation merely by providing patent protection, and is present in plentiful other cases in leading emerging economies. These examples instead project a more proactive role in channeling resources toward local production, preferential government intervention with anticipated gains in local employment, integration with educational institutions, etc.[120]

Russia's relatively low patent propensity rates offer another case in point. For a start, innovation in most sectors of industry in Russia is still highly dependent on governmental spending. In 2008, its business sector performed less than 9 percent of business expenditures on R&D.[121] Second, and most importantly, is Russia's innovation policies in the pharmaceutical sector. Russia most recently has been actively pursuing the aim of becoming a regional leader in the field based on preferential governmental policies.[122] Instead of taking recourse to compulsory licensing as one of the flexibilities available under the TRIPS Agreement, Russia has chosen to put the emphasis on government R&D and to strengthen its domestic pharmaceuticals production based on government procurement preferential policies.[123]

---

referring also to Daniela de Moraes Aviani, Data from National Service for the Protection of Cultivars, available at: www.sbmp.org.br/6congresso/wp-content/uploads/2011/08/1.-Daniela-Aviani-Panorama-Atual-no-Brasil.pdf.

[119] See BRICS, The BRICS Report: A Study of Brazil, Russia, India, China, and South Africa with Special Focus on Synergies and Complementarities, at 108–10.

[120] See Frederick Abbott, Carlos Correa, and Peter Drahos, supra note 108, at 15.

[121] See OECD, OECD Reviews of Innovation Policy: Russian Federation 2011.

[122] See Tetyana Payosova, supra note 108, at 250.

[123] Two of the main growth drivers of the Russian pharmaceutical industry are indeed state procurements in the hospital segment and government financing aimed at providing medicines. See, e.g., Deloitte, Development Trends and Practical Aspects of the Russian Pharmaceutical Industry – 2014 (2014), at 10; NovaMedica, The Pharma Letter, Russian Government to Change Rules on Public Procurement of Drugs (August 28, 2013), available at: http://novamedica.com/media/theme_news/p/631#sthash.lFR4tPlr.dpuf (stating in view of much critique over an excessive governmental preferential take that currently public procurement accounts for over 30% of the Russian pharmaceutical market. In 2012 the value of the segment reached 236 billion rubles ($7.18 billion)); Michael Edwards, R&D in Emerging Markets: A new Approach for a New Era, McKinsey, February 2010 (upholding that as R&D in developed markets fail to meet the pharmaceutical needs of emerging economies, there is need for localized R&D in emerging economies), at: www.mckinsey.com. In consequence, scholars such as Tetyana Payosova add that in the near future Russia is expected to change its profile from a pharmaceutical-importing to a pharmaceutical-exporting country. See Tetyana Payosova, supra note 108, at 250–51.

China, to be sure, witnesses a comparable pattern. As part of China's 12th Five-Year Plan (2011–15) for National Economic and Social Development, China has set itself very ambitious goals in terms of increasing governmental R&D expenditure. China is attempting to raise its R&D expenditure to 2.2 percent as a proportion of GDP and is aiming for 3.3 invention patents for every 10,000 head of population by 2015.[124] In China, however, it is less certain how such preferential policies toward the increase in R&D expenditures comply with a most noticeable parallel increase in local patent applications.

4.4   Theoretical Ramifications

The core empirical findings discussed underline numerous theoretical ramifications which necessitate much additional research. There are possibly three such theoretical lead ramifications.

For a start, to date no empirical finding have assessed South–South positive externalities over basic or scientific research. More particularly, there simply is no clear account of comparable costs and intensity of research-related transactions developing countries bestow on other developing countries. Equally, no findings exist as to the possible externalities bestowed by developing countries on developed ones. To be sure, some literature already depicts the more general narration of Southern innovation adopted in Northern countries directly. Termed "reverse innovation" by Vijay Govindarajan and Chris Trimble in their much-lauded 2012 Harvard Business Review Press book *Reverse Innovation: Create Far from Home, Win Everywhere*, the term explains how in certain cases Northern countries in fact adopt Southern innovative products developed in Southern countries. Yet, there is simply no measurable account for the costs and intensity of archetypal reverse innovation spillovers.[125] There are, in conclusion, no deriving patent policies therein.

These measurements by all means may bear following policy implications for such Southern countries. Basic research investment by one country is surely known to be beneficial to other countries, either indirectly, in the form of technological spillovers from the entry of successfully innovative firms, or directly – that is, as outlined by Nelson,[126] Arrow,[127] and

---

[124] See Legislative Affairs Office of the State Council PRC, "12th five-Year Plan (2011–2015) for National Economic and Social Development of P.R China," March 17, 2011.

[125] See Vijay Govindarajan and Chris Trimble, Reverse Innovation: Create Far From Home, Win Everywhere, *Harvard Business Review Press* (2012), in *Strategy & Innovation*, May 2012, vol. 10(2) (discussing numerous case studies of reverse innovation-based production); Vijay Govindarajan, The Case for "Reverse Innovation" Now, October 26, 2009 (discussing the historical process that have led to reverse innovation).

[126] Richard R. Nelson, supra note 83.   [127] Kenneth Arrow, supra note 38.

Cohen et al.,[128] by access of firms to increases in the knowledge base in another country.[129] The subsequent interaction between basic and applied research was thus crucial on numerous occasions identified to take place within the OECD context. To illustrate, such are the further development and refinement of the nuclear magnetic resonance spectrometer and the transmission electron microscope, respectively.[130] The positive side-effects occur through various channels whose outputs are supply of trained scientists, new scientific and technological instrumentation, start-ups and spin-offs from universities, or prototypes of new products and processes.

Surely there are arguments for and against the coordination of basic research across countries.[131] In the global context of the North–South division, the impact of basic research transactions remains unclear. At a start, when basic research is viewed as a global public good whose output is freely available and whose consumption is non-rivalrous and non-excludable, as discussed following Nelson's 1959 insight, the standard "free-ride" argument could suggest that uncoordinated investment decisions will entail considerable under-investment also across countries. Potentially that is so also by advanced economies toward basic research associated with emerging ones.[132] Basic research may alternatively be viewed as a regional good having emerging economies or middle-income regions in mind, while carrying international spillovers. As a consequence, basic research may induce and increase prospects of innovative yield measured as patent propensity rates for regional firms in these countries.[133] Innovative firms, moreover, may be able to increase the rents generated by these innovations through exports or FDIs. The possibility of capturing rents in foreign markets by taking away business from established firms could thus suggests that basic research investments have negative externalities on other countries, which would cause over-investment also across the archetypical development divide.[134]

---

[128] Wesley M. Cohen, Richard R. Nelson, and John P. Walsh, Links and Impacts: The Influence of Public Research on Industrial R&D, *Management Science*, 48(1), 1 (2002).

[129] Hans Gersbach and Maik T. Schneider, supra note 11, at 5.

[130] Hans Gersbach, Gerhard Sorger and Christian Amon, supra note 5, at 7–8 ("[t]he interdependence of basic and applied research and their mutual intensification is fundamental to the invention, development, and commercialization of new products and technologies"), at 8.

[131] See Hans Gersbach and Maik T. Schneider, supra note 11, at 1.

[132] Cf. Hans Gersbach and Maik T. Schneider, supra note 11, at 1.   [133] Ibid.

[134] Cf: ibid., at 2, referring to Martin Baily and Hans Gersbach, Efficiency in Manufacturing and the Need for Global Competition, Brookings Papers on Economic Activity, *Microeconomics*, 307 (1995) (for discussion on negative and positive externalities with emphasis on advanced economies); Wolfgang Keller and Stephen R. Yeaple, Multinational Enterprises, International Trade, and Productivity Growth: Firm-level Evidence from the United States, NBER Working Papers 9504, National Bureau of

From a patent-policy perspective, more precisely, this ultimately means that there is also no good sense of why basic research performed in emerging economies impacts as it does on patent intensity as a proxy of domestic innovation – that is, against the backdrop of relatively low patent propensity rates associated with these Southern economies. As a result, it is yet uncertain what is the precise impact of emerging econo- mies' inclination to proffer outsized basic research investments assuming possibly sub-optimal patent propensity rates.

A second theoretical ramification follows. To date there still is no measurement as to whether developing countries generate higher rate of patent propensity relative to basic research, given the degree of the basicness of these countries' inventions. In other words, we still cannot simply compare the science-dependent patents, or whether basic research is comparably patentable across the development divide even while accounting for leading patent offices, including the USPTO, EPO, SIPO, and the JPO. To date, we find only scarce studies of this issue, even when advanced economies are accounted for. One such important study by Francis Narin et al. serves as a case in point. The authors explain how, as of 1986, technology reflected in USPTO patents was much more "science-dependent" than ten years earlier.[135] They further show that the time-lags in the citations from patents to other publications are diminishing rapidly, and that science-intensive patents are relatively highly cited.[136] The significance of the lack of such analysis for the Southern economies is, of course, that this would allow us to better model innovation-based growth in emerging economies, by ultimately assessing the efficacy of these relatively lower countries' patent propen- sity rates altogether: if a relatively lower patent propensity incorporates a relatively high patent intensity in respect of basic research, this could mean that fewer patents do not mean less innovation based on basic or scientific research, insofar as patenting policy is concerned.

A third theoretical ramification derives from the previous point. Examined from the vantage point of enhanced basic research rates in

Economic Research, Inc. (2003) (presenting the same argument); Laura Alfaro, Areendam Chanda, Sebnem Kalemli-Ozcan, and Selin Sayek, How Does Foreign Direct Investment Promote Economic Growth? Exploring the Effects of Financial Markets on Linkages, NBER Working Papers 12522, National Bureau of Economic Research, Inc. (2006) (presenting the same argument).

[135] See Francis Narin and J. Davidson Frame, The Growth of Japanese Science and Technology, *Science* 245 (1989) (authors have shown sharply upward trends in the frequency with which US patents, originating in a number of countries, contain citations to publications other than patents: from about 0.2 cites to "other publications" per US patent in 1975, to between 0.9 cites for US patents of US origin – and 0.4 cites for US patents of Japanese origin – in 1986), at 600–4.

[136] Ibid.

emerging economies compared to advanced ones, basic research could possibly be more closely associated with commercial use for future policy orientation in these countries. The OECD's 2012 definition of basic research differs from the so-called Pasteur's Quadrant – a term coined by Princeton University Professor Donald Stokes. In his important book *Pasteur's Quadrant: Basic Science and Technological Innovation*, Stokes characterizes basic research with immediate commercial use in view.[137] Louis Pasteur's research is thought to exemplify this type of method, which bridges the gap between "basic" and "applied" research.[138] The validity of this proposition further necessitates granular empirical analysis if patent policy is to restrain the blocking of downstream research inventions across countries.

Conclusion

To date, differences of R&D by type in terms of the relationship with patent policy across countries have barely been measured. Even less attention, perhaps, has been given to fluctuations in scientific research by comparison to other R&D-related measurements, such as over human resources, institution funding, and performance or internationalized R&D spillovers.

Numerous insights may already suggest why Southern economies generally show higher basic R&D as a percentage of GERD rates than advanced countries. In absolute terms, R&D expenditures on basic and applied research as well as on experimental development are clearly higher in advanced economies by comparison to middle-income or emerging economies. The comparison between the two economy types becomes more subtle when patent activity intensity factors are accounted for in terms of their percentage of GDP and GERD. From a patent-policy standpoint, these measurements may go a long way to explaining why different countries and country groups differ in patent intensity as measured by a plethora of types of R&D rates.

In terms of our three patent clusters, no significant differences can be seen in terms of basic and applied research. The only significant difference found is between the leaders and marginalized clusters, and relates

---

[137] See Donald E. Stokes, *Pasteur's Quadrant – Basic Science and Technological Innovation* (Brookings Institution Press, 1997).

[138] See also Karl Max Einhaäupl, supra note 94 (theorizing on four types of R&D: (1) pure basic research, (2) use-inspired basic research, in reference to Pasteur's approach, (3) pure applied research, and (4) research which is at the beginning neither use-oriented nor done with the aim of generalization); John M. Dudley, Defending Basic Research, *Nature Photonics* 7, 338 (2013) (presenting the same argument).

to experimental development as a percentage of GERD – the more patentable and commercial type of non-basic R&D. This finding, limited though it is, nevertheless supports the conclusion that advanced economies, or more narrowly countries belonging to the leaders patent cluster, are more significantly characterized by added commercial and thus patentable types of R&D activity. The latter finding may ultimately explain why advanced countries closely associated with the leaders patent cluster are characterized by higher patent intensity rates in general. It may also offer insights for emerging economies closely associated with the two lower patent clusters regarding their archetypal industry–academia gap challenge, with its ramifications for innovation and patent policy.

Appendix E   GERD by Type of R&D and Patent Clusters

E.1   Descriptive Statistics

The following plots describe for each sector and cluster the distributions along time:

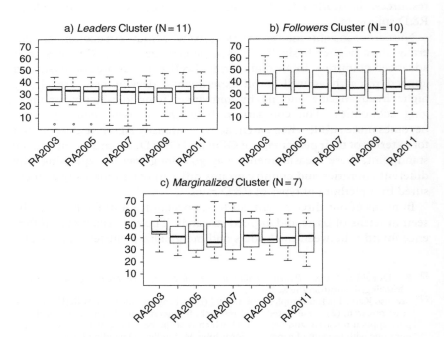

E.1 Applied Research GERD by Patent Clusters (%) (2000–2009)

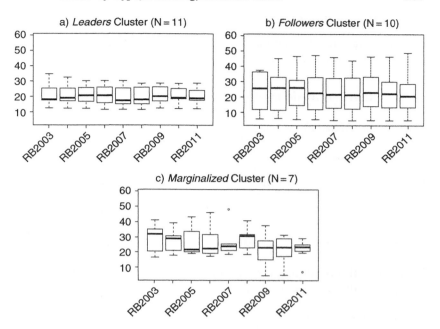

E.2 Basic Research GERD by Patent Clusters (%) (2000–2009)

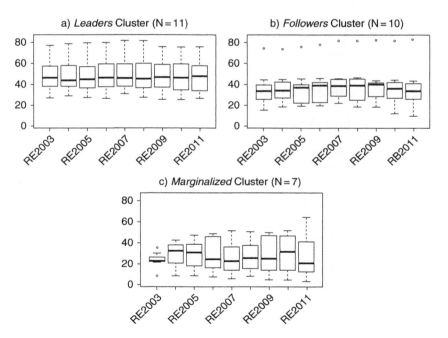

E.3 Experimental Development GERD by Patent Clusters (%) (2000–2009)

E.2   Regression Models

The only significant difference found between the leaders and marginalized clusters was for experimental development.

E.2.1   Regression Model for Applied Research

The Mixed Procedure

Type 3 Tests of Fixed Effects

Effect	Num DF	Den DF	F Value	Pr > F
year	8	200	0.68	0.7099
r_cluster_km	2	25	2.38	0.1130
r_cluster_km*year	16	200	0.82	0.6603

No statistical differences were found between the three clusters and no changes in time were identified.

Differences of Least Squares Means

Effect	r_cluster_km	year	_r_cluster_km	_year	Adj P
r_cluster_km	1		2		0.4413
r_cluster_km	1		3		0.1362
r_cluster_km	2		3		1.0000

E.2.2   Regression Model for Basic Research

The Mixed Procedure

Type 3 Tests of Fixed Effects

Effect	Num DF	Den DF	F Value	Pr > F
Year	8	200	1.44	0.1800
r_cluster_km	2	25	0.94	0.4033
r_cluster_km*year	16	200	0.60	0.8830

No statistical differences were found between the three clusters and no changes in time were identified.

E.2.3   Regression Model for Experimental Development

The Mixed Procedure

Type 3 Tests of Fixed Effects

Effect	Num DF	Den DF	F Value	Pr > F
year	8	200	1.00	0.4371
r_cluster_km	2	25	4.15	0.0277
r_cluster_km*year	16	200	0.77	0.7147

Significant differences were found between the leaders and marginalized clusters, with a higher mean for the former cluster (labeled 1) than the latter (labeled 3).

Differences of Least Squares Means

Effect	r_cluster_km	year	_r_cluster_km	_year	Adj P
r_cluster_km	1		2		0.2942
r_cluster_km	1		3		0.0351
r_cluster_km	2		3		0.6261

5   Patent Intensity by Employment and Human Resources

Introduction

Human capital theory and patent intensity as proxy for domestic innovation are intertwined in two ways.[1] One sphere of connection relates to the effect innovation-based growth has on innovation-related employment, and more specifically on R&D personnel. Valid concern over "jobless growth" has led international organizations, including the International Labor Organization (ILO), the United Nations Industrial Development Organization (UNIDO), the Industrial Development Board (IDB), and the OECD, to tackle this phenomenon, especially as countries recover from the 2008 subprime crisis.[2] This battle has taken the form of an international campaign that ultimately reached state level. This campaign has been funneled most noticeably by the European Commission's formulation of its Europe 2020 strategy with the aim of creating conditions for sustainable growth.[3] At the heart of the controversy lies a clash between two views. One view argues that innovation is also labor saving and directly creates unemployment. A recent study, for example, uses the number of linked patents at the European, Japanese, and US patent offices, known as triadic patents, in 21 industrial countries granted over the period 1985–2009 as a proxy for innovation,

---

[1] The term "human capital" is commonly defined as all the knowledge, education, training, and experience that is embodied in workers. See e.g., Christine Greenhalgh and Mark Rogers, Innovation, *Intellectual Property and Economic Growth* (Princeton University Press, (2010), at 229–31; James S. Coleman, Social Capital in the Creation of Human Capital, *American Journal of Sociology* (1988), at S95–S120; For a cost-based analysis, see e.g., Peer Ederer, Lisbon Council Policy Brief, Innovation at Work: The European Human Capital Index (2006) (defining Human capital as "[t]he cost of formal and informal education expressed in Euros and multiplied by the number of people living in each country"), at 2.

[2] Gustavo Crespi and Ezequiel Tacsir, Effects of Innovation on Employment in *Latin America*, Inter-American Development Bank Institutions for Development (IFD) Technical Note, Inter-American Development Bank, Washington, DC (2012) 1. See also, UNIDO, Industrial Development Report 2013. Sustaining Employment Growth: The Role of Manufacturing and Structural Change. United Nations Industrial Development Organization, Vienna (2013).

[3] European Commission, Communication from the Commission to the European Parliament, the Council, the European Economic and Social Committee and the Committee of the Regions, Europe 2020 Flagship Initiative Innovation Union, COM (2010) 546 final, European Commission, Brussels (2010).

assessing the impact of innovation on the aggregate unemployment rate.[4] The results demonstrate that technological change tends to amplify unemployment, although this effect is not maintained in the long term.[5]

The other view, based on compensation theory, has its origins in the thought of classical economists, including Jean-Baptiste Say,[6] David Ricardo,[7] Karl Marx,[8] and others,[9] who argued that product innovations and indirect income and price effects can offset the direct impact of job destruction which process innovations incorporated in new machineries and equipment brings about.[10] This first of two categories of correlation between patent intensity and human capital thus far has not led to theoretical underpinning regarding the possible effect of patent intensity as a particular indicator for innovation on R&D-related employment.

---

[4] Horst Feldmann, Technological Unemployment in Industrial Countries, Journal of *Evolutionary Economics* 23(5) 1099 (2013).

[5] Ibid.

[6] Jean-Baptiste Say, *A Treatise on Political Economy; or, The production, Distribution and Consumption of Wealth* (Augustus M. Kelley Publishers, 1st edn. 1803), New York: (1964).

[7] David Ricardo, Principles of Political Economy, in *The Works and Correspondence of David Ricardo* (Piero. Saffra, ed.) (Cambridge University Press, 3rd edn., 1821 [1951]).

[8] Karl Marx, *Capital: A Critical Analysis of Capitalist Production* (Foreign Languages Publishing House, 1st edn. 1867 [1961]).

[9] See e.g., Stefan Lachenmaier and Horst Rottmann, Effects of Innovation on Employment: A Dynamic Panel Analysis, *International Journal of Industrial Organization* 29(2) 210 (2011) (using a longitudinal data set on German manufacturing firms over the period 1982–2002, authors find a significantly positive effect for various products but also process innovation variables on labor demand); Michaela Niefer, Patenting Behavior and Employment Growth in German Start-up Firms: A Panel Data Analysis, Discussion Paper No. 05-03, ZEW (2003) (on the varying effect of innovation on corporate employment growth).

[10] For modern day scholarship following Compensation theory, see Marco Vivarelli, Innovation, Employment, and Skills in Advanced and Developing Countries: A Survey of the Economic Literature, *Journal of Economic Issues*, 48(1) (2014) 123 (hereinafter, Vivarelli, Innovation, employment, and skills) (showing the direct labor-saving effect of process innovation and the job-creating impact of product innovation for Italy and the United States between 1960 and 1988); Rupert Harrison, Jordi Jaumandreu, Jacques Mairesse, and Bettina Peters, Does Innovation Stimulate Employment? A Firm-level Analysis Using Comparable Micro-data From Four European Countries. NBER Working Paper No. 14216 (2008) (study concluded that process innovation tends to displace employment based on firm-level data obtained for France, Germany, Spain, and the UK – that is, while product innovation is basically labor-friendly); Mario Pianta, Innovation and Employment, in Jan Fagerberg and David C. Mowery (eds.), *The Oxford Handbook of Innovation* (Oxford University Press, 2005); Mariacristina Piva and Marco Vivarelli, Innovation and Employment: Evidence from Italian Microdata, *Journal of Economics* 86, 65 (2005); Mariacristina Piva and Marco Vivarelli, Technological Change and Employment: Some Micro Evidence from Italy, *Applied Economics Letters* 11, 373 (2004); Vincenzo Spiezia and Marco Vivarelli, Innovation and employment: A critical survey, in Nathalie Greenan, Yannick L'Horty and Jacques Mairesse (eds.), *Productivity, Inequality and the Digital Economy: A Transatlantic Perspective* (MIT Press, 2002), 101; Pascal Petit, Employment and Technological Change, in *Handbook of the Economics of Innovation and Technological Change* (Paul Stoneman, ed.) (Amsterdam, 1995).

Human capital theory has a second inverse effect on innovation and patent-related growth. Human capital adheres to a positive relation between the growth of R&D employment and economic growth based on patent intensity.[11] In other words, deeming R&D employment an important inventive input ultimately enables firms and industries to become more innovative, as reflected in their propensity to patent.[12] As recently acknowledged by WIPO[13] and theoreticians,[14] R&D is largely considered unprofitable for low levels of human capital, such as disadvantaged developing countries. R&D thus becomes cost-effective only when human capital reaches a certain threshold.[15]

Be that as it may, most studies done at the firm level indeed confirm that firms with a higher proportion of scientists and engineers may present superior innovation-based economic growth,[16] correlating to an increase in these firms' patenting rates.[17] Equally importantly, inefficient investment in human capital is considered to result in under- or over-investment in R&D and ultimately to inefficient patenting rates.[18] Over-investment in R&D, in particular, occurs whenever the yield of domestic innovation

---

[11] See J. Vernon Henderson, Ari Kuncoro and Matthew Turner, Industrial Development in Cities, *The Journal of Political Economy* 103, 1067 (1995); Thomas Brenner and Tom Broekel, Methodological Issues in Measuring Innovation Performance of Spatial Units, Papers in Evolutionary Economic Geography, No. 2009–04, Urban & Regional Research Centre Utrecht, Utrecht University (2009).

[12] Frederic M. Scherer, Firm Size, Market Structure, Opportunity, and the Output of Patented Inventions, *American Economic Review* 55, 319 (1965).

[13] See WIPO, The Global Innovation Index 2014: The Human Factor in Innovation, Soumitra Dutta, Bruno Lanvin, and Sacha Wunsch-Vincent (eds.) (2014), at 71, Fig. 2.

[14] See e.g., Anders Sorensen, R&D, Learning and Phases of Economic Growth, *Journal of Economic Growth*, 4, 429 (1999).

[15] Ibid.

[16] Paul Romer, What Determines the Rate of Growth and Technological Change?, working paper #279, The World Bank (September 1989) (finding a positive correlation between the number of scientists and engineers R&D personnel and R&D activities and following economic growth); Manfred von Stadler, Engines of Growth: Education and Innovation, *Jahrbuch für Wirtschaftswissenschaften / Review of Economics* vol. 63(2), 113 (2012) (presenting the same argument).

For related work on migration of R&D personnel, see e.g., Christian Zellner, The Economic Effects of Basic Research: Evidence for Embodied Knowledge Transfer via Scientists' Migration, *Research Policy* 32 1881 (2003) (arguing that a substantial rate of the broader economic benefits to society from publicly funded basic research are related to scientists' migration into the business sector of the innovation system).

[17] See e.g., Kuo-Feng Huang and Tsung-Chi Cheng, Determinants of Firms' Patenting or not Patenting Behaviors, *Journal of Engineering and Technology Management*, vol. 36, 52 (2015), at 57, referring to David J. Storey and Bruce S. Tether, Public Policy Measures to Support New Technology-based Firms in the European Union, *Research Policy* 26, 1037 (1998).

[18] See e.g., Ana Balcão Reis and Tiago Neves Sequeira, Human Capital and Overinvestment in R&D, *The Scandinavian Journal of Economics*, vol. 109(3) 573 (2007) (showing that inefficient investment in human capital results in under- or over-investment in R&D, based on an endogenous growth model). Reis and Neves add that

would potentially remain under-patented, as is at times the case across many countries.[19]

Regrettably, human capital literature focuses persistently on developed or advanced countries, while largely ignoring developing countries and emerging economies. Nevertheless, the evidence from other developing countries and poor regions in developed countries offers two insights. Firstly, there is preliminary evidence of a positive effect of human capital on innovative rates of the type of improvements and follow-up innovation in developing countries.[20] Secondly, and more importantly, Norwegian economist Jan Fagerberg et al. have shown, using data from the 1980s for 64 European regions, that due to a low share of business sector workforce employed in R&D, most poor European regions fail to take advantage of the more advanced technologies available elsewhere, in contrast to the superior growth performance of richer European regions.[21] The authors corroborate their finding on the basis of a positive relation between the number of patent applications of a country and growth.[22] From the standpoint of this chapter's empirical methodology, the salient point is that the authors measure the share of labor force employed in innovation and patenting-related activity, and not merely the total quantity of resources devoted to R&D, either researchers or expenditures.[23] These findings also confirm that firms and industries in developing countries systematically operate below the technology frontier frequently measured by R&D personnel spending, with lower levels of managerial and production skills.[24]

---

over-investment in R&D is thus a negative externality imposed by technological development on human capital accumulation. Ibid. at 598.

[19] The empirical part henceforth offers an account of the connection between human capital inputs and patent intensity. For a more far-reaching view incorporating human capital into research activity per se, see Keith Blackburn, Victor T.Y. Hung, and Alberto F. Pozzolo, Research, Development and Human Capital Accumulation, *Journal of Macroeconomics*, 22(2) 189 (2000) (offering an endogenous growth model that integrates R&D with human capital accumulation, while showing that R&D-related growth is independent of research activity which, itself, is driven by human capital accumulation).

[20] See Micheline Goedhuys, Norbert Janz, and Pierre Mohnen, Knowledge-Based Productivity in "Low-Tech" Industries: Evidence from Firms in Developing Countries, UNU-MERIT Working Paper 2008-007, Maastricht: UNU-MERIT (2008); cf: Beñat Bilbao-Osorio and Andrés Rodríguez-Pose, From R&D to Innovation and Economic Growth in the EU, *Growth and Change*, 35(4), 434 (2004) (showing that positive externalities from higher educational attainment are found in the form a faster technology transfer (and a higher rate of innovation)).

[21] Jan Fagerberg, Bart Verspagen, and Marjolein Canieëls, Technology, Growth and Unemployment Across European Regions, *Regional Studies*, 31, 457 (1997).

[22] Ibid.   [23] Ibid.

[24] See e.g., Micheline Goedhuys, Norbert Janz, and Pierre Mohnen, supra note 20; Laura Barasa, Peter Kimuyu, Patrick Vermeulen, Joris Knoben, and Bethuel Kinyanjui, *Institutions, Resources and innovation in Developing Countries: A Firm Level Approach, Creating Knowledge for Society* (Nijmegen, December 2014); Micheline Goedhuys,

R&D spending includes personnel salaries and wages, material and supplies, durable equipment, land, and buildings, as well as other costs such as energy, water, and maintenance. Of these, it should be noted, the salaries and wages of R&D employees account for approximately one-third of total R&D expenditure, as shown by Luc Soete's invaluable studies as early as 1979.[25] Following Soete's work, and in keeping with this chapter's empirical analysis, salaries and wages of R&D personnel are systematically considered the most important input into the process of knowledge generation, whether patented or not.[26]

5.1   Human Capital and Patent Intensity

5.1.1   The Value of Human Capital for Patent Policy

A question remains: What makes human capital theory valuable for innovation and patent-related policy? A preliminary response to this question may be found in the *2013 EU Survey on R&D Investment Business Trends*,[27] analyzing 172 responses from mainly large firms out of a subsample of 1000 EU-based companies in the *2012 EU Industrial R&D Investment Scoreboard*.[28] The report confirms that among EU countries, quality of R&D personnel is most frequently stated as one of the top three factors determining attractiveness in these countries.[29] This is closely followed by proximity to technology poles and incubators, and the quantity of R&D personnel. The inclusion of quantity of R&D personnel in the labor market is one of the top three factors contrasts given the observed lack of sufficient quantity in Germany.[30] In the emerging economies of China and India, the first ranked factor is the *quantity and cost of R&D personnel*.[31]

Learning, Product Innovation, and Firm Heterogeneity in Developing Countries: Evidence from Tanzania, *Industrial and Corporate Change*, 16(2) 269 (2007).

[25] Luc L. Soete, Firm Size and Inventive Activity: The Evidence Reconsidered, *European Economic Review*, 12(4) 319 (1979).

[26] See J. Vernon Henderson, Ari Kuncoro, and Matthew Turner, supra note 11; Thomas Brenner and Tom Broekel, supra note 11.

[27] See European Commission, 2013 EU Survey on R&D Investment Business Trends, Monitoring Industrial Research: The 2006 EU Survey on R&D Investment Business Trends (Alexander Tübke and René van Bavel, eds.) (2013). This report analyses the 172 responses of mainly large firms from a subsample of 1000 EU-based firms in the 2012 EU Industrial R&D Investment Scoreboard.

[28] Ibid., at 5 and fn. 1 (explaining that the 2012 EU Industrial R&D Investment Scoreboard incorporates 405 EU-based companies of the world top 1500 companies in the 2012 Scoreboard and 595 additional companies from the EU with an R&D investment above 5.26 million Euros in 2011).

[29] Ibid., at 22 and table 2, and 24 and fig. 16.   [30] Ibid.

[31] Ibid., at 26 and fig. 18. In this case the quantity and cost of R&D personnel was chosen alongside market size and growth as determinant for internationalized R&D

A second highly informative survey accounting for the centrality of human capital theory for innovation and patent-related policy was published in 2006 by Georgia Institute of Technology economists Jerry Thursby and Marie Thursby for the US National Academies.[32] The two researchers surveyed high-level R&D executives in 203 MNCs, most of whom were headquartered in the United States or Western Europe. The questions were posed in terms of R&D employment (rather than expenditures) in order to minimize inaccuracies and problems due to currency conversion. The most important driver of location choice for R&D factors was consistent with the above-mentioned 2013 EU Survey on R&D Investment Business Trends, i.e. closeness to highly qualified R&D personnel.[33] There was little significant difference between the US and Western European firms in human capital-related rankings.[34] The second part of Thursby and Thursby's survey focused on the location of the firms' most important R&D facilities, distinguishing between those located in developed countries and those located in emerging economies.[35] For location in the home country and other developed countries, the factors of access to scientists and engineers, both as employees and at universities, along with the factors of intellectual property protection and ownership, were clearly important factors, as summarized in Table 5.1 below.[36]

In recent years, numerous other international organizations have adopted equivalent human capital-related findings, most noticeably

---

attractiveness for MNEs. For the Indian case see also, UNCTAD, World Investment Report, New York and Geneva, United Nations (2005), at 167; New Scientist, Silicon Subcontinent: India is Becoming the Place to be for Cutting-edge Research, (19 February 2005). UNCTAD's findings are herein are based on Prasada Reddy's findings as of 2000, recently reprinted in Prasada Reddy, *Global Innovation in Emerging Economies* (Routledge, 2011), at 139. In a survey conducted at the end of the 1990s, the availability of Indian R&D personnel was ranked by MNEs as the most important reason for locating R&D (4.12 out of 5). Ibid. For the Chinese case, *See* UNCTAD, World Investment Report, Ibid., at 166 and sources therein.

[32] See Jerry G. Thursby and Marie Thursby, *Here or There? A Survey of Factors in Multinational R&D Location* (National Academies Press, 2006).

[33] Ibid., at 21–28. When considering the siting of a new R&D facility, 70% of the respondents foresaw the expansion instead of the relocation of present facilities. See discussion also in John H. Dunning and Sarianna M. Lundan, The Internationalization of Corporate R&D: A Review of the Evidence and Some Policy Implications for Home Countries, *Review of Policy Research*, vol. 26, Numbers 1–2, 13 (2009) (referring to the earlier but somewhat comparable 2006 *EU Survey on R&D Investment Business Trends*), at 21.

[34] See Jerry G. Thursby and Marie Thursby, supra note 32, ibid.

[35] Ibid., at table 7 and figure 9.

[36] Subsidies, tax breaks and the absence of legal requirements were found to be the least important factors in choosing a location, regardless of the development level. Ibid.

Table 5.1 *Factors Considered Important When Locating an R&D Facility*

Factors	Name	Home Country	Developed Economy	Emerging Economy
Country has high growth potential	Growth	NA	3.5	4.3
The R&D facility was established to support sales to foreign customers	SupSales	NA	3.35	3.6
The R&D facility was established to support production for export	SupExport	NA	2.75	2.6
The establishment of an R&D facility was a regulatory or legal prerequisite for access to local markets	LegalReg	NA	1.9	2
There are highly qualified R&D personnel	QualR&D	4.5	4.2	3.75
There is good IP protection	IPProtect	4.25	4.15	3.65
There are university faculty with special scientific or engineering expertise	UnivFac	3.95	3.55	3.2
It is easy to negotiate ownership of IP from research relationship	Ownership	3.85	3.35	3.45
It is easy to collaborate with universities	CollabUniv	3.85	3.5	3.25
There are few regulatory and/or research restrictions in this country	FewRestrict	4.35	2.75	2.8
The cultural and regulatory environment is conducive to spinning off or spinning in new businesses	Spin	3	2.55	2.55
Exclusive of tax breaks and direct government assistance, the cost of R&D are low	Costs	2.75	2.7	3.4
We were offered tax breaks and/or direct government assistance	TaxBreaks	2.5	2.75	2.2

Source: Jerry G. Thursby and Marie Thursby, *Here or There? A Survey of Factors in Multinational R&D Location* (National Academies Press, 2006), at 21–28

UNCTAD.[37] The expansion of R&D in developing countries, albeit still limited, is thus heavily influenced by the availability of knowledge workers.[38] The improved supply of highly skilled people is occurring as a consequence of purposeful and long-term policies to raise educational standards, mainly at the tertiary level, alongside efforts to recruit human resources from abroad.[39]

These findings are pivotal to incorporating R&D personnel-related policies for developing countries, including in absolute terms, given that the EU, the United States and China have almost the same number of researchers. In 2008, there were remarkably 1.5 million full-time equivalent (FTE) researchers in the EU compared to 1.6 million in China, and 1.4 million in 2007 in the United States. Since then, China has passed the EU and the United States in total number of researchers.[40] As human capital affects innovation-based growth, this also affects the propensity to patent as a proxy for innovation, as discussed in the empirical part of this chapter.

5.1.2   R&D Personnel, Linear Growth, and Patenting

The concept of human capital sparked a shift in contemporary economic growth analysis away from a largely physical (and intellectual property-based) view of capital accumulation to its perception as a process that

---

[37] See UNCTAD, World Investment Report 2005, supra note 31, at 203. But see also the Community Innovation Survey 2010: http://unctad.org/en/Docs/wir2005_en.pdf. When firms in the European Union were asked which factors were the most important obstacles to innovation, the lack of qualified personnel as a highly important factor hampering innovation activities on average ranked only 6th for innovative enterprises and 7th for non-innovative enterprises out of 11 factors proposed. Ibid.

[38] Beginning in the 1990s, MNEs set up global R&D facilities in India including Motorola, Microsoft, Oracle, DuPont, Philips, IBM, STMicroelectronics, Daimler-Benz, Pfizer, and many others. Prasada Reddy, supra note 31, at 108.

[39] See UNCTAD, World Investment Report 2005, supra note 31, at 203. On education as an innovation input in developing countries, see also, Edmund Amann and John Cantwell, *Innovative Firms in Emerging Markets Countries* (Oxford University Press, 2012); Paul J. Robson, Helen M. Haugh and Bernard A. Obeng, Entrepreneurship and Innovation in Ghana: Enterprising Africa, *Small Business Economics*, 32(3), 331 (2009) (for a positive relation between education level and innovation in Ghana); *Technology, Learning, and Innovation: Experiences of Newly Industrializing Economies* (Linsu Kim and Richard R. Nelson, eds.) (Cambridge University Press, 2000). Keith Pavitt noticeably upholds that training and skills, in particular, are in fact a more effective innovation input than academic knowledge. See Keith Pavitt, What Makes Basic Research Economically Useful?, *Research Policy*, Volume 20(2), April 1991, 109 (table 1 upholds that in the majority of scientific fields, regardless if they are basic or applied sciences, academic training and skills are more essential than is academic research over a far larger number of industrial technologies), at 114 and table 1.

[40] See European Commission, Innovation Union Competitiveness report 2011 (2011), at 88.

integrally involves the productive quality of human beings. The pioneering work of Nobel Laureates Theodore Schultz[41] and Gary Becker[42] in the 1960s contributed greatly to the shift in emphasis away from archetypical physical capital (and IPR-based) accumulation and pointed the way to a systematic study of the role of human capital theory and related policies.[43]

Theodore Schultz identified human capital narrowly with investment in technological and other education, proposing that "important increases in national income are a consequence of additions to the stock of this form of [human] capital."[44] He went on to argue that investment in education could account in large part for the increase in per capita income in the United States.[45] Indeed, Schultz was the first scholar to raise what later became an overarching UN-level policy focusing on developing countries, whereby assistance to developing countries should redirect its attention from the formation of nonhuman capital to human capital. Gary Becker continued this line, broadening the notion of human capital beyond formal schooling and analyzing the rate of return on human capital investment to include additional aspects of human capital, such as general and specific informal gathering of information and on-the-job training enhancing a worker or researcher's productivity. Becker also advocated additional investments to improve "emotional and physical health."[46] According to Becker, factors that weight return include uncertainty and the nonliquid nature of the investment, as well as capital market imperfections and differences in abilities and opportunities.[47]

---

[41] Theodore William Schultz, Capital Formation by Education, *Journal of Political Economy*, vol. 68, 571 (1960). For the following work of Schultz, see Theodore W. Schulz, *Investment in Human Capital: The Role of Education and of Research*, London: Free Press: Collier-Macmillan (1971), at 47; see generally Theodore Schultz, *The Economic Value of Education* (Columbia University Press, 1963).

[42] See Gary Stanley Becker, Investment in Human Capital: A Theoretical Analysis, *Journal of Political Economy*, vol. 70, 9; Gary Stanley Becker, *Human Capital: A Theoretical and Empirical Analysis with Special Reference to Education* (Columbia University Press, 1964).

[43] See Andreas Savvides and Thanasis Stengos, *Human Capital and Economic Growth* (Stanford University Press, 2009), at 5.

[44] Theodore William Schultz, Capital Formation by Education, supra note 41, at 571.

[45] Lutz Arnold, Keith Blackburn, and others later developed influential endogenous growth models that integrate purposive R&D activity with human capital accumulation, where the engine of growth is the investment in schooling. See Lutz Georg Arnold, Growth, Welfare, and Trade in an Integrated Model of Human Capital Accumulation and R&D, *Journal of Macroeconomics*, 20(1) 81 (1998); Keith Blackburn, Victor T.Y. Hung, and Alberto F. Pozzolo, supra note 19; Alberto Bucci, Monopoly Power in Human Capital-Based Growth, *Acta Oeconomica*, vol. 55(2) 121 (2005).

[46] See Gary Stanley Becker, supra note 42.   [47] Ibid.

Beyond the US firm-level, these factors would in time yield human capital measurements for firms, industries, and countries.[48]

Still in the realm, soon after Schultz and Becker's breakthroughs, Becker and Chiswick,[49] added the insight that different investments in human capital and the corresponding rates of return are also policy considerations that largely determine the distribution of earnings, and hence ultimately influence earnings-based economic growth. Policy regarding what Becker and Chiswick referred to as "institutional factors" could include subsidies for education, inheritance of property income, and ultimately differences in the innovative capabilities and opportunities of firms, industries, and countries. In an explanation that can be used as an analogy for developing countries, the case of Southern US states shows that investment in formal education successfully explained differences in average (white male) wages across both the US South and non-South. The return to education for each schooling level – namely, low, medium, and high number of years in schooling – ultimately was higher in the South. The variance of (the log of) earnings and years of schooling was also larger in the South.[50]

Subsequently, interest in the economic importance of human capital remained dormant for two decades until the 1980s when development occurred both in theory and in UN-level policy, led by the World Bank and followed by WIPO and other UN-level organs. Human capital was

---

[48] For studies also at the individual level, see Mark Gradstein and Moshe Justman, Human Capital, Social Capital, and Public Schooling, *European Economic Review*, nr. 44, 879 (2000) (concluding that employees that are better prepared, have more practical experience, and have invested more time, energy, and resources in completing their skills are better able to obtain welfare for themselves and for the whole society); Javier Gimeno, Timothy B. Folta, Arnold C. Cooper, and Carolyn Y. Woo, Survival of the fittest? Entrepreneurial Human Capital and the Persistence of Underperforming Firms, *Administrative Science Quarterly*, nr. 42, 750 (1997) (human capital as a source of competitive advantage of employees, companies and societies).

For analysis of firm-level innovation inputs in developing countries, including training, information search, communication facilitates, and other human capital inputs, see e.g., Steve W. Bradley, Jeffery S. McMullen, Kendall Artz, and Edward M. Simiyu, Capital Is Not Enough: Innovation in Developing Economies, *Journal of Management Studies*, 49(4), 684 (2012); Gustavo Crespi and Pluvia Zuñiga, Innovation and Productivity: Evidence from Six Latin American Countries, *World Development*, 40(2), 273 (2011); James Tybout, Manufacturing Firms in Developing Countries: How Well Do They Do, and Why?, *Journal of Economic Literature*, 38(March): 11 (2000).

[49] See Gary S. Becker and Barry R. Chiswick, Education and the Distribution of Earnings, *American Economic Review*, vol. 53, 358 (1966).

[50] On the rates of return to education across countries, see also Jason Dedrick and Kenneth L Kraemer, *Asia's Computer Challenge* (Oxford University Press, 1998) (using the number of individuals holding scientific and technological degrees to assess a nation's innovativeness); David J. Storey and Bruce S. Tether, supra note 17 (asserting that supply of postgraduate degrees in science and technology affects innovation in technology-based firms).

first acknowledged by the World Bank in the early 1980s, beginning with the cautious adoption of education as part of an emphasis on manpower economic growth. *The 1980 World Development Report* and *1980 Education Sector Policy Paper* both highlighted the direct productivity benefits of primary education, analyzing rates of return to education and drawing in particular on the studies of George Psacharopoulos, the head of the Research Unit of the Bank's Education Department, providing international comparisons of rates of return on education.[51] From this time forth, emphasis on the analysis of rates of return on education became an inherent component of World Bank policy.[52]

It is not coincidental that the ascendancy of human capital, in conjunction with the application of rates of return, corresponded with the advent of the Washington Consensus. Under the auspices of the World Bank, human capital ultimately provided the opportunity for the neoliberal agenda to be applied to education, allowing the World Bank to continue its involvement and even increase its influence in the education sector across the development divide.[53] The almost universal incorporation of human capital theory into innovation-based economic growth and patent-related policy was due in a large part to the World Bank's immense influence on WIPO.[54] Recently, in its 2014 yearly *Global Innovation Index*

---

[51] Pauline Rose, From Washington to Post-Washington Consensus: The Triumph of Human Capital, in *The New Development Economics: Post Washington Consensus Neoliberal Thinking* (Jomo K.S. and Ben Fine, eds.) (2006) 162, at 165.

[52] See George Psacharopoulos and Harry Partinos, Returns on Investment in Further Update, World Bank Policy Research Working Paper 2881, Washington DC (2002); George Psacharopoulos, *Returns on Education: An International Comparison*, Amsterdam: Elsevier (1973); George Psacharopoulos, Returns on Education: An Updated International Comparison, *Comparative Education*, 17(3) 321 (1981); George Psacharopoulos, Returns to Education: A Further International Update and Implications, *Journal of Human Resources*, 20(4) 683 (1985); George Psacharopoulos, Returns to Investment in Education: A Global Update, *World Development*, 22(9) 1325 (1994).

[53] Some scholars have noticed, however, that in perspective, human capital dialectics had little impact however on the World Bank. See Devesh Kapur, John Lewis and Richard Webb, *The World Bank: Its Half-Century, vol. I: History* (Washington DC: Brookings Institute, 1997) (explaining that such has been the result, given the World Bank's perceived comparative advantage in other areas, the "inherent subjectivities of a soft sector" and the potential influence of political issues), at 168–69.

[54] For an overview of the World Bank's promotion of human capital-related policy, see Pauline Rose, supra note 51, at 174; but see Mark Blaug, *The Economics of Education and the Education of an Economist* (Edward Elgar Publishing, no. 48, 1987) (concluding that the World Bank's human capital research program seems increasingly unconvincing as it does not explain either the patterns of educational finance or public ownership of schools and colleges), at 849; Ben Fine and Pauline Rose, Education and the Post-Washington Consensus, in *Development Policy in the Twenty-First Century: Beyond the Post-Washington Consensus* (Ben Fine, Contas Lapavitas, and Jonathan Pincus, eds.) (Routledge, 2001) (adding that the rise of human capital theory within the World Bank grew out of the more specific application of cost–benefit analysis to calculation of rates of return. Exactly the

(GII) and related yearly GII report entitled *Human Capital in Innovation*, WIPO dedicated the entire report to reviewing the role of human capital in fostering innovation in view of the development divide.[55] WIPO nevertheless perceives human capital as a factor that cannot easily be merged into innovation and patent-related policies for two key reasons.

The first reason concerns the challenge of insufficient empirics. WIPO's 2014 GII report acknowledges that due to relative scarcity of innovation-specific empirical studies, it remains difficult to apply a comprehensive definition of human skills in innovation policy.[56] Equally importantly, it is difficult to measure human capital and innovation outputs and outcomes within innovation and patent-related policies as a whole.[57] The organizations' *2015 GII Report* adds that although innovation literature emphasizes the role of human capital for innovation and development "these innovation input factors seem to be the most difficult of all inputs in which to achieve good scores, both in general and for low-income countries in particular."[58]

The second constraint on the application of WIPO's human capital analysis to its innovation and patent-related policy relates to the organization's self-acknowledged slow development on the ground. As the WIPO 2015 GII report recognized, although the literature emphasizes the important role of human capital in development and innovation, low- and lower-middle-income innovation progresses is slower than would be

same methodology is applicable to any factor with an economic effect regardless of education specifically); Sam Bowles and Herb Gintis, *Schooling in Capitalist America: Educational Reform and the Contradictions of Economic Life* (Routledge and Kegan Paul, 1976) (criticizing the notion of human capital as its focus on costs and benefits is unable to address questions of how and why national education systems emerge, and how and why did they evolve differently); Ben Fine and Ellen Leopold, *The World of Consumption* (Routledge, 1993) (adding that education, training, and skills should be understood to be attached to a "system of provision" with a series of economic and other activities such as building schools to setting curricula and training teachers and are heavily embodied in social structures, relations, and processes).

[55] The GII 2014 is the 7th edition of the GII report co-published by Cornell University, INSEAD and WIPO. See The Global Innovation Index 2014, supra note 13.

[56] Ibid., at 69.   [57] Ibid.

[58] See *The Global Innovation Index 2015: Effective Innovation Policies for Development* (2015) (Soumitra Dutta, Bruno Lanvin, and Sacha Wunsch-Vincent, eds.), at XIX. For EU-centered empirics on human capital, see e.g., Patrick Vanhoudt, Thomas Mathä, and Bert Smid, How Productive Are Capital Investments in Europe?, EIB papers, 5(2) (2000) (Due to the core obstacles concerning data availability for regions, there is little evidence on the effects of human capital at the EU regional level); Tondl further shows that incomes and productivity of Southern EU regions are positively linked to school enrolment. See Gabriele Tondl, The Changing Pattern of Regional Convergence in Europe, *Jahrbuch für Regionalwissenschaft*, 19(1) 1 (1999). For a broader conclusion on the lack of data concerning education across the entire human capital literature, see Andreas Savvides and Thanasis Stengos, supra note 43, at 8.

anticipated from a policy standpoint.[59] These results, WIPO assesses, do not necessarily imply a lack of political will by these developing countries in contrast to other innovation and patent-related policies.[60] WIPO's 2015 report thus concludes that pursuing and excelling in these aspects takes more time than originally thought based on the experience of dozens of countries across the development divide.[61]

In the early 1990s, a third major policy-oriented breakthrough occurred in development theory, based on the work by Indian economist Amartya Sen and Pakistani development theorist Mahbub ul Haq regarding human capability over two decades.[62] Their insights led to the gradual appearance of a human development paradigm, partly manifested in the *Human Development Report* by the United Nations Development Programme (UNDP), and the much used composite index of ranking countries, the Human Development Index (HDI).[63] Within the context of technology and innovation-related policy the HDI has, on balance, been seriously criticized on a number of grounds, including its failure to consider technological development and its lack of attention to development from a global perspective.[64] This index presently has little or no relevancy in the context of human capital-related innovation policies across the UN and the WTO.

The current policy-orientation resurrection with human capital began with the seminal paper by renowned Harvard University macroeconomist Robert Barro, who emphasized the empirical determinants of long-term economic growth.[65] While Barro's paper did not relate explicitly to the function of human capital, it did place human capital (identified with formal education and measured by enrolment rates) at center stage in the economic growth process. Shortly after, economists Greg Mankiw, David Romer, and David Weil provided a theoretical explanation for the vital role of human capital in the growth process.[66] They verified empirically

---

[59] The Global Innovation Index 2015, ibid, at 72.   [60] Ibid.   [61] Ibid.

[62] See Amartya Sen, *Commodities and Capabilities* (North-Holland, 1985); Amartya Sen, Well-being, Agency and Freedom: The Dewey Lectures 1984, *Journal of Philosophy*, 82(4) 169 (1985). See also Sudhir Anand and Amartya Sen, Human Development and Economic Sustainability, *World Development*, Elsevier, vol. 28(12) 2029 (2000).

[63] The HDI index is a composite statistic of life expectancy, education, and income per capita indicators, used to rank countries into four tiers of human development. See UNDP, The Human Development concept, at: http://hdr.undp.org/en/humandev.

[64] See Hendrik Wolff, Howard Chong, and Maximilian Auffhammer, Classification, Detection and Consequences of Data Error: Evidence from the Human Development Index, *Economic Journal* 121 (553): 843 (2011).

[65] Robert J. Barro, Economic Growth in a Cross-Section of Countries, *Quarterly Journal of Economics*, vol. 106, 407 (1991).

[66] N. Gregory Mankiw, David Romer, and David N. Weil, A Contribution to the Empirics of Economic Growth, *Quarterly Journal of Economics*, vol. 107, 407 (1992).

that differences in cross-country income per capita can be understood by dissimilarities in savings, education, and population growth. The Mankiw–Romer–Weil model adheres to human capital as an input into an aggregate production function that presupposes declining returns to the reproducible factors of physical (including IPR) and human capital-based production. Yet even the Mankiw–Romer–Weil calculation does not mitigate WIPO's above-mentioned and self-acknowledged empirical impasse.

A pivotal paper published in 1988 by Nobel Laureate Robert Lucas and entitled *On the Machines of Economic Development* focuses on the reproducible nature of human capital and the possibility of externalities generated by human capital not only on individuals gathering knowledge but also on their coworkers, colleagues, and others.[67] Lucas's model provided the first human capital approach to endogenous growth.[68] Simply put, since human capital accumulation is the "engine" of growth, growth will itself also be endogenous. Lucas's important contribution on the concept of growth based on human capital accumulation was followed two years later by Becker, Murphy, and Tamura,[69] and a year later by Nancy Stokey.[70]

Henceforth, theoreticians began the search for human capital externalities, a research topic that has to date yielded mixed and indecisive evidence, especially for developing countries. One noticeable finding, adopted at the UN-level, concerns the "knowledge trap" facing disadvantaged countries. Many developing countries thus contend with a lack of human capital to confront lack of policy coordination, growing complexity, and the fragmentation of policy-making venues on the transnational and

---

[67] See Robert Emerson Lucas, On the Machines of Economic Development, *Journal of Monetary Economics*, vol. 22, 3 (1988). The model explains that individuals split their time between work and training. So, there is a trade-off, as during training individuals lose part of their work income, yet they raise their future productivity and therefore their future wages.

[68] See Andreas Savvides and Thanasis Stengos, supra note 43, at 6.

[69] Gary S. Becker, Kevin M. Murphy, and Robert Tamura, Human Capital, Fertility, and Economic Growth, *Journal of Political Economy*, vol. 98, no. 5, pages S12–37 (October 1990), reprinted in *Human Capital: A Theoretical and Empirical Analysis with Special Reference to Education* (3rd edn) (1994). Becker, Murphy, and Tamura offer empirical support for Lucas' spillover analysis, explaining that more human capital per person reduces fertility rates, because human capital is more productive in generating goods and additional human capital rather than more children.

[70] Nancy Stokey, Human Capital, Product Quality, and Growth, *Quarterly Journal of Economics*, 106(2) 587 (1991). Stokey's model uniquely assumes that labor is heterogeneous and differentiated by the level of human capital. In the model, goods are differentiated in terms of quality. As aggregate human capital grows, output accordingly consists of adding higher-quality goods to production and dropping lower-quality goods thereof.

international levels. These countries are thus penalized by the knowledge-intensity demanded at the UN-level as well as at the WTO.[71] This notion corresponds with Stephen Redding's insight that countries may become trapped in a "low-skills" equilibrium, characterized by a inadequately trained employees and low product quality.[72]

Another approach to evaluating the contribution of human capital to aggregate growth was formulated by Jess Benhabib and Mark Spiegel,[73] based on the traditional methodology for calculating growth, according to which the growth of output is determined by the gathering of inputs and TFP growth. The point of departure and the original contribution of their approach is to include only physical capital and labor as inputs and to model human capital as contributing to the increase of TFP rather than as an input to aggregate production. From a policy perspective, the contribution of human capital to TFP growth is twofold. Firstly, it determines the speed by which any country across the development divide is able to close the gap between its level of TFP and that of the technological leader or the catch-up effect.[74] As Nelson and Phelps earlier suggested,[75] a larger stock of human capital makes it easier for a country to attract the new products or ideas that have been discovered somewhere else.[76] Therefore, a follower country with more human capital, whether a developed or developing country, tends to grow faster because it catches up more rapidly with the technological leader.

---

[71] See Margaret Chon, Denis Borges Barbosa, and Andrés Moncayo von Hase, Slouching Toward Development in International Intellectual Property, *Michigan State Law Review* 71 (2007), at 89 referring to Sylvia Ostry, After Doha: Fearful New World?, Bridges, Aug. 2006, at 3, available at www.ictsd.org/monthly/bridges/BRIDGES 10-5.pdf. For the WTO, see Gregory Shaffer, Can WTO Technical Assistance and Capacity Building Serve Developing Countries?, *Wisconsin International Law Journal*, vol.23 643 (2006) ("Implementation often requires developing countries, unlike developed countries, to create entirely new regulatory institutions and regimes"), at 645.

[72] As Redding explains, workers invest in human capital or the acquisition of skills, while firms invest in quality-augmenting R&D, the two types of investment exhibit pecuniary externalities and are strategic complements. See Stephen Redding, The Low-Skill, Low-Quality Trap: Strategic Complementarities between Human Capital and R&D, *The Economic Journal*, vol. 106, No. 435, 458 (1996).

[73] See Jess Benhabib and Mark M. Spiegel, The Role of Human Capital in Economic Development: Evidence from Aggregate Cross-Country Data, *Journal of Monetary Economics*, vol. 34, 143 (1994).

[74] For OECD countries, see Dirk Frantzen, R&D, Human Capital and International Technology Spillovers: A Cross-Country Analysis, *The Scandinavian Journal of Economics*, vol. 102(1) 57 (2000) (Human capital is measured to affect productivity growth and there is evidence of interaction with the catch-up process over the period 1965–1991in all 21 OECD countries except New Zealand).

[75] Richard R. Nelson and Edmund S. Phelps, Investment in Humans, Technological Diffusion, and Economic Growth, *American Economic Review*, 56(2) 69 (1966).

[76] Ibid.

Secondly, the contribution of human capital to TFP growth is manifested whenever human capital determines the pace by which any country can adapt and implement foreign technologies domestically.[77] To date, however, Benhabib and Spiegel's work remains under-theorized within the innovation-based economic growth context. Regardless, the theoretical setting discussed adheres to linear innovation-based economic growth.

5.1.3   Non-Linear Human Capital and Endogenous Growth

Within the realm of nonlinear innovation theory examined by this book, several endogenous growth researchers have postulated that the impact of human capital on growth is correspondingly nonlinear.[78] Differences across countries in the timing of the take-off from stagnation to growth have contributed to what Oded Galor refers to as the "Great Divergence" and to the emergence of convergence clubs.[79] Following Galor, this chapter takes a long-term view of development, arguing that the transition from stagnation to growth is a predictable result of the course of development. It acknowledges Galor's view of growth process as characterized by dissimilar stages of development that leads to this chapter's nonlinearities of data concerning countries across the development divide.

Galor's work is based on the studies of economists Steven Durlauf and Paul Johnson,[80] including their division of countries into several groups based on the regression-tree methodology. Durlauf and Johnson show that the growth practice of the groups varies markedly. Kalaitzidakis et al.[81] similarly applied recent nonparametric economic techniques in

---

[77] See Andreas Savvides and Thanasis Stengos, supra note 43, at 7, referring to Jess Benhabib and Mark M. Spiegel, ibid.

[78] The literature review in Grossman and Helpman's endogenous growth analysis cites no fewer than ten potential determinants of long-term growth, alongside human capital investment. These cover physical investment rates, export shares, inward orientation, the strength of property rights, government consumption, population growth, and regulatory pressure. See Gene M. Grossman and Elhanan Helpman, Quality Ladders in the Theory of Growth, *Review of Economic Studies*, LVIII (1991) 43; Gene M. Grossman and Elhanan Helpman, *Innovation and Growth in the Global Economy*, Cambridge: MIT Press (1991).

[79] Oded Galor, From Stagnation to Growth: Unified Growth Theory, in *Handbook of Economic Growth* (Philippe Aghion and Steven N. Durlauf, eds.) (Amsterdam, 2005), 171. Galor's model essentially shows a positive association between human capital accumulation and technological progress. That is, as rising levels of schooling lessen the adverse effect of technological change on human capital aggregation.

[80] See Steven Neil Durlauf and Paul A. Johnson, Multiple Regimes and Cross-Country Growth Behavior, *Journal of Applied Econometrics*, vol. 10, 365 (1995).

[81] See Andreas Savvides and Thanasis Stengos, supra note 43, at 7, referring to Pantelis Kalaitzidakis, Theofanis Mamuneas, Andreas Savvides, and Thanasis Stengos,

order to model the human capital-growth relationship. The advantage of their methodology is that it allows the impact of human capital on growth to differ not only by country, but also by time period, jointly confirming the nonlinear nature of the human capital-growth relationship.[82] The transition to nonlinear human capital modeling of growth, and the shift away from the Solow-Swan model of neoclassical growth, focuses on the recurrence of attention in long-term growth, as manifested in models of endogenous growth as a substitute analytical structure, as the empirics of this chapter entails.[83] The endogenous growth theory ultimately takes human capital as one of the most important inputs in innovation from the macro level into a firm-, industry-, and even country-based analysis.

5.2   The Empirical Analysis

5.2.1   Methodology

Growth theory has thus far reviewed the interdependencies between innovation and employment on three main levels. First are firm-specific skills and knowledge which qualify for human capital.[84] These establish a

---

Measures of Human Capital and Nonlinearities in Economic Growth, *Journal of Economic Growth*, vol. 6, 229 (2001).

[82] See also, Eric A. Hanushek and Dennis Kimko, Schooling Labor Force Quality, and the Growth of Nations, *American Economic Review*, vol. 90 1184 (2000) (using data from six international tests of student achievements in mathematics and science to derive labor quality for 31 countries. Hanushek and Kimko's work was a benchmark in analyzing the quality of education confirming an important effect on the growth of per capita GDP across these countries).

[83] Alongside the endogenous growth theoretical setting for human capital theory, there remain two additional strands of studies that have touched innovation in human capital respect. First, the Schumpeterian studies do include the number skilled workers in their studies, though only as a control variable. See e.g., Zoltan J. Acs and David B. Audretsch, Innovation in Large and Small Firms: An Empirical Analysis, *The American Economic Review*, 678 (1988). A second strand of human capital-related literature is found in management literature. See e.g., Cheng Lin, Ping Lin, Frank M. Song, and Chuntao Li, Managerial Incentives, CEO Characteristics and Corporate Innovation in China's Private Sector, *Journal of Comparative Economics*, 39(2), 176 (2011) (examining World Bank 2002 survey data concluding that CEO education and incentive scheme are positively associated with firm's innovation).

[84] See e.g. Nathalie Greenan and Dominique Guellec, Technological innovation and employment reallocation, *Labour*, 14(4), 547 (2000); Alexander Coad and Rekha Rao, The Employment Effects of Innovations in High-Tech Industries, Papers on Economics & Evolution, #0705, Max Planck Institute of Economics, Jena (2007); Werner Smolny, Innovation, Prices and Employment – A Theoretical Model and an Empirical Application for West German Manufacturing Firms, *The Journal of Industrial Economics*, XLVI(3), 359 (1998); Robert M. Grant, Toward a Knowledge-based Theory of the Firm, *Strategic Management Journal*, Special Issue, nr. 17, 109 (1996).

positive connection between R&D employment and innovation thereof.[85] The second level of human capital analysis is on the industry level, referring to knowledge resulted from experience specific to an industry.[86] The third level of human capital analysis is individual specific human capital, referring to knowledge that can be used for a large range of firms and industries. This can include managerial and entrepreneurial experience,[87] a certain level of education and vocational training,[88] and the total household income.[89]

The analysis in this chapter portrays a country-level analysis instead using six yearly series. These are: (1) GERD per full-time equivalence (FTE) researcher, namely the total domestic intramural expenditure on R&D during a given year divided by the total number of researchers per country;[90] (2) GERD per head count (HC) researcher per country.[91] The analysis of GERD per researcher, was analyzed according to HC, as well

[85] See e.g., Jacques Mairesse and Pierre Mohnen, The Importance of R&D for Innovation: A Reassessment Using French Survey Data, *Journal of Technology Transfer* 30, 183 (2005).

[86] See e.g. Rinaldo Evangelista and Maria Savona, Innovation, Employment and Skills in Services: Firm and Sectoral Evidence, *Structural Change and Economic Dynamics*, 14, 449 (2003); Tommaso Antonucci and Mario Pianta, Employment Effects of Product and Process Innovation in Europe, *International Review of Applied Economics*, vol. 16 (3), 295 (2002); Tito Bianchi, With and Without Co-operation: Two Alternative Strategies in the Food Processing Industry in the Italian South, *Entrepreneurship & Regional Development*, nr. 13, 117 (2001); Martin Kenney and Urs von Burg, Technology Entrepreneurship and Path Dependence: Industrial Clustering in Silicon Valley and Route 128, *Industrial and Corporate Change*, nr. 8, 67 (1999); Robin Siegel, Eric Siegel, and Ian C. Macmillan, Characteristics Distinguishing High-growth Ventures, *Journal of Business Venturing*, nr. 8, 169 (1993).

[87] Johannes M. Pennings, Kyungmook Lee, and Arjen van Witteloostuijn, Human Capital, Social Capital, and Firm Dissolution, *Academy of Management Journal*, nr. 41, 425 (1998).

[88] Thomas Hinz and Monika Jungbauer-Gans, Starting a Business after Unemployment: Characteristics and Chances of Success (Empirical Evidence from a Regional German Labor Market), *Entrepreneurship and Regional Development*, nr. 11, 317 (1999).

[89] Maureen Kilkenny, Laura Nalbarte, and Terry Besser, Reciprocated Community Support and Small Town Small Business Success, *Entrepreneurship and Regional Development*, nr. 11, 231 (1999).

[90] FTE reflects the actual volume of human resources devoted to R&D. One FTE is equal to "one person working full-time on R&D for a period of one year, or more persons working part-time or for a shorter period, corresponding to one person-year." As a result, a person who normally spends 30% of their time on R&D and the rest on other activities should be considered as 0.3 FTE. Similarly, if a full-time R&D worker is employed at an R&D unit for only six months, this results in an FTE of 0.5. On the other hand, HC relates to the total number of persons who are mainly or partially employed in R&D. Thus, it includes staff employed both full-time and part-time. See UNESCO, UNESCO Institute for Statistics (UIS) – Glossary (Gerd per researcher). As the UNESCO report explains, this methodology was adapted from OECD (2002), Frascati Manual, §423 and §331–333.

[91] Headcount data includes the "total number of persons employed in R&D, independently from their dedication. These data allow links to be made with other data series, such as

as FTE. Both were converted according to the PPP of the 2005 dollar prices; (3) Government expenditure per tertiary student as percentage of GDP per capita (in percentage). General government expenditure per student in tertiary education is calculated by dividing total government expenditure on tertiary education by the number of students at tertiary level, expressed as a percentage of GDP per capita.[92] This was defined as the total number of articles divided by the total number of researches; (4) Scientific and technical journal articles per national researcher HC;[93] (5) Scientific and technical journal articles per national researcher FTE; and, lastly, (6) Number of researchers (HC) per labor force per country.

The six series above correspond with the landmark *Human Capital Index* developed by the Lisbon Council, a Brussels-based think tank. The index is set up to model human capital aspects of 13 EU members. The index quantified numerically each countries' ability to develop and nurture its human capital in four different categories collectively comprising the index. This process was based on a novel methodology that is contextually adopted in this chapter.[94] First is the category of *Human Capital Productivity*. This figure measures the productivity of human capital. It is derived by dividing GDP by all of the human capital employed in that country. This diverges from traditional productivity measures in that the figure takes account of how sound educated employed labor is, as a substitute of just how many hours are being worked.[95] The measurement of human capital productivity is accounted for in this chapter by the measurement of scientific and technical journal articles per national researcher HC and Scientific and technical journal articles per national researcher FTE (indicators 4 and 5, respectively).

The second human capital category is *Human Capital Utilization*, measuring national human capital deployment. It differs from traditional employment ratios in that it measures human capital as a proportion of

education and employment data, or the results of population censuses. They are also the basis for calculating indicators analyzing the characteristics of the R&D workforce with respect to age, gender or national origin." UNESCO, UNESCO Institute for Statistics (UIS), Ibid. As the UNESCO report explains, this methodology was adapted from OECD (2002), Frascati Manual, Ibid, §326–27.

[92] Average total (current, capital and transfers) general government expenditure per student in the given level of education, expressed in nominal PPP dollars. UNESCO, UNESCO Institute for Statistics (UIS), supra note 90.

[93] Scientific and technical journal articles refer to the number of scientific and engineering articles published in the following fields: physics, biology, chemistry, mathematics, clinical medicine, biomedical research, engineering and technology, and earth and space sciences. See World Bank Data IBRD-IDA – Glossary, at: http://data.worldbank .org, referring to the National Science Foundation, Science and Engineering Indicators.

[94] See Peer Ederer, supra note 1, at 2. For the Index's methodology, see ibid., at 20.

[95] Ibid.

the overall population.[96] The measurement of human capital productivity is partly accounted for in this chapter by the measurement of the number of researchers (HC) per labor force per country (indicator 6) and by GERD per FTE researcher and GERD per HC researcher per country (indicators 1–2). The third is category is the *Human Capital Endowment*, measuring the cost of all types of education and training in a particular country per employed person.[97] The measurement of human capital endowment is accounted for in this chapter by the measurement of government expenditure per tertiary student as percentage of GDP per capita (indicator 3).

Lastly, human capital is accounted for by the Lisbon Council's Human Capital Index by the *Demography and Employment* category. This category looks at existing economic, demographic, and migratory trends to estimate the number of people that will be employed in the year 2030 in each country.[98] The latter account is discussed in Chapter 6, in relation to the patent intensity characteristics by the three patent clusters concerning domestic/overseas patenting rates measured thereof.

With the exception of the series on government expenditure per tertiary student as percentage of GDP per capita, for which the available data herein starts in 1998, all the series cover the period 1996–2013. Therefore, for all series except the series on government expenditure per tertiary student as percentage of GDP per capita, the criterion for a country to be included in the imputation process was at least one observation for the years 1996–1998 and at least one observation for the years 2011–2013. For the series on government expenditure per tertiary student as percentage of GDP per capita, there was at least one observation for the years 1998–2000, and at least one observation for the years 2011–2013. Kalman smoothing was applied for imputing the series, but when it failed to converge, moving average imputation was used instead.[99]

5.2.2   Findings

5.2.2.1 Patent-Researcher Gap Between Patent Clusters
Descriptive statistics (after imputation) of averages over the midrange of years 2003–2009 are presented below, for each series and for each of the

---

[96] Ibid.   [97] Ibid., at 4.   [98] Ibid.

[99] Before applying the MIXED procedure to analyze each series, there was an investigation into which would be the best Box-Cox transformation for each cluster, so that the required normality assumption for using this procedure be justified. The log transformation was found to be the best for all the series. These transformations were used before the imputations and following that, imputed values were back transformed. When the minimal value was zero, then 0.5 was added to all values of the series before taking the log, to avoid trying to take log of zero.

three patent clusters. The first row corresponds to the original scale and the second to the log scale. In addition, Box Plots of these data are presented for the six indicators measured per clusters 1–3 accounting for the leaders, followers and marginalized patent clusters, respectively, and in Figure Series 1.

The final conclusions for each series are briefly summarized after the display of each of the corresponding descriptive tables. These conclusions are based on the results obtained from the regressions fitted by using the mixed models included in Appendix F: Employment and Human Resources by Patent Cluster.

5.2.2.1.1 *GERD per FTE Researcher*

All clusters together

n	mean	sd	median	min	max
49	149.02	88.49	138.55	35.51	513.92
49	4.82	0.62	4.93	3.56	6.23

By cluster

Cluster 1

n	mean	sd	median	trimmed mean	mad	min	max	range
19	195.61	57.06	190.01	194.61	48.74	89.76	318.51	228.75
19	5.23	0.31	5.25	5.24	0.24	4.50	5.76	1.26

Cluster 2

n	mean	sd	median	trimmed mean	mad	min	max	range
20	108.38	55.46	100.45	102.19	53.24	35.53	230.73	195.19
20	4.55	0.54	4.61	4.55	0.45	3.57	5.44	1.87

Cluster 3

n	mean	sd	median	trimmed mean	mad	min	max	range
10	141.8	143.00	99.06	108.57	75.15	35.51	513.92	478.40
10	4.6	0.81	4.58	4.53	0.82	3.56	6.23	2.67

In conclusion, no significant changes in time were found for all three clusters. A significant difference was found between the leaders and followers patent clusters, with higher values for the leaders cluster.

5.2.2.1.2 GERD per HC Researcher

All clusters

n	mean	Sd	median	min	max
45	81.67	82.90	52.82	4.42	513.92
45	4.01	0.92	3.96	1.46	6.23

By cluster

Cluster 1

n	mean	sd	median	trimmed mean	mad	min	max	range
8	145.05	29.3	142.28	145.05	30.50	109.35	189.79	80.44
8	4.96	0.2	4.96	4.96	0.23	4.69	5.24	0.55

Cluster 2

n	mean	sd	median	trimmed mean	mad	min	max	range
21	58.58	32.11	48.33	53.42	20.84	25.09	152.92	127.83
21	3.95	0.48	3.88	3.92	0.54	3.22	5.02	1.80

Cluster 3

n	mean	sd	median	trimmed mean	mad	min	max	range
16	80.31	124.48	38.13	54.75	42.29	4.42	513.92	509.50
16	3.63	1.23	3.59	3.60	1.24	1.46	6.23	4.76

Significant changes in time were found, but differing among clusters. On average, however, an increase over time has been witnessed, with higher values for the leaders cluster compared with the two other clusters.

5.2.2.1.3 *Government Expenditure per Tertiary Student as Percentage of GDP Per Capita*

All clusters

n	Mean	sd	median	min	max
44	32.99	22.54	26.38	9.05	151.81
44	3.35	0.51	3.27	2.19	5.02

By cluster

Cluster 1

n	mean	sd	median	trimmed mean	mad	min	max	range
15	31.05	11.97	27.78	30.84	10.77	9.05	55.75	46.71
15	3.35	0.44	3.32	3.39	0.35	2.19	4.02	1.82

Cluster 2

n	mean	sd	median	trimmed mean	mad	min	max	range
19	31.90	14.49	26.04	31.15	7.62	12.84	63.66	50.81
19	3.37	0.42	3.26	3.37	0.28	2.55	4.14	1.59

Cluster 3

n	mean	sd	median	trimmed mean	mad	min	max	range
10	38.00	41.82	23.87	27.34	11.26	9.49	151.81	142.32
10	3.31	0.77	3.16	3.22	0.55	2.24	5.02	2.78

Significant changes in time were found for all three clusters with a decrease in time. Also, significant differences in the overall means were found between all three clusters.

5.2.2.1.4 *Scientific and Technical Journal Articles per Researcher HC*

All clusters

n	mean	sd	median	min	max
44	0.19	0.24	0.15	0.02	1.63
44	−2.05	0.83	−1.91	−4.06	0.48

By cluster

Cluster 1

n	mean	sd	median	trimmed mean	mad	min	max	range
8	0.23	0.09	0.21	0.23	0.07	0.13	0.37	0.24
8	-1.51	0.36	-1.58	-1.51	0.35	-2.02	-0.99	1.03

Cluster 2

n	mean	sd	median	trimmed mean	mad	min	max	range
20	0.15	0.07	0.15	0.15	0.07	0.02	0.31	0.29
20	-2.03	0.63	-1.91	-1.96	0.44	-3.92	-1.17	2.75

Cluster 3

n	mean	sd	median	trimmed mean	mad	min	max	range
16	0.21	0.39	0.10	0.12	0.09	0.02	1.63	1.61
16	-2.33	1.08	-2.42	-2.41	1.01	-4.06	0.48	4.54

Significant changes in time were found, showing a decrease in but differing among clusters. Also, significant differences were found between clusters, with higher values for the leaders cluster compared with the two other clusters.

5.2.2.1.5 *Scientific and Technical Journal Articles per Researcher FTE*

All clusters

n	mean	sd	median	min	max
49	0.32	0.23	0.29	0.06	1.63
49	-1.32	0.55	-1.25	-2.82	0.48

By cluster

Cluster 1

n	mean	sd	median	trimmed mean	mad	min	max	range
18	0.34	0.12	0.32	0.33	0.06	0.15	0.60	0.45
18	-1.14	0.36	-1.16	-1.14	0.20	-1.90	-0.51	1.39

Cluster 2

n	mean	sd	median	trimmed mean	mad	min	max	range
21	0.28	0.11	0.27	0.28	0.08	0.06	0.50	0.44
21	-1.38	0.49	-1.35	-1.34	0.42	-2.82	-0.71	2.11

Cluster 3

n	mean	sd	median	trimmed mean	mad	min	max	range
10	0.35	0.46	0.21	0.22	0.09	0.08	1.63	1.55
10	-1.52	0.85	-1.62	-1.64	0.48	-2.56	0.48	3.04

Significant decrease in time was found, but differing among clusters. Higher values were found for the leaders cluster compared with the followers cluster.

5.2.2.1.6 *Number of Researchers (HC) per Labor Force*

All clusters

n	mean	sd	median	min	max
45	0.01	0.01	0.00	0.0	0.02
45	-5.65	1.15	-5.52	-8.4	-3.84

By cluster

Cluster 1

n	mean	sd	median	trimmed mean	mad	min	max	range
8	0.01	0.00	0.01	0.01	0.00	0.01	0.02	0.01
8	-4.50	0.38	-4.49	-4.50	0.21	-5.24	-3.92	1.32

Cluster 2

n	mean	sd	median	trimmed mean	mad	min	max	range
21	0.01	0.00	0.00	0.00	0.00	0.00	0.02	0.02
21	-5.54	0.83	-5.51	-5.47	0.82	-7.33	-4.15	3.19

Cluster 3

n	mean	sd	median	trimmed mean	mad	min	max	range
16	0.00	0.01	0.00	0.00	0.00	0.0	0.02	0.02
16	-6.37	1.27	-6.21	-6.41	1.18	-8.4	-3.84	4.56

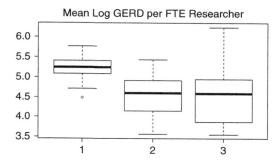

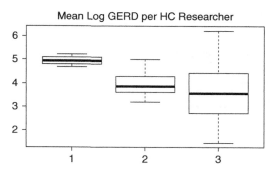

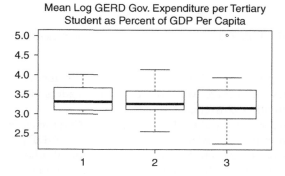

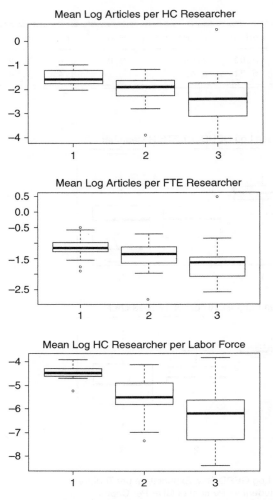

Figure Series 1 Box Plots of the Average of the Log Series (2003–2009)

Significant changes in time but differing among cluster were found. Also, higher values on average were found for the leaders cluster compared with the two other clusters.

The following is a summary of all the conclusions. With the exception of government expenditure per tertiary student as percent of GDP per capita, for which there were no differences among clusters, for all other variables the leaders cluster (labeled 1) had higher mean values compared with the followers patent cluster (labeled 2). For three out of the six series, cluster 1 also had higher mean values compared with the marginalized cluster (labeled 3). These three were: GERD per HC researcher, articles per HC researcher and HC researcher per labor force. For GERD per FTE researcher, there were no significant changes along time. For government expenditure per tertiary student as percent of GDP per capita, there were the same changes in time for all three patent clusters. For all other series, there were changes along time that differed among clusters. The significant difference between clusters were also tested for each year. A consistently significant difference among clusters was found for each of the years for the three series: GERD per FTE researcher, GERD per HC researcher and HC researcher per labor force. For the two articles-related series the difference became non-significant toward the end of the series.

These results loosely correspond with earlier findings. As Yale University economist Samuel Kortum shows, focusing on advanced economies as of 1997, patent per researcher rates are said to decline over time as technological breakthroughs become increasingly hard to come by.[100] As a policy concern, the number of researchers must rise exponentially to generate a constant flow of new patented inventions.[101] This reasoning surely assumes that the growth in research effort produces constant productivity growth if the size distribution of inventions is stationary.[102] The growth in research employment itself is fueled by an increase in the value of patented inventions relative to wages, which is in turn sustained by growth in the labor force.[103]

Samuel Kortum explains that a key implication of the theory is that the value of patented inventions rises over time, causing researchers to expend ever greater resources to discover them.[104] Kortum offers two corroborating patent statistical findings. The first refers to Mark Schankerman and Ariel Pakes' work,[105] showing that in the United Kingdom, France and Germany the age at which patents were allowed to lapse (because the inventor failed to pay a renewal fee) tended to rise over time. From 1965 to 1975 patents per researcher fell sharply in these countries, but the decline was offset by a rise in the average value of a patent estimated on the basis of renewal statistics.[106]

Second, there is a long-term trend for inventions to be patented more internationally, as Kortum explains. In the 1950s, the ratio of US

---

[100] Samuel Kortum, Research, Patenting and Technological Change, Econometrica, vol. 65:6, 1389 (1997), at 1389.
[101] Ibid.   [102] Ibid.   [103] Ibid.   [104] Ibid.
[105] Mark Schankerman and Ariel Pakes, Estimates of the Value of Patent Rights in European Countries During the Post-1950 Period, *The Economic Journal*, 96, 1052 (1986).
[106] Ibid.

inventions seeking patent protection in the United Kingdom to US inventions seeking patent protection in the United States was about 10 percent. By the 1970s this ratio had climbed to almost 20 percent, and by the early 1990s it was 25 percent. The increase has been even more dramatic for US inventions seeking patent protection in Germany and Japan.[107]

In support of the first corroborating finding Kortum mentions, this chapter's analysis thus adds that when the significant difference between clusters was tested yearly, a consistent significant GERD per researcher-related expenditure per patent propensity difference among clusters was found for each of the years for the three series: GERD per FTE researcher, GERD per HC researcher and HC researcher per labor force. For the two articles' series the difference became non-significant toward the end of the series as said – all indicating a differing GERD per researcher-related expenditure by patent propensity gap between the three patent clusters for each of the years. This was further corroborated with a finding of a significant patent-researcher gap mostly between the leaders cluster and the two other clusters of all series except for government expenditure per tertiary student as percentage of GDP per capita, as explained.

5.2.2.2 Human Capital Input Deficit in Developing Countries

In one key instance, Christopher Freeman and World Bank researchers explain that within developing countries, such as in the case of Latin America, the private sector has expressed concern that the skills generated by universities do not match its needs.[108] Two out of three Latin American country researchers work in the public sector, mostly in universities, and only one in ten are employed in the business sector. With the exception of Costa Rica – where around 25 percent of researchers work in the business sector – the analogous figure does not exceed 12 percent in any Latin American country. In terms of R&D spending, experimental development (as opposed to basic or applied research) accounts for less than 30 percent in Latin American countries, compared to more than 60 percent in countries such as the Republic of Korea and the United States.[109] Thus, there appears to be a discrepancy linking the policies

---

[107] Samuel Kortum, supra note 100, at 1389.
[108] David de Ferranti, Guillermo Perry, Indermit Gill, William Maloney, Jose Luís Guasch, Carolina Sanchez-Paramo, and Norbert Schady, *Closing the Gap in Education and Technology* (World Bank, 2003), at 228; Christopher Freeman, The National System of Innovation in Historical Perspective, *Cambridge Journal of Economics*, 19, 5 (1995).
[109] See Lea Velho, Science and Technology in Latin America and the Caribbean: An Overview, Discussion Paper, 2004-4 (Maastricht: UNU-INTECH) (2004), at 17.

taken to promote skills and the demand from the private sector, moderately reflecting the existing industrial specialization toward natural resources and assembly operations based on low labor costs.[110]

This early finding possibly relates to the Human Capital Index's *Human Capital Utilization* category measuring national human capital deployment. As seen, in continuation, the measurement of the number of researchers (HC) per labor force per country (indicator 6, above), GERD per FTE researcher and GERD per HC researcher per country (indicators 1–2, above) all indicate a significant deficit from the perspective of the two lower patent clusters. There is still a need to further corroborate Christopher Freeman et al.'s above finding concerning the optimal institutional input by GERD type also by patent clusters. This chapter's findings nevertheless shed light on an underlying human capital input gap across the development divide largely represented by the three patent clusters.

5.3   Theoretical Ramifications

Economic theory still lacks a clear measurement for the impact of innovation on employment, and there remains a strong need for empirical analyses capable of testing this connection.[111] Prominent studies, such as those of Harvard University economist Lant Pritchett,[112] and Jess

---

[110] Ibid.; Mario Cimoli, João Carlos Ferraz, and Annalisa Primi, Science and technology policies in open economies: the case of Latin America and the Caribbean, Paper presented at the first meeting of ministers and high authorities on science and technology, Lima, Peru, 11 12 November (similarly referring to Latin American and Caribbean mismatch between demand-side needs and human capital supply side offering) (2004), at 11.

[111] For findings upholding no significant link or outright negative impact of new technology on jobs, see e.g., Vincent Van Roy, Daniel Vertesy, and Marco Vivarelli, Innovation and Employment in Patenting Firms: Empirical Evidence from Europe, IZA Discussion Paper No. 9147, (June 2015) (using a unique longitudinal database of approximately 20,000 patenting firms from 22 European countries between 2003 and 2012, testing the effect of innovation activity on job creation), at 3 and Sec. 2; Tor Jakob Klette, and Svein Erik Førre, Innovation and Job Creation in a Small open, Economy: Evidence from Norwegian Manufacturing Plants 1982–92, *Economics of Innovation and New Technology* 5, 247 (1998) (examining 4,333 Norwegian manufacturing firms over the period 1982–1992 and finding no significant relationship between R&D intensity and net job creation); Erik Brouwer, Alfred Kleinknecht, and Jeroen O. N. Reijnen, Employment Growth and Innovation at the Firm Level, *Journal of Evolutionary Economics* 3, 153 (1993) (finding a negative relation between employment and aggregate R&D expenditures. Authors also find a positive relationship when only product innovations were considered in a cross-sectional study of 859 Dutch manufacturing firms).

[112] Lant Pritchett, Where Has All the Education Gone, World Bank working papers, no. 1581 (1996).

Benhabib and Mark Spiegel,[113] similarly suggest that it is the level of educational attainment that influences growth, not its change, as human capital's reverse effect on innovation remains imprecise. More recent studies also show considerable measurement errors, arguing how human capital analysis can lead to particularly erratic observations when looking at human capital growth.[114] Others confirm that many potential policy influences operate not only directly on growth, but also obliquely via the enlistment of resources for fixed investment.[115] Economists Daniel Cohen and Marcelo Soto, in their revealingly titled article *Growth and Human Capital: Good Data, Good Results*, further confirm these disturbing findings.[116] As a result, the impact of human capital empirical analysis on the propensity to patent across countries deserves additional research due to a few core concerns.

To begin with, human capital literature largely assumes that the training of employees requires that they be offered concentrated, long-term training and learning experiences.[117] That, goes the theory, would assumingly put off dependence on easy imitation or substitution in the short term.[118] There nevertheless remains a distinction between what New York University economist William Baumol describes as the differing competencies needed by inventors and entrepreneurs as opposed to skills needed by professionals conducting such research activities and underlying education and training.[119] Inventors and entrepreneurs often have only basic formal technological and scientific education and training.[120]

---

[113] Jess Benhabib and Mark Spiegel, supra note 73.

[114] Alan B. Krueger, Mikael Lindahl, Education for Growth: Why and for Whom, NBER working paper, no. 7591(2000); Javier Andrés, Ángel de la Fuente, and Rafael Doménech, Human Capital in Growth Regressions: How Much Difference Does Data Quality Make?, CEPR discussion paper, no. 2466, London: Centre for Economic Policy Research (2000).

[115] Andrea Bassanini, Stefano Scarpetta, and Philip Hemmings, Economic Growth: The Role of Policies and Institutions, Panel Data Evidence from OECD Countries, OECD working paper, STI 2001/9 (2001).

[116] Daniel Cohen and Marcelo Soto, Growth and Human Capital: Good Data, Good Results, CEPR discussion paper, no. 3025, London: Centre for Economic Policy Research (2001).

[117] Wesley M. Cohen and Daniel A. Levinthal, Absorptive Capacity: A New Perspective on Learning and Innovation, *Administrative Science Quarterly* 35(1), 128 (1990); Michael D. Michalisin, Robert D. Smith, and Douglas M. Kline, In Search of Strategic Assets, *The International Journal of Organizational Analysis*, 5(4) 360 (1997).

[118] Richard Reed and Robert J. DeFillippi, Casual Ambiguity, Barriers to Imitation and Sustainable Competitive Advantage, *Academic Management Review* 15(1) 88 (1990).

[119] William Baumol, Education for Innovation, in *Innovation Policy and the Economy* (Adam Jaffe, Josh Lerner, and Scott Stern, eds.), vol. 5, MIT Press (2005).

[120] Petr Hanel, Skills Required for Innovation: A Review of the Literature, Note de Recherche (2008), at 9, referring to Bernard Bonin and Claude Desranleau, *Innovation Industrielle et Analyse économique* (HEC, 1987).

On the other hand, R&D-based firms employ highly educated researchers with extant knowledge and high academic degrees. According to Baumol, "rigorous education plays a critical role in support of technical progress, and R&D expenditure of giant corporations together with the efforts of the independent entrepreneurs-innovators provide a crucial contribution to the process."[121] The two types of human capital are thus characteristically distinct from one another and play complementary roles – that is, as they jointly contribute to the propensity to patent across firms, industries and countries. These differences could potentially impact upon the propensity to patent in firms, industries and states rather distinctively. Moreover, as there are no curricula for education of innovators, such objectives would be quite different from those pursued by engineering and scientific education.[122] As Baumol further explains:[123] "Education designed for technical competence and mastery of available body of analysis and education designed to stimulate originality and heterodox thinking tend to be substitutes rather than complements." Baumol's absorbing examination nevertheless remains debatable and requires granular conceptual refinement. It is sufficient to recall researchers such as Bernard Bonin and Claude Desranleau,[124] who reviewed a time-series sample of major innovations in numerous industries and countries. They concluded, possibly contrary to the above position, that the majority of inventors alongside innovators reviewed were similarly technically or scientifically trained.[125] Accordingly, future research should carefully account for two mechanisms through which R&D employment could affect patent propensity rates. The first is R&D personnel that can produce innovation directly using resources in relevant R&D firm's departments. The second mechanism is applied whenever R&D facilitates non-R&D personnel, mainly engineers and managers, in improving innovation. Human capital theory already acknowledges that the existence of skilled labor is a more significant factor in the transmission of tacit knowledge than university or industry research.[126] In both cases, R&D-related personnel constitute a complementary form of human

---

[121] William Baumol, supra note 119, at 38.

[122] See e.g., Belton M. Fleisher, Yifan Hu, Haizheng Li, and Seonghoon Kim, Economic Transition, Higher Education and Worker Productivity in China, *Journal of Development Economics*, 94(1), 86 (2011) (classifying employees into highly educated and less educated by averaging the workers' schooling codes for each occupation level). The authors conclude that highly educated employees mainly consists of "engineering and technical personnel" and "managerial personnel (including sales)," at 9.

[123] William Baumol, supra note 119, at 35.   [124] Petr Hanel, supra note 120, at 226.

[125] Ibid.

[126] David B. Audretsch and Maryann P. Feldman, R&D Spillovers and the Geography of Innovation and Production, *American Economic Review* vol. 86, 630 (1996).

capital to be accounted for in terms of its impact on patentable innovation.[127]

A second core theoretical ramification of this chapter relates to the effect of human capital on unpatentable innovation. Innovation theory identifies four types of unpatentable innovation that should be addressed by human capital theory. Firstly, there are imitative activities, including reverse engineering, which typically do not require R&D-literate personnel, as they are dependent mainly on the firm's technical personnel and engineers.[128] Secondly, firms can make small modifications or incremental non-patentable changes to products and processes while relying on engineers. This reality relates to the fact that innovation processes in low- and medium-technology sectors are related to adaptation and learning by doing, based on design and process optimization, rather than to R&D and R&D-literate personnel.[129] A third type of unpatentable innovation to be addressed by human capital theory relates to the preference of firms to combine existing knowledge bases – that is, particularly industrial design and engineering projects and related personnel, which, as non-R&D forms of innovation, are also often unpatentable.[130] Lastly, firms sometimes keep their share of R&D employees on total employment stable in order to ensure future growth or to maintain an innovative image, even if this is uncorroborated by patent intensity.[131]

The third theoretical ramification raised by this section relates to the unchallenged aspects of human capital across differing industries. Only a handful of studies have searched for and found differences in the employment/job-creation effect of innovation across different industry groups.

---

[127] See Christian Rammer, Dirk Czarnitzki, and Alfred Spielkamp, Innovation Success of Non-R&D-performers: Substituting Technology by Management in SMEs, *Small Business Economics* 33, 35 (2009) (arguing that "SMEs may opt for refraining from R&D and relying more on innovation management tools in order to achieve innovation success"), at 35; Xiuli Sun, Firm-level Human Capital and Innovation: Evidence from China, Partial Doctor of Philosophy thesis in the School of Economics, Georgia Institute of Technology (2015) ("innovation includes not only R&D innovation but also non-R&D innovation"), at 18.

[128] See Technology, Learning, and Innovation: Experiences of Newly Industrializing Economies, supra note 39.

[129] Povl A. Hansen and Göran Serin, Will Low Technology Products Disappear?: The Hidden Innovation Processes in Low Technology Industries, *Technological Forecasting and Social Change*, 55(2) 179 (1997).

[130] Christoph Grimpea and Wolfgang Sofka, Search Patterns and Absorptive Capacity: Low and High Technology Sectors in European Countries, *Research Policy*, 38(3) 495 (2009).

[131] See Matthias Buerger, Tom Broekel, and Alex Coad, Regional Dynamics of Innovation: Investigating the Coevolution of Patents, Research and Development (R&D), and Employment, *Regional Studies*, vol. 46(5) 565 (2012), at 572.

Notably, Greenhalgh et al.[132] explored a panel of United Kingdom firms over the period 1987–1994. Their fixed effects cumulative estimates showed a reserved, but positive effect of R&D expenditures on employment. Interestingly, when the researchers divided the panel into high- and low-tech sectoral groups, they found that the positive impact of R&D on employment was confined to high-tech sectors. Recently, Buerger et al.[133] have studied the co-evolution of R&D expenditures, patents and employment. Their main result, similar to that of Greenhalgh et al., was that patents positively and significantly impact employment in two high-tech sectors, namely medical/optical equipment and electrics/electronics, but are not significant in two more traditional technology sectors, namely chemicals and transport equipment.[134] Human capital theory is insufficient regarding the reverse impact of human capital on industry-specific patentable innovation. In time, policy-makers will have to ensure that the education system delivers the kind of skills that are most in demand for patentable innovation-based economic growth.

One way to address this challenge is to use state coordinative regulation, through what Green et al. term a "skills coordinator."[135] To accelerate skills formation in relevant innovative technological industries, governments need an informed view of the skills that are in demand. South Korea, Singapore, and Taiwan offer illustrative working examples of this. In Singapore, for example, the Ministry of Trade and Industry, the Economic Development Board, and the Council for Professional and Technical Education jointly supervise prospect skills needs, drawing on inputs from foreign and local investors as well as from education and

---

[132] Christine Greenhalgh, Mark Longland, and Derek Bosworth, Technological Activity and Employment in a Panel of UK Firms, *Scottish Journal of Political Economy* 48, 260 (2001).

[133] Matthias Buerger, Tom Broekel, and Alex Coad, supra note 131 (authors use data concerning four manufacturing sectors across German regions over the period 1999–2005).

[134] For an analogous study on employment impact of R&D expenditures, across industry types, see also, Francesco Bogliacino, Mariacristina Piva and Marco Vivarelli, R&D and Employment: An Application of the LSDVC Estimator using European Data, *Economics Letters* 116 (1) 56 (2012) (detecting a significant R&D expenditures impact on employment only in services and high-tech manufacturing but not in the more traditional manufacturing sectors. Authors use a panel database covering 677 European manufacturing and service firms over 19 years (1990–2008)).

See also Vincent Van Roy, Daniel Vertesy and Marco Vivarelli, supra note 111 (estimating a labor-friendly nature of innovation measured in terms of forward-citation weighted patents, based on a panel dataset analysis covering approximately 20,000 European patenting firms between 2003 and 2012. This positive impact of innovation is statistically significant only for firms in the high-tech manufacturing sectors, while not significant in low-tech manufacturing and services.).

[135] See Francis Green, David Ashton, Donna James, and Johny Sung, The Role of the State in Skill Formation: Evidence from the Republic of Korea, Singapore, and Taiwan, *Oxford Review of Economic Policy*, 15, 1 (1999), at 82–96.

training institutions. This information is matched against national policy objectives and used to build targets for a range of universities, schools, polytechnics, and the Institute for Technical Education.[136]

A fourth theoretical ramification remains. It refers to the archetypal gender gap in innovation and patenting. This partly accounted for gender gap connotes a strong negative relationship between a country's expenditure on R&D and the proportion of women in science.[137] The deeply rooted EU campaign against gender inequality in the sciences serves as a first-rate point of departure.[138] To begin with, the European Commission's 2011 *Innovation Union Competitiveness Report* grimly concludes that the highest proportions of women are found in those countries with the lowest R&D expenditure on R&D personnel. On balance, the lowest proportions of women are in the sectors with the uppermost R&D spending on R&D personnel.[139]

[136] See UNCTAD, World Investment Report 2005, supra note 31, at 203, referring to Francis Green, David Ashton, Donna James and Johny Sung, ibid, at 88.

[137] For an account of the gender research productivity gap, see Yu Xie and Kimberlee A. Shauman, Sex Differences in Research Productivity: New Evidence about an Old Puzzle, *American Sociological Review*, 63 (6), 847 (1998) (explaining that the representation of women in science and engineering in the United States of research productivity increased between 1969 and 1993. The gap however can largely be explained in terms of differences in personal characteristics, structural positions and marital status). Yu Xie and Kimberlee A. Shauman, *Women in Science: Career Processes and Outcomes* (Harvard University Press, 2003) (if the differences in the distribution of resources such as space, equipment, and time are taken into account, the research productivity gap between men and women remain negligible); J. Scott Long, *From Scarcity to Visibility: Gender Differences in the Careers of Doctoral Scientists and Engineers* (National Academy Press, 2001) (while females are earning an increasing proportion of the doctorates in science and engineering, they are not participating in the S&E workforce at commensurate levels). For studies indicating women academicians' relatively lower employment rates, see also Diana Bilimoria and Xiangfen Liang, State of Knowledge about the Workforce Participation, Equity, and Inclusion of Women in Academic Science and Engineering, in *Women, Science, and Technology: A Reader in Feminist Science Studies* (Mary Wyer, Mary Barbercheck, Donna Cookmeyer, Hatice Ozturk, Marta Wayne, eds.) (Routledge, 3rd edn., 2014) 21 [hereafter, Women, Science and Technology] (accounting for the relatively lower presence of women with science and engineering doctorate degrees holding academic positions along the faculty ranks), at 34–34 and figure 3.3; Michael T. Gibbons, Engineering by the Numbers, ASEE (2011) (women doctorate holders by academic positions constituted 12.7% of tenured or tenure-track faculty in 2009).

[138] In 1998 the European Commission's Research Directorate-General set up a first expert group on women in science. The resulting European Technology Assessment Network (ETAN) report, titled Science policies in the European Union – Promoting excellence through mainstreaming gender equality (2000), commenced the official European program to combat gender inequality in science and technology. On national policies to combat the scientific gender bias, see discussion at European Commission, Innovation Union Competitiveness report 2011, supra note 40, at 214 and table II.3.1.

[139] See European Commission, Innovation Union Competitiveness report 2011, supra note 40, at 213; See European Commission, Waste of Talents: Turning Private Struggles into a Public Issue; Women and Science in the ENWISE Countries, A report

The gender gap certainly exists in emerging economies and other developing countries. In European ENWISE countries (Central and Eastern European countries and the Baltic States), women constitute the majority of teaching staff (54 percent), but tend to be concentrated in the more junior academic positions.[140] Insufficient funding, poor infrastructure, and outdated equipment in these countries are all factors that obstruct the development of research communities, particularly in areas with low expenditure on R&D. Since these areas tend to be those where a large proportion of female scientists are employed, women scientists ultimately face a higher risk of missing out on research and patenting opportunities.[141] Women's patenting rates are low given their representation in science, technology, engineering, and mathematics and their authorship of scientific papers (the proportion of women academic researchers varies widely across EU countries, from 45 percent to 11 percent).[142]

The gender gap over patenting rates was stressed across the Atlantic in the 2011 Leahy-Smith American Invents Act, which mandated the Director of the USPTO to "establish methods for studying the diversity of patent applicants, including those applicants who are minorities, women, or veterans." In a seminal 2015 survey conducted by Indiana University researchers, the survey examined 4.6 million utility patents issued from 1976 to 2013,[143] concluding that women's rate of filing patents with the US Patent and Trade Office across 185 countries has increased from 2.7 percent of total patent intensity to 10.8 percent over the nearly forty-year period.[144] Adjusting for co-authorship, similar

to the European Commission from the ENWISE Expert Group on Women Scientists in Central and Eastern European Countries and in the Baltic States, Luxembourg (2003) [hereafter, The ENWISE Report]. The ENWISE (Enlarge Women In Science to East) report by the European Commission in 2003 analyzes the situation facing women scientists in Bulgaria, the Czech Republic, Estonia, Hungary, Latvia, Lithuania, Poland, Romania, the Slovak Republic, and Slovenia.

[140] Ibid., The ENWISE Report, Ibid., at 58 and table 2.3.   [141] Ibid., at 8.

[142] See European Commission, She Figures 2015: Gender in Research and Innovation, Brussels: European Commission (2015) ("The evolution of the proportion of women in grade A academic positions between 2010 and 2013 confirms that women continue to be vastly underrepresented in top positions within the Higher Education Sector"), at 1 and figure 1; Kjersten Bunker Whittington and Laurel Smith-Doerr, Women Inventors in Context: Disparities in Patenting across Academia and Industry, *Gender and Society*, 22(2), 194 (2008).

[143] See Cassidy R. Sugimoto, Chaoqun Ni, Jevin D. West, and Vincent Larivière, The Academic Advantage: Gender Disparities in Patenting (May 27, 2015), at 5 and figure 2.

[144] For additional confirming figures, see Jennifer Hunt, Jean-Philippe Garant, Hannah Herman, and David J. Munroe, Why Don't Women Patent?, IZA Discussion Paper No. 6886 (September 2012), at 1, referring to the National Survey of College Graduates 2003 (showing that American women patent at only 8% of the male rate) and the US

estimates show that women account for 8.2 percent of patents filled by Americans at the EPO in 2005.[145]

The Indiana University report also indicates that the rate of women filing patents has risen fastest within academia compared to all other sectors of the innovation economy. As the report indicates, from 1976 to 2013 the overall percentage of patents filled by women rose from an average of 2–3 percent across all areas to 18 percent in academia, 12 percent in individuals, and 10 percent in industry.[146]

Of the various explanations offered to this gender gap, one of the most telling was formulated by sociologist Kjersten Bunker Whittington and colleagues, who explain that as fewer women are interested in becoming involved with commercial work with their scientific endeavors, they also bear fewer commercial applications and are less exposed to knowledge of how the commercial process works.[147] European Commission researchers Naldi and Vannini Parenti offer an analogous explanation, suggesting that women are more intensely involved in producing publications than in commercially patenting them.[148]

Bureau of Labor Statistics 2011 (earn 81% of male full-time weekly earnings). In scientific disciplines female patenting rates are higher, as in life science. See W. W. Ding, F. Murray and T. E. Stuart, Gender Differences in Patenting in the Academic Life Sciences, *Science*, 313 (5787), 665 (2006) (evaluated a random sample of 4227 life scientists over a 30-year period, showing that female academic scientists patent at approximately 40% of the rate of men); G. Steven McMillan, Gender Differences in Patent Activity: An Examination of the US Biotechnology Industry, 80 *Scientometrics* 683 (2009) (finding a substantive patenting gender gap in the field of biotechnology even though the field is attractive to women for employment), at 684.

[145] Rainer Frietsch and others add that the highest shares were for Spain and France (12.3% and 10.2%, respectively). The lowest shares were for Austria and Germany (3.2% and 4.7% respectively). See Rainer Frietsch, Inna Haller, Melanie Funken-Vrohlings, and Hariolf Grupp, Gender-specific Patterns in Patenting and Publishing, *Research Policy*, 38 590 (2009).

[146] But see studies on rates of female patenting within localized specific scientific fields arguing for under-representation of women in specific fields of science. Waverly W. Ding, Fiona Murray, and Toby E. Stuart, Gender Differences in Patenting in the Academic Life Sciences, *Science*, 313, 665 (2006); Jerry G. Thursby and Marie Thursby, Gender Patterns of Research and Licensing Activity of Science and Engineering Faculty, *Journal of Technology Transfer* 30, 343 (2005); Kjersten Bunker Whittington and Laurel Smith-Doerr, Gender and Commercial Science: Women's Patenting in the Life Sciences. *Journal of Technology Transfer*, 30, 355 (2005).

[147] Kjersten Bunker Whittington, Gender and Scientific Dissemination in Public and Private Science: A Multivariate and Network Approach, Department of Sociology, Stanford University; Kjersten Bunker Whittington and Laurel Smith-Doerr, Women and Commercial Science: Women's Patenting in the Life Sciences, *Journal of Technology Transfer*, 30, 355 (2005) (the gender discrepancies in the propensity to publish or patent in academia compared with in industry is explained in terms of women's fewer opportunities to publish as they are less encouraged or because they choose less exploitable research areas), at 365, 366.

[148] See Fulvio Naldi and Ilaria Vannini Parenti, Scientific and Technological Performance by Gender (European Commission, 2002).

As part of the EU's intense efforts in the field, the 2011 European Commission's Innovation Union Competitiveness report named six hypotheses that set the foundations for future empirical findings.[149] Explaining the negative link between the proportion of women in R&D and the level of development of the country's innovative yield, these hypotheses could also offer considerable insight for future empirical analysis across the North–South divide. These hypotheses are the lower salaries of women researchers, lower-paid sectors of R&D, "feminine" subdivisions of R&D, a male "brain drain," higher overall levels of employment for women, and combinations of these factors.[150] Moreover, the lack of a strong social network has repeatedly been cited as another core explanation for the suppressed commercialization activities of women.[151]

One way in which European and American universities have responded to this is through the creation of Technology Transfer Offices (TTOs), which were established in the American instance in order to meet the demands of the US Bayh–Dole Act in promoting the commercial exploitation of inventions that result from government-funded research.[152] Since the Bayh–Dole Act, there has been an exponential increase in the number of TTOs at universities.[153] This important yet insufficient regulatory mechanism may explain the increase in the proportion of women inventorship over recent decades.[154]

---

[149] See European Commission, Innovation Union Competitiveness report 2011, supra note 40, at 213, referring to European Commission, Benchmarking policy measures for gender equality in science (2008).

[150] Ibid.

[151] Maria Abreau and Vadim Grinevich, The Nature of Academic Entrepreneurship in the UK: Widening the Focus on Entrepreneurial Activities, *Research Policy*, 42, 408 (2013); Anat BarNir, Starting Technologically Innovative Ventures: Reasons, Human Capital, and Gender, *Management Decision*, 50(3), 399 (2012); Dora Gicheva, Albert N. Link, Leveraging Entrepreneurship through Private Investments: Does Gender Matter? *Small Business Economics*, 40, 199 (2013); Jennifer Hunt, Jean-Philippe Garant, Hannah Herman, and David J. Munroe, Why Are Women Underrepresented amongst Patentees? *Research Policy*, 42, 831 (2013).

[152] Maria Abreau and Vadim Grinevich, The Nature of Academic Entrepreneurship in the UK: Widening the Focus on Entrepreneurial Activities, *Research Policy*, 42, 408 (2013).

[153] Inmaculada de Melo-Martin, Patenting and the Gender Gap: Should Women be Encouraged to Patent More? *Science and Engineering Ethics*, 19, 491 (2013).

[154] See Cassidy R. Sugimoto, Chaoqun Ni, Jevin D. West, Vincent Larivière, supra note 143, at 8, referring to Jordi Duch, Xiao Han T. Zeng, Marta Sales-Pardo, Filippo Radicchi, Shayna Otis, Teresa K. Woodruff and Luís A. Nunes Amaral, The Possible Role of Resource Requirements and Academic Career-Choice Risk on Gender Differences in Publication Rate and Impact, *PLoS ONE* 7(12): e51332 (2012) (adding that TTOs lead to reduced gender-related perceptions of risk); Waverly W. Ding, Fiona Murray and Toby E. Stuart, Gender Differences in Patenting in the Academic Life Sciences, *Science*, 313, 665 (2006) (arguing that TTOs incentivize academic

In conclusion, the magnitude of the gender gap in patent intensity raises the concern that, rather than reflecting comparative advantage or differing tastes by gender, the gap reflects gender inequity and an inefficient use of female innovative capacity.[155] That is especially so given the disparity between low female patenting rates and the relatively high rates of female with science or engineering degrees,[156] or with scientific publications.[157] This surely means that rates of return to female education are not only fragile, but should not serve as the sole basis for measuring and ultimately closing the gap.[158] However, there is relatively little evidence based on cross-country comparisons, and especially with an emphasis on disadvantaged countries and their endogenous growth indicators.[159]

Conclusion

Most studies done at the firm level confirm that firms with a higher proportion of scientists and engineers may show superior innovation-based economic growth which in correlation also increases these firms' patenting rates. Human capital literature, regrettably, continuously

---

entrepreneurship based on the facilitation the construction of collegial networks and organizational support).

[155] Jennifer Hunt, Jean-Philippe Garant, Hannah Herman and David J. Munroe, Why Don't Women Patent?, IZA Discussion Paper No. 6886 (September 2012), at 1.

[156] Ibid. ("Only 7 percent of the gender gap in commercialized patents is owing to women's under-representation in S&E, compared to 78 percent owing to the patenting rate gap among holders of S&E degrees"), at 15.

[157] Rainer Frietsch, Inna Haller, Melanie Funken-Vrohlings, and Hariolf Grupp, supra note 145 (the share of women penning scientific publications in 2005 stood at 24.1% in the United States and 19.2% in Germany, yet women's share in patent applications for the same year was much lower at 8.2% in the United states and 4.9% in Germany), at 17 and table 4.

[158] See John E. Knodel and Gavin W. Jones, Post-Cairo Population Policy: Does Promoting Girl's Schooling Miss the Mark?, *Population and Development Review*, 22(4): 683 (1996) (adding that rates of return are extremely fragile basis on which to justify investment in female education, quite apart from providing little or no guidance on how to succeed in achieving expansion of, and parity in, educational provision); Sally Baden and Anne Marie Goetz, Who needs [Sex] when you can have [Gender]? Conflicting Discourses on Gender at Beijing, Feminist Review 56 (Summer): 3 (1997) ("Tenuous evidence on the relationship between female education and fertility decline, or female education and productivity, can easily be challenged, weakening the justification for addressing gender issues, with a danger that resources will be withdrawn"), at 10.

[159] For a case study of Malaysia's cultural differences in women in sciences, see Ulf Mellström, The Intersection of Gender, Race and Cultural Boundaries, or Why Is Computer Science in Malaysia Dominated by Women?, *Social Studies of Science*, vol. 39(6) 885 (2009) (offering a critique of a western bias in gender and technology studies, based on the Malaysian case study, arguing for more cultural context sensitivity).

focuses on developed or advanced countries, while the literature largely overshadows developing countries including emerging economies.

The analysis in this chapter portrays a country-level analysis using six yearly series. These are: (1) GERD per FTE researcher, (2) GERD per HC researcher per country, (3) government expenditure per tertiary student as percentage of GDP per capita (in percentage), (4) scientific and technical journal articles per national researcher (HC), (5) scientific and technical journal articles per national researcher FTE, and, lastly (6) number of researchers (HC) per labor force per country. Except for the series on government expenditure per tertiary student as a percentage of GDP per capita, for which the available data starts in 1998, all the series cover the period 1996–2013. The findings of this chapter show that except for government expenditure per tertiary student as a percentage of GDP per capita for which there were no differences among clusters, for all other variables the leaders cluster had higher mean values compared with the followers patent cluster.

For three out of the six series, the leaders cluster had also higher mean values compared with the marginalized cluster. These three were GERD per HC researcher, articles per HC researcher, and HC researcher per labor force. For GERD per FTE researcher, there were no significant changes over time. For government expenditure per tertiary student as a percentage of GDP per capita, there were the same changes in time for all three patent clusters. For all other series, there were changes along time that differed among cluster. The significant difference between clusters were also tested for each year. A consistently significant difference among clusters was found for each of the years for the three series: GERD per FTE researcher, GERD per HC researcher, and HC researcher per labor force. For the two articles' series the difference became non-significant toward the end of the series. These findings contribute to the understanding of how the three patent clusters differ also in terms of their patent intensity in relation to human capital measurements.

Appendix F Employment and Human Resources by Patent Cluster

F.1 Regression Models

It should be emphasized that the models were fitted to the log transformed data. For each series, a graphical display of the predicted values for each cluster (on the log scale) is presented, based on the fitted model. These graphs describe the relationship of the dependent variable to cluster and time.

F.1.1   Log GERD per FTE Researcher

No significant changes in time for all three clusters. Significant difference was found between leaders and followers patent clusters. Higher values exist for the leaders cluster.

The Mixed Procedure

Type 3 Tests of Fixed Effects

Effect	Num DF	Den DF	F Value	Pr > F
Year	15	690	0.57	0.8991
r_cluster_km	2	46	13.70	<.0001
r_cluster_km*year	30	690	0.53	0.9816

Least Squares Means

Effect	r_cluster_km	Estimate	Standard Error	DF	t Value	Pr > |t|
r_cluster_km	1	5.2356	0.06698	46	78.16	<.0001
r_cluster_km	2	4.5243	0.1292	46	35.02	<.0001
r_cluster_km	3	4.5556	0.2757	46	16.52	<.0001

Differences of Least Squares Means

Effect	r_cluster_km	_r_cluster_km	Estimate	Standard Error	DF	t Value	Pr > |t|
r_cluster_km	1	2	0.7113	0.1455	46	4.89	<.0001
r_cluster_km	1	3	0.6800	0.2837	46	2.40	0.0207
r_cluster_km	2	3	−0.03127	0.3045	46	−0.10	0.9186

Differences of Least Squares Means

Effect	r_cluster_km	_r_cluster_km	Adjustment	Adj P
r_cluster_km	1	2	Bonferroni	<.0001
r_cluster_km	1	3	Bonferroni	0.0620
r_cluster_km	2	3	Bonferroni	1.0000

Though according to the graph there is a slight increase in time for clusters 2 and 3, and there also seems to be a difference between cluster 1 and 3, these were not strong enough to be statistically significant.

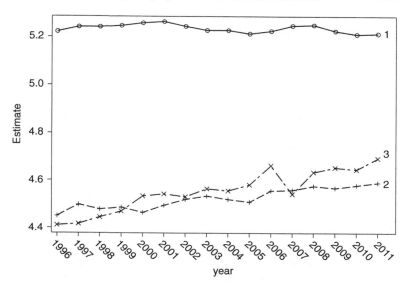

F.1.2   Log GERD per HC Researcher

There are significant changes in time differing among clusters, and on average an increase with time. There is a significant difference between clusters with higher values for cluster 1 compared with clusters 2 and 3.

The Mixed Procedure

Type 3 Tests of Fixed Effects

Effect	Num DF	Den DF	F Value	Pr > F
year	15	630	2.25	0.0044
r_cluster_km	2	42	39.84	<.0001
*r_cluster_km*year*	30	630	2.02	0.0012

Least Squares Means

Effect	r_cluster_km	Estimate	Standard Error	DF	t Value	Pr > |t|
r_cluster_km	1	4.9544	0.07110	42	69.68	<.0001
r_cluster_km	2	3.9003	0.1051	42	37.09	<.0001
r_cluster_km	3	3.5324	0.3283	42	10.76	<.0001

Differences of Least Squares Means

Effect	r_cluster_km	_r_cluster_km	Estimate	Standard Error	DF	t Value	Pr > |t|
r_cluster_km	1	2	1.0541	0.1269	42	8.30	<.0001
r_cluster_km	1	3	1.4219	0.3359	42	4.23	0.0001
r_cluster_km	2	3	0.3678	0.3447	42	1.07	0.2920

Differences of Least Squares Means

Effect	r_cluster_km	_r_cluster_km	Adjustment	Adj P
r_cluster_km	1	2	Bonferroni	<.0001
r_cluster_km	1	3	Bonferroni	0.0004
r_cluster_km	2	3	Bonferroni	0.8761

The graph shows the difference between the clusters in the changes along time, as well as in their level, with a significantly higher mean in cluster 1, compared with clusters 2 and 3.

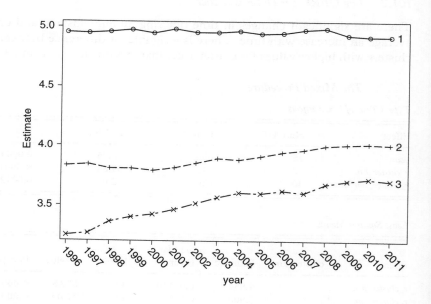

F.1.3   Log Expenditure per Tertiary Student as Percentage of GDP Per Capita (in Percentage)

There are significant changes in time for all 3 clusters with a decrease in time. There is a significant difference between the three clusters.

The Mixed Procedure

Type 3 Tests of Fixed Effects

Effect	Num DF	Den DF	F Value	Pr > F
Year	13	533	4.45	<.0001
r_cluster_km	2	41	0.05	0.9482
r_cluster_km*year	26	533	0.68	0.8879

Least Squares Means

Effect	r_cluster_km	Estimate	Standard Error	DF	t Value	Pr > |t|
r_cluster_km	1	3.3740	0.1216	41	27.74	<.0001
r_cluster_km	2	3.4208	0.1026	41	33.35	<.0001
r_cluster_km	3	3.3668	0.2292	41	14.69	<.0001

Differences of Least Squares Means

Effect	r_cluster_km	_r_cluster_km	Estimate	Standard Error	DF	t Value	Pr > |t|
r_cluster_km	1	2	−0.04673	0.1591	41	−0.29	0.7705
r_cluster_km	1	3	0.007192	0.2595	41	0.03	0.9780
r_cluster_km	2	3	0.05392	0.2511	41	0.21	0.8311

Differences of Least Squares Means

Effect	r_cluster_km	_r_cluster_km	Adjustment	Adj P
r_cluster_km	1	2	Bonferroni	1.0000
r_cluster_km	1	3	Bonferroni	1.0000
r_cluster_km	2	3	Bonferroni	1.0000

The graph shows the decrease in time for all the clusters, with no significant difference between the clusters.

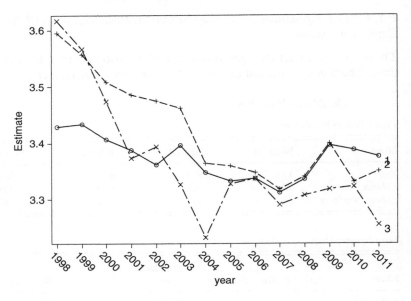

F.1.4   Log Scientific and Technical Journal Articles per Researcher HC

There were significant changes in time for a decrease on average, but differing among clusters. Higher values for cluster 1 compared with clusters 2 and 3.

The Mixed Procedure

Type 3 Tests of Fixed Effects

Effect	Num DF	Den DF	F Value	Pr > F
year	15	615	55.12	<.0001
r_cluster_km	2	41	8.30	0.0009
r_cluster_km*year	30	615	5.06	<.0001

Least Squares Means

Effect	r_cluster_km	Estimate	Standard Error	DF	t Value	Pr > |t|
r_cluster_km	1	−1.6522	0.1332	41	−12.41	<.0001
r_cluster_km	2	−2.2851	0.1325	41	−17.24	<.0001
r_cluster_km	3	−2.6379	0.2760	41	−9.56	<.0001

Differences of Least Squares Means

Effect	r_cluster_km	_r_cluster_km	Estimate	Standard Error	DF	t Value	Pr > |t|
r_cluster_km	1	2	0.6329	0.1879	41	3.37	0.0017
r_cluster_km	1	3	0.9858	0.3064	41	3.22	0.0025
r_cluster_km	2	3	0.3528	0.3061	41	1.15	0.2558

Differences of Least Squares Means

Effect	r_cluster_km	_r_cluster_km	Adjustment	Adj P
r_cluster_km	1	2	Bonferroni	0.0050
r_cluster_km	1	3	Bonferroni	0.0076
r_cluster_km	2	3	Bonferroni	0.7674

The graph shows the difference between the clusters in the changes along time, as well as in their level, with a significantly higher mean in cluster 1, as compared with clusters 2 and 3. The graph shows the increase in time which looks stronger in cluster 3. When we examine the difference between clusters for each year, we can see in the graph that the difference between clusters gets smaller in time. Tests of these differences (slices according to the MIXED procedure) indicated a non-significant difference among clusters in the end of the series (2009−2011, P>.05).

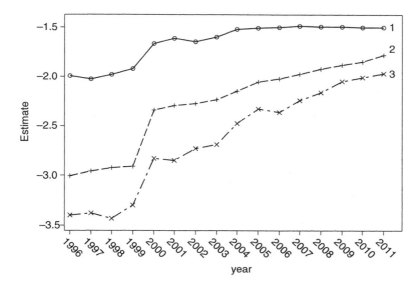

F.1.5   Log Scientific and Technical Journal Articles per Researcher FTE

There are significant changes in time of a decrease, but differing among clusters. There are higher values for cluster 1 compared with cluster 2.

The Mixed Procedure

Type 3 Tests of Fixed Effects

Effect	Num DF	Den DF	F Value	Pr > F
year	15	690	41.00	<.0001
r_cluster_km	2	46	4.08	0.0233
r_cluster_km*year	30	690	4.26	<.0001

Least Squares Means

Effect	r_cluster_km	Estimate	Standard Error	DF	t Value	Pr > |t|
r_cluster_km	1	-1.2708	0.08821	46	-14.41	<.0001
r_cluster_km	2	-1.6398	0.1113	46	-14.73	<.0001
r_cluster_km	3	-1.7437	0.2694	46	-6.47	<.0001

Differences of Least Squares Means

Effect	r_cluster_km	_r_cluster_km	Estimate	Standard Error	DF	t Value	Pr > |t|
r_cluster_km	1	2	0.3691	0.1420	46	2.60	0.0125
r_cluster_km	1	3	0.4730	0.2834	46	1.67	0.1020
r_cluster_km	2	3	0.1039	0.2915	46	0.36	0.7231

Differences of Least Squares Means

Effect	r_cluster_km	_r_cluster_km	Adjustment	Adj P
r_cluster_km	1	2	Bonferroni	0.0376
r_cluster_km	1	3	Bonferroni	0.3059
r_cluster_km	2	3	Bonferroni	1.0000

The graph of the predicted values according to the fitted model shows that the difference between the clusters changes over time, as well as their level. Though the mean difference is larger between clusters 1 and 3 compared with the difference between clusters 1 and 2, the significance is only reached for the latter, since the standard error of the estimated means is much larger for cluster 3 compared with that of cluster 2. Tests for each year of the difference between clusters (slices according to the MIXED procedure) demonstrate what we can see in the graph: that the difference gets smaller along time. The significant difference occurs until 2005 ($P<0.05$), and then gradually decreases till 2011 when its p–value is 0.92.

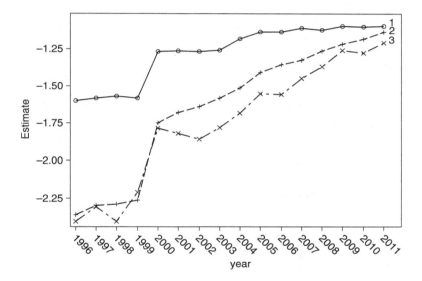

F.1.6   Log Number of Researchers (HC) per Labor Force

There are significant changes in time but differing among clusters with a slight increase on average. There are higher values for cluster 1 compared with clusters 2 and 3.

The Mixed Procedure

Type 3 Tests of Fixed Effects

Effect	Num DF	Den DF	F Value	Pr > F
Year	15	630	12.84	<.0001
r_cluster_km	2	42	19.46	<.0001
r_cluster_km*year	30	630	1.74	0.0090

Least Squares Means

Effect	r_cluster_km	Estimate	Standard Error	DF	t Value	Pr > |t|
r_cluster_km	1	−4.5958	0.1323	42	−34.74	<.0001
r_cluster_km	2	−5.6007	0.1874	42	−29.89	<.0001
r_cluster_km	3	−6.3735	0.3065	42	−20.79	<.0001

Differences of Least Squares Means

Effect	r_cluster_km	_r_cluster_km	Estimate	Standard Error	DF	t Value	Pr > |t|
r_cluster_km	1	2	1.0049	0.2294	42	4.38	<.0001
r_cluster_km	1	3	1.7777	0.3338	42	5.32	<.0001
r_cluster_km	2	3	0.7728	0.3592	42	2.15	0.0373

Differences of Least Squares Means

Effect	r_cluster_km	_r_cluster_km	Adjustment	Adj P
r_cluster_km	1	2	Bonferroni	0.0002
r_cluster_km	1	3	Bonferroni	<.0001
r_cluster_km	2	3	Bonferroni	0.1118

The graph shows the difference between the clusters in the changes along time, as well as in their level, with a significantly higher mean in cluster 1 compared with clusters 2 and 3. There is an increase in time in cluster 1.

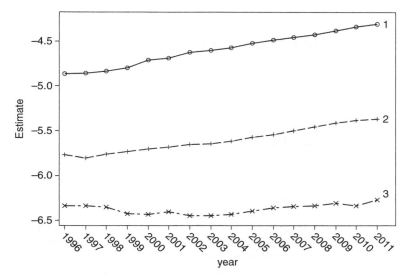

In conclusion, except for Government expenditure per tertiary student as a percentage of GDP per capita, for which there were no differences among clusters, for all other variables cluster 1 had higher mean values compared with cluster 2. For 3 out of the 6 series, cluster 1 had also higher mean values compared with cluster 3. These three were GERD per researcher HC, articles per researcher HC, and researcher HC.

For GERD per researcher FTE, there were no significant changes along time. For Government Expenditure per Tertiary student as percent of GDP per capita, there were the same changes in time for all three clusters. For all other series, there were changes along time that differed among clusters. The significant differences between clusters were also tested for each year. A consistently significant difference was observed among clusters for each of the years for the three series: GERD per researcher FTE, GERD per researcher HC, and Researcher HC per Labor Force. For the two articles series, the difference became non-significant toward the end of the series.

6　Spatial Agglomeration of Innovation and Patents

Introduction

Cluster analysis of patenting intensity is also characterized spatially. Recent years have seen growing interest among economic geographers in the uneven spatial patterns of innovation.[1] As this intellectual trail still lacks a comprehensive theoretical framework, it is characterized by shared themes revolving around the inability of conventional geographies to explain fully the complex character of innovation-based growth patterns, particularly beyond the urban and country levels.

Innovation theory offers two spatial research strands as a strategic decision to access and absorb relevant location-based externalities. The first focuses on the determinants of the location choice to start new innovative firms. This usually appears in econometric models as a distinct dependent variable, namely the location of a new firm, positioned against a set of explanatory variables that aim to describe location determinants, which may include R&D activities. The second tradition appeals to the

---

[1] See, e.g., Peter Thompson and Melanie Fox-Kean, Patent Citations and the Geography of Knowledge Spillovers: A Reassessment, *American Economic Review*, vol. 95(1) 450 (2005); Maryann P. Feldman, The New Economics of Innovation, Spillovers and Agglomeration: A Review of Empirical Studies, *Economics of Innovation and New Technology* 8, 5 (1999), at 6–7; David Audretsch and Maryann P. Feldman, Knowledge Spillovers and the Geography of Innovation and Production, Discussion Paper 953, London: Centre for Economic Policy Research (1994); Adam B. Jaffe, Manuel Trajtenberg and Rebecca Henderson, Geographical Localization of Knowledge Spillovers as Evidenced by Patent Citations, *Quarterly Journal of Economics* 108, 577 (1993).

On the broader context of economic geography, see also Paul Krugman, *Geography and Trade* (Leuven University Press, 1991); Paul Krugman, On the Relationship between Trade Theory and Location Theory, *Review of International Economics 1*, 110 (1993); Paul Krugman, The "New Theory" of International Trade and the Multinational Enterprise, in *The Multinational Corporation in the 1980s* (Charles Kindleberger, ed.) 57 (MIT Press, 1983); Michael E. Porter, Location, Competition and Economic Development: Local Clusters in a Global Economy, *Economic Development Quarterly* 14(1) 15 (2000); Michael E. Porter, Clusters and Competition: New Agendas for Companies, Governments, and Institutions 197, in Michael E. Porter, *On Competition* (Harvard Business School Press, 1998); Michael Storper, *The Regional World: Territorial Development in a Global Economy* (Guildford Press, 1997); Michael J. Piore, Charles F. Sabel, *The Second Industrial Divide* (Basic Books, 1984).

concept of geographically mediated spillovers and includes a geographic dimension to the determinants of innovation as an output. Usually, it draws on the production function approach and uses some measure of innovation as the dependent variable against a set of possible explanatory variables. This chapter similarly uses this production function approach, employing this book's three-patent-cluster framework as the dependent variables.

The entire body of literature on economic geography and regional innovation system theory has to date focused on advanced countries, while traditionally assuming equal policy ramifications across the development divide. In place of this approach, this chapter offers an analytical framework of regional agglomeration of innovation and patenting including developing countries. It does so based on two measurements of spatial innovation: the degree of newness, and related patent quality across patent clusters.

6.1 The Positive Framework: Patenting Between Economic Geography and Innovation Theory

Economic geographers regularly measure trade according to location-specific externalities. In particular, these follow the prolific exponent of the new trade theory Paul Krugman in assigning key significance to the part that the domestic national geography may play in determining its trade pattern.[2]

Economic geographers' efforts have also followed a second avenue concerning the competitive advantage between trading nations. The latter is most distinctively related to the renowned business economist Michael Porter, who foretold the role of national geographical clustering of industries on international competitive advantage.[3] Krugman and others argue, therefore, that to understand trade it is necessary to measure the processes leading to the local and regional concentration

---

[2] See Paul Krugman, Geography and Trade, supra note 1; For a critical assessment of Krugman's insight, see also: Ron Martin and Peter Sunley, Paul Krugman's Geographical Economics and Its Implications for Regional Development Theory, *Economic Geography*, vol. 72(3) 259 (1996); Anthony G. Hoare, Review of Paul Krugman's "Geography and Trade," *Regional Studies* 26, 679 (1992). Martin and Sunley add that Krugman's new trade theory offers a rapprochement between trade theory and location theory, in reference to Paul Krugman, The Current Case for Industrial Policy 160 in *Protectionism and World Welfare* (Dominick Salvatore, ed.) (Cambridge University Press, 1993), at 263.

[3] Michael E. Porter, The Competitive Advantage of Nations (Macmillan, 1990); Michael E. Porter, The Role of Location in Competition, *Journal of the Economics of Business* 1(1) 35 (1994). Porter, like Krugman, argues that economic geography should become a "core discipline in economics." Ibid., at 790; Cf. Paul Krugman, Geography and Trade, supra note 1, at 33.

of production. To this end, Krugman, like other economic geographers, draws on a range of geographical economic ideas, most of which were derived from Sir Alfred Marshall's 1890 illustrious book *Principles of Economics*.[4]

Writing at the end of the nineteenth century, Marshall can surely not have foreseen the immense impact of his work on innovation theory. His work focused strongly on traditional traded goods and has been influential in regional and urban studies.[5] Most noticeably, his work has led to the concept of "Marshallian externalities" – knowledge spillovers accounting for the benefits associated with cluster formation between firms, innovative or not.

The now-standard classification of Marshallian externalities is attributed to economist Edgar Malone Hoover in *Location Theory and the Shoe and Leather Industries*, published in 1936 by Harvard University Press.[6] Hoover distinguishes between two types of location-related Marshallian externalities, namely localization economies and urbanization economies.[7] Localization economies reflect the benefits generated by the proximity of firms producing similar goods.[8] Simply put, within a geographic region, localization economies are exterior to a firm but interior to an industry. Black and Henderson introduced Marshall's idea that agglomeration economies are the result of positive spillovers between firms that share the same location, namely localization economies.[9]

Analogously, in 1992, Edward Glaeser, Hedi Kallal, José Scheinkman, and Andrei Shleifer elaborated the Marshall-Arrow-Romer (MAR)

---

[4] Alfred Marshall, *Principles of Economics* (Macmillan, 8th edn., 1920) (Original work published 1890). Marshall accounts for concepts as localization economies, through traditional location theories, to notions of cumulative causation which remain outside the scope of this book. On the relevancy today of Marshall's pioneering work on British industry, see also, Jacques-Laurent Ravix, Localization, Innovation and Entrepreneurship: An Appraisal of the Analytical Impact of Marshall's Notion of Industrial Atmosphere, *Journal of Innovation Economics & Management* 2014/2 (n°14), 63 (2014), at 69–70; Fiorenza Belussi and Katia Caldari, At the Origin of the Industrial District: Alfred Marshall and the Cambridge School, *Cambridge Journal of Economics, 33*, 335 (2009), at 338–339. See also discussion herein.

[5] See, e.g., Alfred Marshall, supra note 4, Book IV, Ch. X, §3, at 225.

[6] Edgar Malone Hoover, *Location Theory and the Shoe and Leather Industries* (Harvard University Press, 1936), chapter 6 (titled: *Economies of concentration*), at 89–115, at 90.

[7] Ibid.

[8] Localization economies relate to economies of intra-industry specialization that allow a finer division of function among firms, labor market economies that reduce search costs for firms seeking workers with specific training, and communication economies that can speed up adoption of innovations. Ibid.

[9] Duncan Black and J. Vernon Henderson, A Theory of Urban Growth, *Journal of Political Economy*, vol. 107(2) 252 (1999).

knowledge spillover between firms in an industry.[10] According to the MAR spillover view, the proximity of firms within a common industry often affects how well knowledge passes among firms to ease innovation and growth.[11] The exchange of ideas largely occurs from employee to employee through imitation.[12] It also travels by means of espionage and through rapid inter-firm movement of highly skilled employees from different firms.

As MIT economists Eric von Hippel,[13] AnnaLee Saxenian,[14] and later also Maryann Feldman and Frank Lichtenberg[15] explain, such forms of tacit knowledge have a higher degree of uncertainty than tangible knowledge. This is also what makes face-to-face interaction and communication so important when geographic proximity is accounted for.[16] In other words, the less codified and tangible the knowledge, the greater the degree of centralization in geographic organization. These MAR knowledge spillovers also serve as the focal point of innovation and patenting-related literature, as this chapter explains.

The second type of Marshallian externalities are known as the urbanization economies, which are defined by all the advantages associated with the overall level of activity prevailing in a particular urban area among technologies and sectors. These archetypal inter-industry externalities reflect the benefits from operating in large population centers with correspondingly large overall labor markets and large diversified service sectors

---

[10] See Edward Glaeser, Hedi Kallal, José Scheinkman, and Andrei Shleifer, Growth in Cities, *The Journal of Political Economy*, vol. 100(6) Centennial Issue 1126 (1992).

[11] See Edward Glaeser, Hedi Kallal, José Scheinkman, and Andrei Shleifer, ibid., at 1127. See also Gerald A. Carlino, Knowledge Spillovers: Cities' Role in the New Economy, *Business Review*, Q4 17 (2001).

[12] Carlino. supra note 11, at 1127. Carlino adds that the closer the firms are to one another, the greater the MAR spillover.

[13] See Eric Von Hipple, Sticky Information and the Locus of Problem Solving: Implications for Innovation, *Management Science 40*, 429 (1994).

[14] See AnnaLee Saxenian, *Regional Advantage: Culture and Competition in Silicon Valley and Route 128* (Harvard University Press, 1994). Saxenian describes how gathering places, such as the Wagon Wheel Bar located only a block from Intel, Raytheon, and Fairchild Semiconductor, served as informal recruiting centers as well as listening posts. In her book, Saxenian offers other examples, such as the Route 128 corridor in Massachusetts, the Research Triangle in North Carolina, and suburban Philadelphia's biotechnology research and medical technology industries.

[15] See Maryann P. Feldman and Frank R. Lichtenberg, The Impact and Organization of Publicly-Funded Research and Development in the European Community, NBER Working Paper 6040) (1997) (Using data on the results of publicly supported R&D projects in the European Community to construct several indicators of tacitness, authors indicate that the more tacit the knowledge generated by the R&D, the more geographically centralized R&D activities become), referring also to Nathan Rosenberg, *Perspectives on Technology* (Cambridge University Press, 1976) (concerning the tacitness of technological knowledge), at 78.

[16] See Eric Von Hipple, supra note 13, at 442.

to interact with manufacturing.[17] Urbanization economies, in other words, are defined as scale effects associated with city size or density.

Historical data have consistently shown that patent originations have indeed been concentrated to a large extent in cities and metropolises. In his Harvard University Press book *The Spatial Dynamics of US Urban Industrial Growth, 1800–1914*, published in 1966, Allen Pred examined US patent data for the mid-nineteenth century, showing that patent activity in the three-principal cities just then was four times greater than the national average. In 1971 Robert Higgs showed that the number of US-granted patents between 1870 and 1920 was positively related to urbanization intensity.[18]

In a more recent seminal study, Maryann Feldman and David Audretsch used the US Small Business Administration's innovation data-base, which is composed of innovations collected from new product announcements in manufacturing trade journals. The authors showed that in 1982 only 150 of the innovations (4 percent) covered by their data set occurred outside of metropolitan areas. Almost half of all innovations occurred in four metropolitan areas, namely New York (18.5 percent), San Francisco (12 percent), Boston (8.7 percent), and Los Angeles (8.4 percent).[19]

The empirical performance of innovation in geographic locations as one source of sustained growth is still incomplete.[20] A few studies on balance have shown that, paradoxically, not locating in a cluster may actually hold some advantages, by allowing firms to introduce new products earlier than their competitors and preserve their privacy.[21]

---

[17] Edgar Malone Hoover, supra note 6, chap. 6.

The analysis of externalities, known also as knowledge spillovers, reside alongside two additional topics. The first of these is Marshall's "economies of specialization", whereby a localized industry can support a greater number of specialized local suppliers of industry-specific intermediate inputs and services. In this way it could obtain a greater variety at a lower cost. This book's empirics do not offer an industry-specific analysis, given data coverage constraints due to an emphasis on developing countries. The second Marshallian geographic account is known as "labor market economies." It foretells how localized industries attract and create pools of workers with similar skills, smoothing the effects of the business cycle (both on unemployment and wages) through the effects of large numbers. This argument loosely corresponds with human capital analysis which is discussed in Chapter 5 of this volume.

[18] See Allen R. Pred, *The Spatial Dynamics of US Urban Industrial Growth, 1800–1914* (Harvard University Press, 1966).

[19] See Maryann P. Feldman and David B. Audretsch, Innovation in Cities: Science Based Diversity, Specialization and Localized Competition, *European Economic Review* 409 (1999).

[20] See, e.g., Morgan Kelly and Anya Hageman, Marshallian Externalities in Innovation, *Journal of Economic Growth*, vol. 4(1) 39 (1999), at 39.

[21] Ray P. Oakey and Sarah. Y. Cooper, High Technology Industry, Agglomeration and the Potential for Peripherally Sited Small Firms, *Regional Studies*, 23, 347 (1989); Luis

In particular, Suarez-Villa and Walrod found that non-clustered electronic establishments spent on average 3.6 times more on R&D and employed 2.5 times more R&D personnel than clustered ones.[22]

In direct relation with innovation theory, there are two strands of research that focus on location choice as a strategic decision to access and absorb location-based externalities.[23] The first focuses on the determinants of the location choice to start new innovative firms. Usually, it appears to econometric models with a distinct depend variable, namely the location of a new firm against a set of explanatory variables that aim to describe location determinants, which may include R&D activities. Most works focus on new firms, whereas its relevance may vary with firms' R&D intensity.[24] Most importantly, because geographic characteristics are exogenous it is possible to disentangle effects.[25]

The second tradition appeals to the concept of geographically mediated spillovers. It includes a geographic dimension to the determinants of innovation as an output. Usually, it draws on the production function approach and uses some measure of innovation as the dependent variable against a set of possible explanatory variables. The knowledge production function also implies that innovative activity should cluster in regions where knowledge-generating inputs are largest and thus where knowledge spillovers are most common.[26]

The groundbreaking work in this field was pioneered by Adam Jaffe,[27] who modified Zvi Griliches'[28] knowledge production function to include

Suarez-Villa and Wallace Walrod, Operational Strategy, R&D and Intra-metropolitan Clustering in a Polycentric Structure: The Advanced Electronics Industries of the Los Angeles Basin, *Urban Studies*, 34, 1343 (1997).

[22] Luis Suarez-Villa and Wallace Walrod, supra note 22.

[23] See Isabel Mota and António Brandão, Modeling Location Decisions – The Role of R&D Activities, European Regional Science Association in its series ERSA conference papers No. ersa05p612 (2005), at 2. Cf. Maryann P. Feldman, supra note 1, at 6.

[24] This approach remains outside the scope of this book's analysis. For literature on this account, see, e.g., David B. Audretsch, Erik E. Lehmann, and Susanne Warning, University Spillovers: Strategic Location and New Firm Performance, Discussion Paper 3837, CORE (2003) and Douglas Woodward, Octávio Figueiredo, and Paulo Guimarães, Beyond the Silicon Valley: University R&D and High-Technology Location, *Journal of Urban Economics*, vol. 60(1) 15 (2006).

[25] Maryann P. Feldman, supra note 1, at 6–7.   [26] Ibid., at 8.

[27] See Adam B. Jaffe, Real Effects of Academic Research, *American Economic Review* 79(5), 957 (1989)). But see more preliminary work as of 1962 by Wilbur R. Thompson, Locational Differences in Inventive Efforts and Their Determinants 253, in Richard R. Nelson (ed.), *The Rate and Direction of Inventive Activity: Economic and Social Factors* (Princeton University Press, 1962).

[28] Zvi Griliches, Issues in Assessing the Contribution of R&D to Productivity, *Bell Journal of Economics*, vol. 10 92 (1979).

a geographical dimension by measuring local patent counts[29] or other forms of R&D and innovation-related counts.[30] Aiming to assess the real effects of local academic research, Jaffe first reclassified patents into a restricted number of technological areas, and then showed that the number of patents for each US state for each technological area is a positive function of the R&D performed by local universities. The relationship between patents and university R&D is then interpreted as a sign of the existence of some localized knowledge spillovers from academic institutions into the local business sector.[31]

---

[29] For studies measuring local patent counts, see Anthony Arundel and Aldo Geuna, Does Proximity Matter for Knowledge Transfer from Public Institutes and Universities to Firms?, Working Paper 73, SPRU (2001); Attila Varga, Local Academic Knowledge Transfers and the Concentration of Economic Activity, *Journal of Regional Science*, vol. 40 (2) 289 (2000); Bart Verspagen and Wilfred Schoenmakers, The Spatial Dimension of Knowledge Spillovers in Europe: Evidence from Patenting Data,' paper presented at the AEA Conference on Intellectual Property Econometrics, Alicante, 19–20 April (2000); Morgan Kelly and Anya Hageman, Marshallian Externalities in Innovation, *Journal of Economic Growth*, 4 (March), 39 (1999) (using US patent counts at the state level, showing that patent intensity exhibits strong spatial clustering independently of the distribution of employment, and that 'knowledge spillovers' are important determinants of a state's innovative performance); Botolf Maurseth and Bart Verspagen, Knowledge Spillovers in Europe: A Patent Citation Analysis, paper presented at the CRENOS Conference on Technological Externalities and Spatial Location, University of Cagliari, September 24–25 (1999); Paul Almeida and Bruce Kogut, Localization of Knowledge and the Mobility of Engineers in Regional Networks, *Management Science* 45, 905 (1999) (finding evidence that knowledge spillovers from university research to firms are highly localized per semiconductor patent citations). See, notably, Adam B. Jaffe, M. Trajtenberg, and R. Henderson, supra note 1. Using patent citations, Jaffe and others track direct knowledge flows from academic research into corporate R&D. They find that innovative firms are more likely to quote research from a co-localized university that conducts relevant research than from equivalent universities located somewhere else.

[30] Acs and others build upon this last point and replicate Jaffe's exercise by substituting patents with innovation counts, derived from the Small Business Innovation Data Base (SBDIB). The authors show that innovation counts may capture the effect of geographical coincidence that eluded patents. See Zoltan J. Acs, David B. Audretsch, and Maryann P. Feldman, Real Effects of Academic Research: Comment, *American Economic Review*, 82, 363 (1992). The authors offer two distinct innovation production functions, namely for large firms and for small ones. They find that "geographical coincidence" is significant only for small firms, and suggest that this is due to university R&D alternatives for firms' in-house R&D, which is too costly for small firms. See also Isabel Mota and António Brandão, supra note 23 (evaluating the importance of R&D for firms' decision about location while using micro-level data for the Portuguese industrial sector and focus on the location choices made by new starting firms during 1992–2000 within 275 municipalities); Maryann P. Feldman, and David B. Audretsch, R&D Spillovers and the Geography of Innovation and Production, *American Economic Review*, vol. 86(3) 630 (1996) (shows that the geographical concentration of the innovation output is positively related to the R&D intensity), updating Feldman's previous work at Maryann P. Feldman, *The Geography of Innovation* (Kluwer, 1994).

[31] See Adam B. Jaffe, Real Effects of Academic Research, *American Economic Review* 79(5), 957 (1989).

Geographical knowledge spillovers were thus far measured mostly twofold: firstly, in terms of studies on the linkages between patent citations, defined as paper trails; secondly, in terms of studies that measure the mobility of skilled labor, or more simply innovation based on the notion that knowledge spillovers are transmitted through people.

The first area is addressed by studies on the geographical linkages between patent citations, defined as paper trails, and the location of inventors. Perhaps because, as Paul Krugman has argued, "knowledge flows ... are invisible; they leave no paper trail by which they may be measured and tracked,"[32] the measurement of knowledge spillovers has proved a most challenging task. The challenge was taken up most prominently by Jaffe, Trajtenberg, and Henderson,[33] and others that followed,[34] who all pointed out that knowledge spillovers may well leave a paper trail in the citations to prior art incorporated in patents. Furthermore, as patents refer to the country of residence of the inventors, patents become a valuable measurement tool of geographic knowledge flows. Jaffe, Trajtenberg, and Henderson undertook the considerable task of constructing a large dataset of patents and matching the locations of their inventors to the locations of inventors of all patents that subsequently cited them as prior art. In a study that remains seminal, the authors trace the pattern of patent citations to explore both the temporal and geographic span of knowledge spillovers.

The authors show that knowledge spillovers are indeed localized.[35] Cited patents were up to two times more likely to come from the same state, and up to six times more likely to come from the same metropolitan area.[36] Building on Manuel Trajtenberg's approach of linking patent

---

[32] Paul R. Krugman, *Geography and Trade* (MIT Press, 1991), at 53.

[33] Adam B. Jaffe, Manuel Trajtenberg, and Rebecca Henderson, supra note 1. Starting with 1,450 patents that originated in 1980, Jaffe, Trajtenberg, and Henderson trace the characteristics of approximately 5,200 citations to these originating patents from 1980 to 1989. Peter Thompson and Melanie Fox-Kean offered a methodological critique of the Jaffe et al. study. See Peter Thompson and Melanie Fox-Kean, supra note 1 (concluding that there is no guarantee that the control patent has any industrial similarity either to the citing or to the originating patent) at 3.

[34] See, Bronwyn H. Hall and Rosemarie H. Ziedonis, The Patent Paradox Revisited: An Empirical Study of Patenting in the US Semiconductor Industry, 1979–95, RAND *Journal of Economics* 32, 101 (2001) (for an examination of the patenting patterns of 95 US semiconductor firms between 1979 and 1995 showing that patenting is largely a metropolitan phenomenon). Bronwyn and Ziedonis find that during the 1990s, 92% of all patents were granted to residents of metropolitan areas, although only about three-quarters of the US population resides in metropolitan areas. See also the sources in Fn. 29.

[35] Ibid., (concluding that patented inventions as a form of knowledge flows "do sometimes leave a paper trail"), at 578.

[36] Ibid.

applications to other patents that reference or cite them,[37] the three authors upheld that citations are significantly localized by each geographic measurement unit and by different organizational types, such as universities, top corporations, and other corporations.

There are other factors that condition localization upon fostering innovation. Noticeably, the effect of citing patents fades with time as citations show less geographic effects as knowledge diffuses. In a follow-up study, Jaffe and Trajtenberg found that electronics, optics, and nuclear technology have high instant citation but, due to quick obsolescence, the speedy desertion of citations can be observed over time. More generally, the two researchers highlight the conditions on spillovers as the frequency and duration of citations depends on the scientific turf.[38]

The second method of measuring geographical knowledge spillovers relates to studies that measure the geographical mobility of skilled labor on innovation. Individuals, and especially star scientists with the skill, knowledge, and know-how to engage in technological advances, embody and move ideas between locales. Zucker and Darby summarize a series of papers that examine the role of archetypal star scientists as a source of intellectual capital driving the transformation of bioscientific knowledge into commercial applications.[39] Simply put, this category of geographical knowledge spillovers focuses on the human capital of key persons instead of on the average human capital in a local labor market. As Zucker and

---

[37] Manuel Trajtenberg, A Penny for your Quotes, *Rand Journal of Economics* 21(1) 172 (1990).

[38] See Adam Jaffe and Manuel Trajtenberg, Flows of Knowledge from Universities and Federal Labs: Modeling the Flows of Patent Citations over Time and Across Institutional and Geographic Boundaries, NBER Working Paper 5712) (1996) (Authors add that patent citations are more likely to be localized in the first year following the patent).

[39] See Lynne G. Zucker and Michael R. Darby, Star Scientists and Institutional Transformation: Patterns of Invention and Innovation in the Formation of the Biotechnology Industry – Proceedings of the National Academy of Science 93 (November): 12709 (1996). See also: Lynne G. Zucker, Michael R. Darby, and Jeff Armstrong, Geographically Localized Knowledge: Spillovers or Markets?, Economic Inquiry, *Western Economic Association International*, vol. 36(1) 65 (January 1998) (the authors find that the positive effect of research universities on neighboring firms concerns identifiable market exchange between particular university star scientists and firms as opposed to generalized knowledge spillovers); Lynne G. Zucker, Michael R. Darby, and Marilynn B. Brewer, Intellectual Human Capital and the Birth of US Biotechnology Enterprises, *American Economic Review*, 87(1) (1997) (demonstrating boundary spanning between universities and New Biotech Entities (NBEs) via star scientists at universities who have made scientific breakthroughs). Zucker, Darby, and Brewer measure this linkage in terms of the numbers of star scientists and their collaborators in a given area, as predictive of NBEs in controlling for presence of universities and federal funds. See also Lynne G. Zucker and Michael R. Darby, Costly Information in Firm Transformation, Exit, or Persistent Failure, NBER Working Papers 5577, National Bureau of Economic Research, Inc. (1996) (showing that pharmaceutical firms in the United States mostly are more likely to invest in top scientists at large).

Darby demonstrate for the biotech industry, localized intellectual capital generates externalities that are inclined to be geographically restricted to the region where these scientists reside.

Within the context of innovation-based growth, the related term Regional Innovation Systems (RIS) has become increasingly popular among economic geographers and innovation theorists.[40] The term has also been discussed by the OECD in a first study acknowledging regional innovation clusters within OECD countries.[41] Unfortunately, the term RIS has no specified definition. It is instead loosely understood as a network of technologically specialized and locally situated firms, institutions, and research agencies that are involved in a process of the generation, use, and dissemination of knowledge.[42] These regional knowledge externalities are said to permit companies operating near important knowledge sources to introduce innovation at a faster rate than rival firms located elsewhere.[43]

[40] See Thomas Brenner and Tom Broekel, Methodological Issues in Measuring Innovation Performance of Spatial Units, Papers in Evolutionary Economic Geography, No. 04–2009, Urban & Regional Research Centre Utrecht, Utrecht University (2009); Bjørn T. Asheim and Meric S. Gertler, The Geography of Innovation – Regional Innovation Systems, in *The Oxford Handbook of Innovation* 291 (Jess Fagerberg, D. C. Mowery, and Richard R. Nelson, eds.) (Oxford University Press, 2006), at 298, referring to Michael Storper and Anthony J. Venables, Buzz: Face-to-Face Contact and the Urban Economy, *Journal of Economic Geography*, Oxford University Press, vol. 4(4) 351 (2004); David Doloreux, Innovative Networks in Core Manufacturing Firms: Evidence from the Metropolitan area of Ottawa, *European Planning Studies*, vol. 12(2) (2004) 178; David Wolfe, Clusters Old and New: The Transition to a Knowledge Economy in Canada's Regions (Queen's School of Policy Studies, 2003); Bjørn T. Asheim and Arne Isaksen, Regional innovation systems: The integration of local 'sticky' and global 'ubiquitous' knowledge, *Journal of Technology Transfer*, 27, 77 (2002); Philip Cooke, Regional Innovation Systems, Clusters, and the Knowledge Economy, *Industrial and Corporate Change*, 10(4), 945 (2001) (concluding that Europe's innovative gap in comparison with the United States lies in a European regional firm-level market failure); John de la Mothe and Gilles Paquet (eds.), Local and Regional Systems of Innovation (Kluwer Academics Publishers, 1998); Philip N. Cooke, P. Boekholt, and Franz Tödtling, *The Governance of Innovation in Europe* (Pinter, 2000); Vernon Henderson, Ari Kuncoro and Matt Turner, Industrial Development in Cities, *The Journal of Political Economy* 103, 1067 (1995).

[41] OECD, Innovative Clusters: Drivers of National Innovation Systems (OECD Publications, 2001) (explaining that regional clusters in every country or region have unique clusters blends, and emphasizing that regional clusters may transcend geographical levels).

[42] See David Doloreux, Regional innovation systems in the periphery: The case of the Beauce in Québec (Canada), *International Journal of Innovation Management*, 7(1) 67 (2003).

[43] See Stefano Breschi and Francesco Lissoni, Knowledge Spillovers and Local Innovation Systems: A Critical Survey, Industrial and Corporate Change, *Oxford University Press*, vol. 10(4) 975 (2001), at 975.

The term "region" in the context of innovation has similarly been applied to such divergent territories and jurisdictions as urban industrial districts,[44] cities,[45] country regions,[46] countries,[47] and transnational regions.[48] With a few exceptions, contributions in this line of research, however, are still

[44] See, e.g., Bjørn T. Asheim and Arne Isaksen, supra note 40; Erik Brouwer, Hana Budil-Nadvornikova and Alfred Kleinknecht, Are Urban Agglomerations a Better Breeding Place for Product Innovations? An Analysis of New Product Announcements, *Regional Studies*, 33, 541 (1999) (finding that firms that are located in centralized Dutch regions are inclined to produce a higher number of new products in comparison with firms located in more peripheral regions).

[45] See, e.g., James Simmie (ed.) Innovative Cities (Spon Press, 2001); Anthony G. Hoare, Linkage Flows, Locational Evaluation, and Industrial Geography: A Case Study of Greater London, *Environment and Planning* A 7:41–58 (1975). As von Hagen and Hammond argue, the metropolitan rather than the state or broad regional level is the most meaningful geographic unit for analyzing geographical differences in industrial concentrations in the United States. See Jürgen von Hagen and George Hammond, Industrial Localization: An Empirical Test for Marshallian Localization Economies. Discussion paper 917 (Centre for Economic Policy Research, 1994).

[46] See, e.g., Bjørn T. Asheim, Lars Coenen, and Martin Svensson-Henning, Nordic SMEs and Regional Innovation Systems (Nordisk Industrifond, 2003) (analyzing Nordic SMEs that mainly draw on innovation through science-driven R&D, as in biotechnology); Daniel Latouche, Do Regions Make a Difference? The Case of Science and Technology Policies in Quebec, in Hans-Joachim Braczyk, Philip Cooke, and Martin Heidenreich (eds.) *Regional Innovation Systems: The Role of Governances in a Globalized World* (UCL Press, 1998) (for the province of Quebec); Meric S. Gertler and David A. Wolfe, Dynamics of the Regional Innovation System in Ontario, in John de la Mothe and Gilles Paquet (eds.), *Local and Regional Systems of Innovation* (Kluwer Academic Publishers, 1998) (for the Canadian province of Ontario). For the broader economic geography context, see Paul Krugman and A. Venables, *Integration and the Competitiveness of Peripheral Industry 56, in Unity with Diversity in the European Community* (Cambridge University Press 1990) (Christopher Bliss and Jorge Braga de Macedo, eds.) (1990) (on the peripheral regions in the European Union); Paul Krugman, The Lessons of Massachusetts for EMU, in *Adjustment and Growth in the European Monetary Union* (Francisco Torres and Francesco Giavazzi, eds.) 241 (Cambridge University Press, 1993) (on the experience of the New England region and the state of Massachusetts in economic integration); Peter J. Sunley, Marshallian Industrial Districts: The Case of the Lancashire Cotton Industry in the Inter-war Years, Transactions of the Institute of British Geographers n.s. 17, 306 (1992); Guy P. F. Steed, Internal Organization, Firm Integration and Locational Change: The Northern Ireland Linen Complex, 1954–64, *Economic Geography* 47 371 (1971) (on Northern Ireland's linen manufacturing industry); Paul Krugman, History and Industrial Location: The Case of the Manufacturing Belt, *American Economic Review* (Papers and Proceedings) 81, 80 (1991) (explaining the rise of the manufacturing belt in the Northeastern United States during the nineteenth century).

[47] Adam Jaffe, The Real Effects of Academic Research, *American Economic Review* 79, 957 (1989) (arguing that the geographic innovation production function on the US local state level shows geographic spillovers from academia to industry measured by patents); Peter Maskell, Learning in the Village Economy of Denmark: The Role of Institutions and Policy in Sustaining Competitiveness, in Hans-Joachim Braczyk, Philip Cooke and Martin Heidenreich (eds.), supra note 46.

[48] See, e.g., Paul Krugman, Increasing Returns and Economic Geography, *Journal of Political Economy* 99, 483 (1991). On the basis of this location model, Krugman argues that large-scale regions are more significant economic units than nation-states. He writes that a nighttime satellite image of the world shows regional agglomerations rather than national concentrations. Ibid., at 483–484.

based on evidence collected in firms and regions in developed countries.[49] The evidence gathered in a small number of emerging economies recognizes that the majority of inventors are clustered in the most economically advanced regions. This has been shown in studies focusing on China,[50] India,[51] and Latin America, including Brazil, Chile, and Mexico,[52] where FDI is also concentrated.

6.2   Indigenous Patenting and the Degree of Innovativeness

6.2.1   Overview

Given that literally the entire body of literature on economic geography and regional innovation system theory focuses on advanced countries, while traditionally assuming equal policy ramifications across the development divide, a question remains: Do regional agglomerations of innovation and patenting measured through knowledge spillovers differ across geographic regions, or possibly innovation clusters?

At the outset, a geographic estimation of the knowledge production function in comparing regions or country-group clusters remains empirically challenging. There seem to be two core quantitative and qualitative factors affecting the regional agglomeration of innovation. Focusing on the income rate of countries as opposed to this book's related cluster analysis which follows, these factors correspond with South Korean economist Keun Lee's observation,[53] whereby high-income countries tend to show higher localization of knowledge creation and diffusion

---

[49] See Monica Plechero and Cristina Chaminade, From New to the Firm to New to the World: Effect of Geographical Proximity and Technological Capabilities on the Degree of Novelty in Emerging Economies, Paper no. 2010/12 (2010), at 2. But see also Marco Ferretti and Adele Parmentola, *The Creation of Local Innovation Systems in Emerging Countries: The Role of Governments, Firms and Universities* (Springer, 2015).

[50] Yu Zhou and Tong Xin, An Innovative Region in China: Interaction Between Multinational Corporations and Local Firms in a High-Tech Cluster in Beijing, *Economic Geography*, 79(2) 129 (2003) (For a cluster analysis of China's leading ICT service cluster in Zhongguancun, Beijing).

[51] Lee Branstetter, Guangwei Li and Francisco Veloso, The Rise of International Coinvention 135, in (Adam B. Jaffe and Benjamin F. Jones, eds.), *The Changing Frontier: Rethinking Science and Innovation Policy* (University of Chicago Press, 2015) (for the case of India and China).

[52] See *Upgrading to Compete: Global Value Chains, SMEs and Clusters in Latin America* (Carlo Pietrobelli and Roberta Rabellotti, eds.) (Harvard University Press, 2007), chapters 3, 4, and 7, respectively.

[53] Keun Lee further differs methodologically as he measures localization through international patent self-citation. See Keun Lee, *Schumpeterian Analysis of Economic Catch-up: Knowledge, Path-creation and the iddle Income Trap* (Cambridge University Press, 2013), at 218. Lee follows Adam B. Jaffe, Manuel Trajtenberg, and Rebecca Henderson, Geographic Localization of Knowledge Spillovers as Evidenced by Patent Citations,

and a higher degree of innovation.[54] The first quantitative factor is the relatively higher domestic or indigenous patenting rates by these countries' residing inventors – that is, in comparison with relatively lower rates of patenting by oversea inventors. Indigenous patenting rates may further interrelate with technology export rates. Against the backdrop of inter-regional and inter-cluster discrepancies concerning this ratio, developing countries thus also witness equivalent lower technology export rates, as a possible proxy for their altogether higher rate of indigenous or local innovativeness.

A second, more qualitative feature affecting the localized agglomeration of innovation concerns the degree of innovativeness in countries and clustered country groups across the development divide. The arguably lower quality of innovation witnessed in developing countries thus serves as a second attribute of indigenous patent intensity, notwithstanding region–cluster theoretical intricacies.

6.2.2   Indigenous Patenting and the Technology Trade Ratio

As Keun Lee recently noted, trade liberalization continuously adopts a covert assumption that local firms are sufficiently competitive to compete against foreign companies or imported goods. Lee et al. critique this standpoint with regard to many case studies, relating particularly to middle-income countries. Given the questionable nature of this assumption, naive trade liberalization may lead to monopoly by foreign goods or the destruction of local industrial bases.[55] In what Keun Lee and Hochul Shin refer to as a superior opening strategy, they suggest the alternative term "asymmetric opening." In South Korean realpolitik during the period 1967–1993, the two explain, latecomer or catching-up economies liberalized the import of capital goods for the production of final or consumer goods, while nevertheless favoring the protection of their local consumer goods industries by imposing high tariffs on imported goods.[56]

Quarterly Journal of Economics, 108, 577 (1993) in using the patent citation methodology for measuring localization. See Adam B. Jaffe, Manuel Trajtenberg and Rebecca Henderson, supra note 1 (comparing the geographic location of patent citations with that of the cited patents, as an indication of the degree to which knowledge spillovers are geographically localized).

[54] See, e.g., Keun Lee, supra note 53, at 49, 69.

[55] Keun Lee, supra note 53, ibid., at 148 and table 6.3 for the South Korean case of charging asymmetric tariffs, ibid.

[56] Hochul Shin and Keun Lee, Asymmetric Trade Protection Leading not to Productivity but to Export Share Change, The Economics of Transition, The European Bank for Reconstruction and Development, vol. 20(4) 745 (2012).

Analogous examples of middle-income countries illustrate this point. Regarding Indonesia's electronics industry, economists Yohanes Kadarusman and Khalid Nadvi[57] show that when local firms invest in building up their internal technological capabilities, they are able to increase the efficiency of their competitiveness in both the domestic Indonesian market and through export to the Middle East and the ASEAN region. In contrast, the authors show that other Indonesian companies in the same industry have increased their own technological capabilities sub-optimally whenever they were involved in captive relationships with global leading companies and produced for developed countries. Kadarusman and Nadvi conclude that firms were able to be more innovative whenever they focused their innovative efforts on introducing new products adapted to domestic and regional markets.[58] Another such illustration examines the case of Brazil's shoe and furniture industries. As Lizbeth Navas-Alemán analogously adds, the Brazilian Sinos Valley shoe and furniture clusters, where most innovative firms in the fields of design and product development are located, are also oriented toward the domestic market.[59]

However, the best-documented example of the role of domestic and regional trade and its connection with patent intensity by domestic inventors is China. The Chinese government's focus on the promotion of domestic innovation has spurred a patent boom in China.[60] In 2011 alone, the SIPO issued nearly 1 million patents of various kinds, the majority of which were awarded to domestic applicants.[61] By 2012, nearly 60 percent of Chinese invention patents were for inventions initially

---

[57] See Yohanes Kadarusman and Khalid Nadvi, Competitiveness and Technological Upgrading in Global Value Chains: Evidence from the Indonesian Electronics and Garment Sectors, *European Planning Studiesg*, 21(7) 1007 (2013), at 14. Kadarusman and Nadvi conclude that in the case of the Indonesian electronics and garment sectors, "in emphasizing the central position of global lead firms, the Global Value Chain (GVC) framework fails to recognize the potentially key role of local agency." Ibid., at 18. See also, generally, Yohanes Kadarusman, Knowledge Acquisition: Lessons from Local and Global Interaction in the Indonesian Consumer Electronics Sector, *Institutions and Economies*, 4(2) 65 (2012) (on the transition Indonesia is undergoing within the consumers electronics manufacturing from imitation to local innovation).

[58] Yohanes Kadarusman and Khalid Nadvi, supra note 57, at 13.

[59] Lizbeth Navas-Alemán, The Impact of Operating in Multiple Value Chains for Upgrading: The Case of the Brazilian Furniture and Footwear Industries, *World Development*, 39(8), 1386 (2011).

[60] See James McGregor, China's Drive for 'Indigenous Innovation': A Web of Industrial Policies, US Chamber of Commerce (July 26, 2010), at chapter 5 (discussing China's policy focus on indigenous innovation).

[61] See Lee Branstetter, Guangwei Li, and Francisco Veloso, supra note 51 (examining Chinese invention patents, using SIPO microdata on Chinese grants over 1985–2012 period), at 150. For similar earlier SIPO findings, see e.g., SIPO, 2009 Annual Report (2010) at 35.

created domestically before also being patented in China.[62] Sun et al.[63] report that 18 percent of the Chinese firms in their survey have on average been granted 2.33 domestic patents each. Dennis Wei et al.[64] similarly show that 21 percent of the firms interviewed have filed patents, and over half of them have an in-house R&D facility.

China's patent strategy of promoting domestic patenting is just one part of a fifteen-year innovation strategy. Summarized in a 2006 document entitled the *National Medium and Long Term Plan for Science and Technology Development*, this strategy not only specifies domestic invention patent objectives, but more generally calls for China to develop into an "innovation-oriented society" by 2020.[65] The approach suggested by the plan includes *zuzhu chuangxin* – a rather vague concept often translated as "indigenous innovation."[66] It prioritizes certain areas of frontier technology, the development of an "integrated national system of institutions supportive of R&D," increased funding for R&D, and assimilation of foreign technology.[67] The 2006 Plan identifies 17 specific, large-scale science and engineering "megaprojects" that are to receive special attention and funding. However, under the Plan, no industry has been overtly excluded from the goal of raising domestic innovation levels.[68]

In this process, the total R&D expenditure in China grew from 7.4 billion Chinese Yuan Renminbi (RMB) in 1987 to 35 billion RMB in 1995 and 300.3 billion RMB in 2006, with an average annual growth rate

---

[62] See Lee Branstetter, Guangwei Li, and Francisco Veloso, supra note 51 (clarifying that patents granted to MNE Chinese subsidiaries and joint ventures in China are classified as domestic grants by the SIPO), at 150.

[63] Yifei Sun, Yu Zhou, George C. S. Lin, and Yehua H. Dennis Wei, Subcontracting and Supplier Innovativeness in a Developing Economy: Evidence from China's Information and Communication Technology Industry, *Regional Studies*, 47(10) 1766 (2013).

[64] Yehua H. Dennis Wei, *Ingo Liefner, and Changhong Miao*, Network Configurations and R&D Activities of the ICT Industry in Suzhou Municipality, China, Geoforum, 42(4) 484 (2011).

[65] See US Int'l Trade Comm'n (USITC), Pub. No. 4199, China: IP Infringement, Indigenous Innovation Policies and Framework for Measuring the Effects on the US Economy (2010), sec. 5–2; Cong Cao, Richard P. Suttmeier, and Denis F. Simon, China's 15-year Science and Technology Plan, 59 *Physics Today* 38 (2006), at 38, 39.

[66] See Arti K. Rai, US Executive Patent Policy, Global and Domestic 85, in Patent Law in Global Perspective, Ruth L. Okediji and Margo A. Bagley, eds.) (Oxford University Press, 2014), at 94, referring to Peter Yu, Five Oft-Repeated Questions About China's Recent Rise as a Patent Power, *Cardozo Law Review De Novo* 78 (2013).

[67] Arti K. Rai, ibid. (commenting that thus far the US executive branch particularly objected to the government procurement for Chinese firms as part of the Chinese "indigenous innovation" strategy), at 94.

[68] See Cong Cao, Richard P. Suttmeier and Denis F. Simon, supra note 65, at 43, box 2 (for the complete list of key areas, frontier technologies, and megaprojects).

of 21 percent.[69] The goal of promoting Chinese intellectual property in view of domestic and regional innovation promotion was reinforced in China's 2008 National Intellectual Property Strategy (NIPS). The NIPS urges the government to "guide and support [Chinese] market entities to create and utilize intellectual property" in the course of a range of policies linked to indigenous innovation.[70] Similarly, recent guidance from the country's Supreme People's Court on the implementation of indigenous innovation policies instructs the courts to support and promote indigenous innovation. The Court further gave guidance for increasing the level of protection of indigenous IPRs in key technologies.[71] Seen as part of a comprehensive innovation policy, the Chinese policy increasingly addresses areas beyond R&D funding, including industrial research, IPRs, and venture capital.[72]

In adherence with the Chinese indigenous patenting policies, a study of China's Information and Communications Technology (ICT) industry undertaken by Sun et al. found that the impact of Global Value Chain (GVC) on local technological innovation indeed depends in the Chinese case on local suppliers' absorptive capacity.[73] According to these findings, internal R&D efforts are (or at the very least are becoming) the most important source of innovation for firms in China; involvement in GVC

---

[69] See Xiaolan Fu, China's Path to Innovation (Cambridge University Press, 2015) ("High-technology exports from indigenous Chinese firms are very limited"), at 94, 113, 150; Xiaolan Fu and Yundan Gong, Indigenous and Foreign Innovation Efforts and Drivers of Technological Upgrading: Evidence from China, *World Development*, vol. 39(7) 1213 (2011), at 1215, referring to MOST (Ministry of Science and Technology of China), 2010, Statistics of Science and Technology at: www.most.org.cn.

[70] USITC, supra note 65, referring to Government of China, Outline of the National Intellectual Property Strategy, June 2008, Article III.2(11).

[71] USITC, supra note 65, referring to Opinions on the Provision of Judicial Support and Service, Supreme People's Court, www.chinacourt.org/flwk/show.php?file_id=144434 (link to text in Chinese), June 29, 2010.

[72] Cong Cao, Richard P. Suttmeier, and Denis F. Simon, supra note 65, at 38; James McGregor, China's Drive for "Indigenous Innovation," US Chamber of Commerce – Global Regulatory Cooperation Project (July 2010), sec. titled "A Rambling Plan of Breathless Ambition," at 14–15.

[73] In development studies, the concept of a GVC includes the entire spectrum of global level activities which firms and employees conduct to bring a product from its conception to end use. See, e.g., Gary Gereffi and Karina Fernandez-Stark, Global Value Chain Analysis: A Primer, Center on Globalization, Governance & Competitiveness (CGGC) Duke University (May 31, 2011), at 4–5.

On the impact of GVCs on economic growth, see, generally, Raphael Kaplinsky, The Role of Standards in Global Value Chains and their Impact on Economic and Social Upgrading, Policy Research Paper 5396, World Bank (2010); Gary Gereffi, John Humphrey, and Timothy Sturgeon, The Governance of Global Value Chains, *Review of International Political Economy*, vol. 12(1) (2005).

does not per se boost firms' innovation capacity.[74] This and other studies in support of China's indigenous innovation policy[75] tentatively correspond with a 2013 UNCTAD's World Investment Report 2013 entitled *Global Value Chains: Investment and Trade for Development 2013* on GVCs and their contribution to development. The report concluded that despite the fairly positive impact GVCs have on growth, developing countries nevertheless face the risk of operating in permanently low-value-added activities as there is no automatic process that guarantees diffusion of technology through GVCs.[76] Developing countries must therefore carefully assess the particularities of GVC participation and of proactive policies to promote GVC-led development strategies.[77]

A large portion of China's exports is noticeably innovative. Topping the list of the country's leading exports in 2014 were computers ($208B), broadcasting equipment ($157B), telephones ($107B), and integrated circuits ($61.5B).[78] Thus, from an economic geography perspective, China's case of building on indigenous innovation mostly based on domestic patent filings potentially also witnesses trade-related regional or geographic knowledge spillovers. Thus, 31.9 percent of China's total exports by value in 2014 were indeed delivered to other East Asian

---

[74] Yifei Sun, Yu Zhou, George C. S. Lin, and Y. H. Dennis Wei, Subcontracting and Supplier Innovativeness in a Developing Economy: Evidence from China's Information and Communication Technology Industry, *Regional Studies*, 47(10), 1766 (2013) ("The process is mediated by the internal R&D efforts and absorptive capabilities and only in firms with strong internal R&D did subcontracting show positive impacts on firm innovation"), at 1782.

[75] For a supportive illustration of China's indigenous innovation policy in the computer electronics industry, see Qiwen Lu, *China's Leap into the Information Age: Innovation and Organization in the Computer Industry* (Oxford University Press, 2000) (detailing four business histories leading Chinese computer electronics enterprises from their founding in the mid-1980s to the late 1990s amidst the reform of the nation's S&T indigenous policies). See also, generally, Qing Wang, Zongxian Feng, and Xiaohui Hou, A Comparative Study on the Impact of Indigenous Innovation vs. Technology Imports for Domestic Technological Innovation, Science of Science and Management of S.& T (2010) (exploring the impact of indigenous innovation and technology import, export and FDI-for domestic innovation using Chinese province level data between 2000 and 2007. In comparing the eastern, middle, and western parts of China, the authors show that in general indigenous innovation accelerated the development of technological innovation in China).

[76] The United Nations Conference on Trade and Development (UNCTAD), World Investment Report 2013 – Global Value Chains: Investment and Trade for Development 2013.

[77] Ibid. ("Promoting GVC participation implies targeting specific GVC segments and GVC participation can only form one part of a country's overall development strategy"), at 175.

[78] Ibid. China's South-East Asian regional import rate is also illustrative and stands on 39.78%, followed by Europe (19%) and the North America (11%).

regional trade partners.[79] North American importers purchased 18 percent of Chinese shipments, while 20.1 percent arrived in European countries, including the Russian Federation (2.1 percent).[80] To recall, China's cluster country counterparts also extend beyond the South-East Asian region, shaping an intriguing relationship for future region–cluster comparison.

Evidence of support for localized innovation policy is also found in the case of South Korea and Taiwan. Keun Lee shows how the degree of localization of these two countries increased steadily starting in the mid-1980s and eventually caught up with the level of the G5.[81] Lastly, there is the example of India, where research, design, and development of new and renewable energy are similarly designed to be indigenous. In its *11th Plan Proposal*, the Ministry of New and Renewable Energy (MNRE) highlighted the importance of ensuring that domestic innovation prevails.[82] Lastly, as early as 1982, Richard Nelson, in his historical volume *Government and Technical Progress*, argued that the bulk of new technology-based companies in the United States in the 1970s similarly resulted from regional and national markets demanding innovations and accepting risk.[83] These results seem to be broadly consistent with other rather skeptical views, particularly those expressed by Dani Rodrik[84] and others,[85] regarding the incoherent impact FDI-assisted technology

---

[79] See The Observatory of Economic Complexity (OEC), at: http://atlas.media.mit.edu/en/visualize/tree_map/hs92/export/chn/show/all/2014/. The OEC is an online resource for international trade data and economic complexity indicators powered by the MIT Media Lab.

[80] Ibid.   [81] Keun Lee, supra note 53, at 218 and discussion in chapter 3, fig. 3.1.

[82] See Strategic Plan for New and Renewable Energy Sector for the Period 2011–17, Ministry of New and Renewable Energy, The Government of India (February 2011) (aspiring "to achieve the comfort level of indigenous manufacturing base in five years period"), at 78.

[83] See Richard R. Nelson, Government and Technical Progress: A Cross-Industry Analysis (Pergamon Press, 1982) (on the regional pattern of invention thereof), at 243.

[84] On the basis of very mixed evidence, several scholars – notably, Dani Rodrik – warn that although the current policy literature on FDI in host developing countries is overflowing with claims of positive spillovers from FDI, evidence of their existence is less profuse. See, e.g., Dani Rodrik, The New Global Economy and Developing Countries: Making Openness Work, MD Policy Essay No. 24., Baltimore, Overseas Development Council (1999).

[85] See, e.g., Xiaolan Fu and Yundan Gong, supra note 69 (arguing that in recent years a growing number of developing countries questioned the effectiveness of a FDI-led technology upgrading strategy as they shift to indigenous innovation policies), at 1213. See also 1214–1215 (reviewing the preconditions required by developing countries for efficient international technology transfer absorption); Beata S. Javorcik and Mariana Spatareanu, Disentangling FDI Spillovers Effects: What do Firm Perceptions tell us?, In Does Foreign Direct Investment Promote Development? New Methods, Outcomes and Policy Approaches, (Theodore H. Moran, Edward M. Graham and Magnus Blomstrom, eds.) (Washington, Institute for International Economics, 2005), 45; Holger Görg and

transfer has on the development of host developing countries. The variable of localization of knowledge creation and diffusion is, to be sure, more about the question of who is in charge of the catching-up process by developing countries.[86] In the case of numerous middle-income countries, the role of the state seems to have been a more crucial factor than in lower-income countries.

The significance of indigenous innovation policies may ultimately allow such middle-income countries to rise above the perennial middle-income trap, whereby a country that reaches a certain level of earnings becomes stuck at that level. Since the early 2010s, China has faced the possibility of falling into the middle-income country trap following three decades of rapid growth credited to the Beijing Consensus,[87] with its emphasis on increasing indigenous specialization in short-cycle technology sectors.[88] Similarly, India's success with IT services can be regarded as another short-term technology-based sector, since it applies short-cycle technologies, namely IT, to servicing clients.[89]

Given the discrepancy regarding innovation and patenting-based regions and clusters, as this book details, the overall degree of indigenous innovativeness outside highly innovative countries and relevant clusters remains uneven. Evidence from Central and Eastern European countries (CEECs) makes a case in point. CEEC is an OECD term for a group of countries from more than one of this book's patent clusters, and includes Albania, Bulgaria, Croatia, the Czech Republic, Hungary, Poland, Romania, the Slovak Republic, Slovenia, Estonia, Latvia, and Lithuania.[90] British-Croatian economist Slavo Radoševic shows that the relatively higher degree of dependence on external sources of knowledge in CEECs, given the insubstantial innovation capabilities of their domestic firms, may be inconsistent with middle-income countries' call for indigenous innovativeness.[91] As Radoševic explains, weak innovation

David Greenaway, Much Ado about Nothing? Do Domestic Firms really Benefit from Foreign Direct Investment?, *World Bank Research Observer*, vol. 19, 171 (2004).

[86] Keun Lee, supra note 53, at 212.

[87] Keun Lee, supra note 53, at 178. The term "Beijing Consensus" stands for the political and economic policies instigated by the Chinese government after the death of Mao Zedong and the rehabilitation of Deng Xiaoping in 1976. It is considered to have contributed to China's eightfold growth in gross national product over two decades. See, e.g., Joshua Cooper Ramo, The Beijing Consensus, The Foreign Policy Centre (May 2004) (Within the context of innovation policy "Change, newness and innovation are the essential words of power in this consensus"), at 4.

[88] Keun Lee, supra note 53, at 178.   [89] Ibid.

[90] See The OECD, Agricultural Policies in OECD Countries: Monitoring and Evaluation 2000: Glossary of Agricultural Policy Terms, OECD (defining CEECs) (2001).

[91] See Slavo Radoševic, Domestic Innovation Capacity – Can CEE Governments Correct FDI-driven Trends through R&D Policy? 135, In *Closing the EU East-West Productivity Gap: Foreign Direct Investment, Competitiveness, and Public Policy* (David A. Dyker, ed.)

capabilities of local firms and the breach between the "old" S&T systems and new supply of information for firms may have led to growing dependence on overseas technologies.[92]

In conclusion, while international technology transfer offers potential gains, the scope of the advantages may be inadequate due to the inappropriateness of overseas technology to local conditions and the preconditions for effective FDI-assisted technology transfer. The comparative significance of indigenous and foreign innovation varies according to the technological concentration of different industries, the development level of the host country, and the distinction in factor endowments between overseas and host countries.[93]

6.2.3   The Degree of Innovativeness and Patent Activity

A second complementary factor also influences the regional agglomeration of innovation. This relates to the degree of innovativeness across regions, particularly when comparing countries across the development divide and the patent clusters.

The lower overall innovation quality seen in developing countries in itself constitutes a second factor limiting patent intensity in these countries. The degree of innovativeness, measured by patenting activity as proxy for domestic innovation, has for decades been evaluated by examining the proportion of domestic patents for which foreign patent protection is sought, and the number of foreign jurisdictions in which patent protection is sought for a given invention.[94] This dual examination highlights the relatively low patenting rates by residents of developing countries in Northern patent offices, and particularly in the EPO and USPTO.

Regarding Northern patent offices, another important insight explains low-quality patenting and innovativeness. In advanced economies, an

(Imperial College Press, 2006), at 149 (Hereinafter, "Radoševic, Domestic innovation capacity"). Radoševic indicates that the correlation coefficient between payment for licenses and FDI inflows for the CEECs for which data are available is positive and high (0.076): see 149 and fig. 6.12, ibid. See also Slavo Radoševic, Patterns of Preservation, Restructuring and Survival: Science and Technology Policy in Russia in the post-Soviet Era, *Research Policy*, vol. 32(6) 1105 (2003) (Hereinafter, "Radoševic, Patterns of preservation") (offering a seminal post-Soviet R&D model relevant for Russia and other eastern European countries. This growth-based model is based on preservation of existing S&T potential, its restructuring, and its subsequent survival strategies, based mostly on FDI-induced technology transfer).

[92] Radoševic, Domestic innovation capacity, ibid., at 149. Radoševic indicates that the correlation coefficient between payment for licenses and FDI inflows for the CEECs for which data are available is positive and high (0.076), at 149 and fig. 6.12. See also Radoševic, Patterns of preservation, ibid.

[93] Xiaolan Fu and Yundan Gong, supra note 69, at 1214–1215.

[94] Lee Branstetter, Guangwei Li, and Francisco Veloso, supra note 51, at 151.

in-depth investigation of patent examination procedures confirms that lower quality of the examination process and lower costs for patents could lead to a much higher propensity to patent. This in turn could further reduce the quality of the examination process and thus result in lower-quality patents and innovativeness. This vicious cycle was highlighted by Jaffe and Lerner[95] for the United States, and by Guellec and van Pottelsberghe[96] for Europe. It is theoretically illustrated by Caillaud and Duchêne,[97] who suggest that if additional low-quality patents are filed, less funding can be assigned to their examination, which makes it easier to have a patent granted. The lower degree of innovativeness thus serves as a further obstructing patent intensity, especially for developing countries, as this section will discuss.

6.2.3.1 Of Patent Novelty: Between New-to-the-Firm and New-to-the-World

Innovation in developing countries consciously lags behind the technological frontier associated with developed countries.[98] The developing South is still mainly imitative and thus more dependent on the acquisition of technology developed and adapted to the local needs.[99] Unsurprisingly, WIPO's *Global Innovation Index Rankings of 2015* ranks only advanced

---

[95] Adam B. Jaffe and Josh Lerner, *Innovation and Its Discontents: How Our Broken Patent System Is Endangering Innovation and Progress, and What to Do About It* (Princeton University Press, 2004), at 176.

[96] Dominique Guellec and Bruno van Pottelsberghe de la Potterie, *The Economics of the European Patent System* (Oxford University Press, 2007), at 211, 217.

[97] Bernard Caillaud and Anne Duchêne, Patent Office in Innovation Policy: Nobody's Perfect, Paris School of Economics, Working Paper, 2009–39 (2009).

[98] Martin Bell and Keith Pavitt offer a seminal article on the particularities of technological accumulation in developing countries; see Martin Bell and Keith Pavitt, Technological Accumulation and Industrial Growth: Contrasts between Developed and Developing Countries, *Industrial Corporate Change*, 2(2) 157 (1993). Linsu Kim and Richard Nelson offer an important analysis of the rapid growth of South-East Asian Newly Industrialized Industries (NIEs) in the three decades since the 1970s in view of other middle-income countries within the developing world; see Linsu Kim and Richard. R. Nelson, Introduction, in *Technology, Learning and Innovation: Experience of Newly Industrializing Economies* 1, Linsu Kim and Richard R. Nelson (eds.) (Cambridge University Press, 2000) (in referring to NIEs imitative innovation authors explain that creative imitation and innovation to catch-up and challenge advanced countries in relevant industries), at 4.

See also: Martin Srholec, A Multilevel Analysis of Innovation in Developing Countries, UNU-MERIT TIK working paper on Innovation Studies 20080812 (2008) (for a quantitative multilevel model of innovation-measuring framework conditions affecting innovativeness of firms using a large sample of firms from many developing countries); Minyuan Zhao, Conducting R&D in Countries with Weak Intellectual Property Rights Protection, *Management Science*, 5, 1185 (2006) (on China and India's incremental innovation orientation).

[99] See, e.g., Martin Bell and Keith Pavitt supra note 98.

economies in its first 28 places. China is ranked 29th, thus constituting the most highly ranked developing country on the list.[100] On average, middle-income or emerging economies have accumulated a pool of knowledge and skills that distinguish their factor endowment from those of the LDCs, as well as from the industrialized or advanced countries. Emerging economies are thus more likely to generate "intermediate" innovations with middle-level technology intensity.[101] These middle-income countries can collect the gains from investment in such technologies through the sale of patents, payment of royalties, or South–South direct investment in other developing countries.[102] Moreover, for the same relative factor prices, the gain from introducing new technologies increases for these countries in direct proportion to the volume of demand.[103]

Despite the lower average technical efficiency of indigenous firms in comparison to foreign-invested firms, many indigenous firms are nevertheless found to be the leading force on the technological frontier in the low- and medium-technology industries, though foreign-invested firms enjoy a clear lead in the high-technology sector.[104] This reality can be effectively illustrated by reference to the findings of another 2015 Report by UNIDO, entitled *Local Innovation and Global Value Chains in Developing Countries – Inclusive and Sustainable Industrial Development*.[105] The report presents a cluster analysis of types of innovative firms by innovative intensity in developing countries, showing that more than half the cases analyzed fall into the cluster of the *weak innovators*

---

[100] See World Intellectual Property Organization (WIPO), Global Innovation Index rankings 2015, Geneva, at xxx. China received the score of 47.47%.

[101] Xiaolan Fu and Yundan Gong, supra note 69, at 1214.   [102] Ibid.

[103] Ibid., referring to R. Findlay, Relative Backwardness, Direct Foreign Investment and the Transfer of Technology: A Simple Dynamic Model, Quarterly of Journal of Economics, 92(1) 1 (1978).

[104] Xiaolan Fu, supra note 69 ("Indigenous technology will be more efficient than foreign technology in sectors that intensively use factors that developing countries have in abundance"), at 136. See also 124.

In the Chinese case, Fu and Gong add that the industries in which foreign firms have obvious dominance include electronics and telecommunications, instruments and measuring equipment, culture, educational and sports goods, as well as garments and leather products where Hong Kong/Taiwan/Macau (HKTM) firms have a clear lead. Indigenous firms have dominant presence in the low- and low–medium-technology industries, such as food processing, paper-making, and smelting and processing of ferrous and nonferrous metals. See, Xiaolan Fu and Yundan Gong, supra note 69, at 2018 and fig. 3, also at 1223.

[105] See United Nations Industrial Development Organization (UNIDO), Valentina De Marchi, Elisa Giuliani, and Roberta Rabellotti, Local Innovation and Global Value Chains in Developing Countries – Inclusive and Sustainable Industrial Development, Working Paper Series WP 05 (2015).

(54 percent),[106] while a third (28 percent) are *independent innovators*,[107] and only one-fifth of the cases (18 percent) fall in the group of *GVC-led innovators*.[108]

Few empirical studies have investigated the endogenous and exogenous indicators that may positively affect the degree of novelty of innovation, and few of these studies have investigated the phenomenon in developing countries.[109] The most widely used definition of the degree of novelty, at least in surveys, is that of the OECD. The *2005 OECD/ Eurostat Oslo Manual on Innovation-Related Statistics* proposes a principled distinction between three degrees of novelty or more lenient newness of innovation. These are new-to-the-firm; an intermediate level that is new-to-the-market or industry; and new-to-the-world, which constitutes the highest possible degree of novelty.[110] An innovation is new-to-the-world if the firm has introduced a new or significantly improved good or service into the global market before its competitors.[111] It is new-to-the-market or

---

[106] The cluster titled "Weak innovators" agglomerates "firms that show low to moderate innovation records [] draw selectively on some of the knowledge sources available within the GVC, while they poorly use other sources of learning." Ibid., at 22–23 and the cluster analysis presented in the appendix therein, at table A-5.

[107] The cluster titled "Independent innovators" consists of "firms that are also highly innovative, but their learning sources come mainly from outside the GVC, while the latter plays only a marginal role in the transfer of knowledge." Ibid.

[108] The cluster titled "GVC-led innovators" consists of "local firms that are highly innovative and use intensively knowledge sources from within the GVC, as well as a selection of other sources outside the GVC." Ibid.

[109] See Monica Plechero and Cristina Chaminade, supra note 49, at 3. But see María Jesús Nieto and Luís Santamaría, The Importance of Diverse Collaborative Networks for the Novelty of Product Innovation, *Technovation* 27(6–7) 367 (2007) (on the collaboration with firms from external sources as a condition for higher degree of novelty); Bruce Tether, Who co-operates for Innovation, and why: An Empirical Analysis, *Research Policy*, 31 947 (2002) (presenting the same argument); John Humphrey and Hubert Schmitz, Developing Country Firms in the World Economy: Governance and Upgrading in Global Value Chains, INEF Report, No. 61, Duisburg: University of Duisburg (2002) (for the effect of international linkages and the transfer of knowledge on the degree of novelty); Andrea Morrison, Carlo Pietrobelli, and Roberta Rabellotti, Global Value Chains and Technological Capabilities: A Framework to Study Industrial Innovation in Developing Countries, Quaderni SEMeQ n° 03/2006 (2006) (presenting the same argument).

[110] Another viewpoint, Kleinschmidt and Cooper's early study of 195 new products, leads to a more generic triad categorization. Kleinschmidt and Cooper's typology distinguishes between "high," "moderate," and "low" innovativeness. See Elko J. Kleinschmidt and Robert G. Cooper, The Impact of Product Innovativeness on Performance, *Journal of Product Innovation Management*, 8, 240 (1991). For a more critical view of innovation typologies and innovativeness terminology, see Rosanna Garcia and Roger Calantone, A Critical Look at Technological Innovation Typology and Innovativeness Terminology: A Literature Review, *The Journal of Product Innovation Management* 19(2) 110 (2002).

[111] OECD and Eurostat (2005), Oslo Manual: Guidelines for Collecting and Interpreting Innovation Data (Paris: OECD) (Hereinafter, 'Oslo Manual'), at 57, sec. 210.

industry if the firm is the first in that specific market or industry to have implemented it.[112] Lastly, innovation is new-to-the-firm if it was already available from competitors in its market.[113] In the latter type of newness, wider-reaching business networks accelerate the spread of new goods and services, diminishing what growth economists call the "stepping on toes" effect. This growth effect occurs when competing innovators at either the firm or the country level expend time and effort on a problem someone else has already solved. In such cases, originally modeled at the firm level but analogous with country-level phenomena, particularly in developing countries, congestion or network externalities arise, as the payoffs to the adoption of innovations are substitutes, as argued by Dasgupta and Maskin,[114] or complements as otherwise argued by David[115] and Katz and Shapiro.[116] Lastly, although the majority of research takes a firm's perspective toward novelty, others also consider the firm-level categories of new-to-the-adopting-unit[117] and new-to-the-consumer.[118]

The new-to-the-firm innovation, ubiquitous in developing countries, broadly refers to imitative innovations. These may incorporate existing and patented technologies, whereas new-to-the-world innovations include new technologies that may be patentable.[119] What is true at the firm level is also true at the country level. Thus, countries, and in particular developing ones, conducting imitative and new-to-the-firm innovations may have a lower propensity to patent compared to those that develop new-to-the-world innovations.[120] The fact that a great extent of new-to-the-firm or new-to-the-industry innovation is associated

---

[112] Oslo Manual, at 57, sec. 209. Garcia and Calantone offer a secondary distinction between Newness to the market and newness to the industry. An industry may comprise several different markets, such as the computer industry and its mainframe, laptop, and home computer markets. See Rosanna Garcia and Roger Calantone, supra note 110, at 124.

[113] Oslo Manual, at 57, sec. 207.

[114] Partha Dasgupta and Eric Maskin, The Simple Economics of Research Portfolios, *The Economic Journal*, vol. 97, 581 (1987).

[115] Paul David, Clio and the Economics of QWERTY, *American Economic Review*, vol. 75 332 (1985).

[116] Michael L. Katz and Carl Shapiro, Systems Competition and Network Effects, *Journal of Economic Perspectives*, vol. 8(2) 93 (1994).

[117] John E. Ettlie and Albert H. Rubenstein, Firm Size and Product Innovation, *Journal of Product Innovation Management*, 4, 89 (1987)

[118] Kwaku Atuahene-Gima, An Exploratory Analysis of the Impact of Market Orientation on New Product Performance: A Contingency Approach, *Journal of Product Innovation Management*, 12, 275 (1995).

[119] Erik Brouwer and Alfred Kleinknecht Innovative Output, and a Firm's Propensity to Patent, An Exploration of CIS Microdata, *Research Policy* 28 (6) 615 (1999).

[120] Cf. Kuo-Feng Huang and Tsung-Chi Cheng, Determinants of Firms' Patenting or not Patenting behaviors, *Journal of Engineering and Technology Management*, vol. 36, 52 (2015), at 59.

with developing countries' imitation does not necessarily imply the need for patent law leniency or the moderation of the patent law regime.

Imitation ranges from illegal duplicates of popular products to truly innovative new products that are merely inspired by a pioneering brand. Steven Schnaars,[121] following the seminal 1966 article *Innovative Imitation* by Harvard Business School economist Theodore Levitt,[122] categorizes several distinct imitations. These are counterfeits or product pirates,[123] knockoffs or clones,[124] design copies,[125] creative adaptations,[126] technological leapfrogging,[127] and adaptation to another industry.[128] Counterfeits and knockoffs are duplicative imitations – illegal and legal, respectively. Counterfeits are copies that sell under the same brand name as the original, often of lower quality, and thereby depriving the innovator of owed profits.[129]

By contrast, knockoffs or clones are usually legal products in their own right, copying directly the pioneering products in the absence or expiration of patents, copyrights, and trademarks but selling with their own brand names at much lower prices. Clones often exceed the original in quality.[130] From a development economics perspective, duplicate imitation conveys no sustainable competitive advantage to the imitator in a technological sense, but it sustains a competitive edge in price if the imitator's wage cost is significantly lower than the originator's. For this reason, as Linsu and Nelson explain, duplicative imitation, if legal, is an astute strategy in the early industrialization of low-waged, catching-up developing countries.[131] The technology involved is usually mature and obtainable, and duplicative imitation of mature technology is comparatively simple to carry out.[132] However, Linsu and Nelson note that duplicative imitation alone is insufficient if newly industrialized economies (NIEs) are to achieve further industrialization.[133] Both creative imitation and innovation are required, not only to catch up with existing industries, but also to challenge advanced countries in new industries.[134]

6.2.3.2 Patent Quality and the Measurement of Newness
The relationship between low quality and innovativeness or newness at the level of indigenous or domestic innovation is again best documented

---

[121] See Steven Schnaars, Managing Imitation Strategy: How Later Entrants Seize Markets from Pioneers (Free Press, 1994).
[122] Theodore Levitt, Innovative Imitation, Harvard Business Review (Sept.–Oct. 1966).
[123] See Steven Schnaars, supra note 121, at 5.   [124] Ibid., at 6.   [125] Ibid., at 7.
[126] Ibid.   [127] Ibid., at 8.   [128] Ibid.
[129] Ibid., at 5; see also Linsu Kim and Richard. R. Nelson, supra note 98, at 4.
[130] See Steven Schnaars, supra note 121, at 6; Linsu Kim and Richard. R. Nelson, supra note 98, at 4.
[131] See Linsu Kim and Richard. R. Nelson, supra note 98.   [132] Ibid.   [133] Ibid.
[134] Ibid.

by the case of China. It includes potential regional knowledge spillovers within the Chinese and South-East Asian technology markets in general. This section cautiously seeks to draw on the Chinese example for other developing countries, given the limited empirics on patent quality for the latter.

Neither foreign nor indigenous domestic Chinese firms dominate the technology frontier in China. Instead, as Oxford University professor Xiaolan Fu describes in her 2015 book *China's Path to Innovation*, these are the low- and medium-technology sectors in which more indigenous firms are located on the frontier.[135] More concretely, international R&D spillovers facilitated by the FDI spillover effect on Chinese indigenous technical change are mostly insignificant or negative. The exception to this is the medium–low-technology sector, where technical change is negligible.[136]

The relatively lower degree of innovativeness in such NIEs, or developing countries more broadly, has been further documented with reference to the Chinese patenting pattern. Branstetter, Li, and Veloso have identified a qualitative difference between higher-quality Chinese patents granted to foreign inventors and lower-quality Chinese patents granted to domestic inventors.[137] These concerns are strongly reinforced by the low propensity for Chinese inventors to apply for and obtain patent protection for their Chinese inventions in patent authorities outside China. Against the background of these considerations, it is interesting to observe that the top 100 US patent applicants seek to protect nearly 30 percent of their domestic patents in at least one major foreign market, such as Japan or Europe. In striking contrast, authors find that the top 100 Chinese domestic applicants seek patent protection for less than 6 percent of their inventions in the United States, only 4 percent in Europe, and only 1 percent in Japan.[138] Eberhardt, Helmers, and Yu[139] further analyze domestic and foreign patenting by Chinese firms, and conclude that the only Chinese firms engaging in real innovation are those also taking out significant numbers of patents outside China.

---

[135] Xiaolan Fu, supra note 69, at 135. For earlier findings see Xiaolan Fu and Yundan Gong, supra note 69 (using a Chinese firm-level panel dataset of 56,125 Chinese firms between 2001 and 2005, Fu shows that indigenous firms lead on the technological frontier in the low- and medium-technology industries, while foreign-invested firms enjoy lead in the high-technology sector).

[136] Ibid. (for empirical findings between 2001 and 2005) and fig. 4. See also Xiaolan Fu and Yundan Gong, supra note 69, at 1222.

[137] Lee Branstetter, Guangwei Li and Francisco Veloso, supra note 51, at 150–151.

[138] Ibid., at 151.

[139] Markus Eberhardt, Christian Helmers and Zhihong Yu, Is the Dragon Learning to Fly? An Analysis of the Chinese Patent Explosion, CSAE Working Paper no. 2011–15, Centre for the Studies of African Economies (2011).

Numerous explanations have been offered for the Chinese patent quality deficit. Hu and Jefferson,[140] who undertook an early quantitative study of the impressive increase in China's domestic patenting, suggest that it was primarily driven by an increase in the propensity to patent, rather than by an increase in actual innovative efforts.[141] Lei, Sun, and Wright[142] note the widespread government subsidies for domestic patenting in China, showing that increased subsidies appear to increase the number of patents, but not the quantity of innovation.[143] They further explain that Chinese applicants break up inventions into small bites for the purpose of filing multiple applications on one invention in order to maximize subsidies.[144] Meanwhile, Huang found that for invention patents initially applied for during the period 1987–89, domestic applicants allowed their patents to expire earlier than foreign applicants by failing to pay maintenance fees over the full legal life of the patent.[145]

China's above-mentioned patent boom inevitably concerns the ease of obtaining a Chinese patent. Encaoua et al.[146] have addressed this phenomenon on the basis of a hypothesis known as the "laxity of patent offices." Their argument concerning the negative correlation between the ease of obtaining a patent and its quality is echoed in Gallini[147] and Bessen and Meurer,[148] who suggest that the increase in patenting in the United States can be attributed to lower examination standards at the USPTO. This posture may also explain a significant amendment to Chinese patent law in 2008. According to article 22(2) of the Patent Law of the People's Republic of China,[149] the novelty requirement was amended in December

---

[140] Albert G. Hu and Gary H. Jefferson, A Great Wall of Patents: What is Behind China's Recent Patent Explosion?, *Journal of Development Economics* 90(1) 57 (2009).

[141] Ibid.

[142] Zhen Lei, Zhen Sun and Brian Wright, Patent Subsidy and Patent Filing in China, unpublished working paper (2012) (using a panel of more than 3,000 patentees between July 2004 and December 2007 identifying a significant increase in the number of invention patent applications from domestic firms in the Suzhou Municipality after the city increased the patent subsidy by a considerable amount), at 30.

[143] Ibid., at 27.   [144] Ibid., at 29.

[145] Can Huang, Estimates of the Value of Patent Rights in China, UNU-MERIT Working Paper no. 004, Maastricht Economic and Social Research Institute on Innovation and Technology (2012).

[146] David Encaoua, Dominique Guellec, and Catalina Martínez, Patent Systems for Encouraging Innovation: Lessons from Economic Analysis, *Research Policy*, 35(9) 1423 (2006), at 1430.

[147] Nancy Gallini, The economics of patents: Lessons from recent US patent reform, *Journal of Economic Perspectives*, 16(2) 131 (2002).

[148] James Bessen and Michael J. Meurer, *Patent Failure: How Judges, Lawyers and Bureaucrats Put Innovators at Risk* (Princeton University Press, 2008).

[149] See Order of the President of the People's Republic of China No.8, The Decision of the Standing Committee of the National People's Congress on Amending the Patent Law of the People's Republic of China, adopted at the 6th Meeting of the Standing Committee

2008 to adopt the absolute standard for novelty, rather than relative novelty as stipulated in the previous version of the law. Under relative novelty, as long as an invention or technology is new in China, it can be patented in the country. During the initial drafting of the Chinese Patent Law in the 1970s–1980s, it seems impossible to check or prove whether the international novelty of inventions or utility was applied.[150]

Patent quality concerns across the development divide were not only theoretically modeled separately, but were also considered for regulation accordingly, outside the realm of international patent harmonization initiatives.[151] Van Pottelsberghe defines two competing patent quality regulatory concerns as substantive legal standards and their administrative operational design.[152] One of the most controversial proposals to regulate archetypal legal standards of USPTO patent quality, later to be considered by the Obama camp, was the idea of a "gold-plated patent." Stanford law professor Mark Lemley and colleagues presented this proposal in a journal published by the libertarian Cato Institute.[153] In what is effectively the mirror image of the archetypal minimum-standard approach typically adopted in the South regarding quality patenting, the authors propose an essentially two-tiered patent system, which they claim

of the Eleventh National People's Congress on December 27, 2008, Art. 22 ("For the purposes of this Law, existing technologies mean the technologies known to the public both domestically and abroad before the date of application.")

[150] Shoukang Guo, Some Remarks On the Third Revision Draft of the Chinese Patent Law 713, in *Patents and Technological Progress in a Globalized World* (Wolrad Prinz zu Waldeck und Pyrmont, Martin J. Adelman, Robert Brauneis, Josef Drexl, Ralph Nack) (Springer, 2009), at 716.

[151] The OECD, EPO and the USPTO hold a record of patent quality initiatives. The USPTO established an administrative initiative named the Enhanced Patent Quality Initiative (EPQI) for enhancing patent quality and in particular appropriate quality metrics target examination issues. See USPTO, Enhanced Patent Quality Initiative pillars, at: www.uspto.gov/patent/enhanced-patent-quality-initiative-pillars. Polk Wagner independently proposed quality-adjusted patent measure initiative, named the Patent Quality Index (PQI). This proposed numeric index is designed to represent the quality of USPTO patent documents. See https://www.law.upenn.edu/blogs/polk/pqi/faq.html. At the OECD level, numerous initiatives include the OECD, Knowledge Networks and Markets (KNM) "Expert Workshop on Patent Practice and Innovation" (May 2012), and the Patent Quality Workshop Report, EPO-Economic and Scientific Advisory Board (ESAB) (May 2012), at: www.epo.org/about-us/office/esab/workshops.html.

[152] See Bruno van Pottelsberghe de la Potterie, The Quality Factor in Patent Systems, ECARES working paper 2010–027 (2010). Van Pottelsberghe draws a comparison over patent quality between the EPO, JPO, and USPTO using a two-layer analytical framework encompassing "legal standards" and their "operational design." The comparison shows substantive operational design discrepancies suggesting that the EPO provides higher-quality services than the USPTO, while the JPO is in an intermediate position.

[153] Mark A. Lemley, Douglas Lichtman, and Bhaven N. Sampat, What to do About Bad Patents, *Regulation*, vol. 28(4) 10 (Winter 2005–2006).

would dramatically enhance the quality of economically significant USPTO patents.

In order to enhance the quality of qualified patents, the authors explain, such patents would be distinctively subject to more rigorous examination by the USPTO. Gold-plated patents would also apply a stronger presumption of validity for issued patents.[154] The present presumption of a validity hurdle would be replaced with a higher level of deference analogous to the evidence presumption currently given to trademarks and copyrights. The benefit is that the applications will be subjected to heightened scrutiny and the public will receive more information about which patents enjoy higher quality and importance.[155] In most cases in advanced economies, it should be added, the regulation of patent quality advanced at the USPTO,[156] EPO,[157] and by numerous academics, concerns merely the second concern of regulatory operational design.[158]

---

[154] Ibid., at 12–13, reprinted in Douglas G. Lichtman and Mark A. Lemley, Presume Nothing: Rethinking Patent Law's Presumption of Validity 300, in *Competition Policy and Patent Law under Uncertainty: Regulating Innovation* (Geoffrey A. Manne and Joshua D. Wright, eds.) (Cambridge University Press, 2011) (adding that the PTO incentive and resource problem with gold-plated patents could be mitigated by having a separate examiner unit, as the PTO has done for re-examination, and by making it an independent revenue-neutral unit of the PTO), at 326. See also ibid., at 123.

[155] Mark A. Lemley, Douglas Lichtman, and Bhaven N. Sampat, supra note 153 (the authors add that the golden patents also would be less likely to be challenged in court), at 12.

Another archetypal legal standard patent-quality initiative predating Lemley's gold-plated patent initiative came into force in March 2000. Responding to the chorus of quality-related criticisms, the USPTO began a patent-quality improvement initiative for business method patent applications, known as the Second Pair of Eyes Review (SPER) program, adding a second-level examination to allowed applications with a main classification of 705, which is the greatest single concentration of business method patents.

[156] The USPTO in part focused its Enhance Patent Quality Initiative on progress on patent dependency. See William New, IP-Watch, USPTO Acting Director Discusses Patent Quality, Pendency, Harmonization (3/03/2015), at: www.ip-watch.org/2015/03/03/uspto-acting-director-discusses-patent-quality-pendency-harmonisation/ (referring to Deputy Director of the USPTO Michelle Lee in relation to the USPTO's Enhance Patent Quality Initiative, emphasizing that the focus on patent quality therein comes from progress on patent pendency – the time it takes them to process patent applications.).

[157] See, European Patent Office (EPO), EPO Economic and Scientific Advisory Board (ESAB), Recommendations for Improving the Patent System: 2012 Statement (February 2013) (for the first study and set of recommendations concerning patent thickets, quality, and fees at the EPO-level, finding that many problems in the patent system could be addressed by better patent quality).

[158] See, e.g., Mark Lemley and B. Sampat, Examiner Characteristics and the Patent Grant Rate, mimeo: www.nber.org/~confer/2008/si2008/IPPI/lemley.pdf (2008)) (investigating how examiner characteristics affect the outcome of the examination process). For literature offering to reform the US Patent Act while improving patent quality, see Joseph Farrell and Robert P. Merges, Implementing Reform of The Patent System: Incentives to Challenge and Defend Patents: Why Litigation Won't Reliably Fix Patent

A question remains: What explains this limited legal standard intervention while measuring patent quality,[159] if the realpolitik of patent offices is to be overlooked? Two important answers come to mind, relating to the difficulty in regulating quality patenting across firms and technological industries. Firstly, it remains questionable whether enhancing patent quality truly qualifies at the firm level. Jean Lanjouw and Mark Schankerman's study on this account comes to mind. The two researchers model the determinants of the decline in measured research productivity (the patent/R&D ratio) using panel data on US manufacturing firms. The patent data covers US patents applied for during the period 1975–1993 and issued by the beginning of 2000.[160] The study focuses on three research productivity factors: level of demand, quality of patents, and technological exhaustion. The authors come to a startling conclusion. Against the backdrop of much policy-related patent quality initiatives, the two found that differences in average patent quality across firms are strongly associated with the market valuation of firms, with an especially large effect in pharmaceuticals. However, these relationships do not hold up in the time series dimension at the firm level.[161]

A second explanation of the difficulty in legally standardizing patent quality is best associated with the critique made by John Alison et al.[162] in reaction to the USPTO's industry-specific Second Pair of Eyes Review (SPER) program. This program began as a patent quality improvement initiative for business method patent applications by opting for second-level examination as of March 2000. Alison and colleagues found that patent quality problems are systemic rather than localized in selected technological industries, further complicating quality patent regulation.

Office Errors and Why Administrative Patent Review Might Help, 19 *Berkeley Technology and Law Journal* 943 (2004); Bronwyn H. Hall and Dietmar Harhoff, Implementing Reform of The Patent System: Post-Grant Reviews in the US Patent System – Design Choices and Expected Impact, 19 *Berkeley Technology and Law Journal* 989 (2004).

[159] OECD researchers have recently completed what is the most thorough contribution to the measurement and definition of patent quality. See Mariagrazia Squicciarini, Hélène Dernis, and Chiara Criscuolo, Measuring Patent Quality: Indicators of Technological and Economic Value (OECD France) (June 6, 2013).

[160] See Jean O. Lanjouw and Mark Schankerman, Research Productivity and Patent Quality: Measurement with Multiple Indicators, CEPR. Discussion Papers in its series CEPR Discussion Papers with number 3623 (October 2002).

[161] Instead, the authors explain, the patent quality accounted for a constructed patent quality index is most useful when one averages quality either over time for firms, or over firms for a given year. Ibid., at 5.

[162] See John R. Allison and Starling David Hunter, On the Feasibility of Improving Patent Quality One Technology at a Time: The Case of Business Methods, *Berkeley Technology Law Journal*, 21(2) 729 (2015) (offering a justification for avoiding industry-specific policies as the USPTO "issues substantial numbers of low quality patents in practically all fields"), at 62.

In conclusion, notwithstanding region–cluster theoretical intricacies, the explanations affecting the localized agglomeration of innovation in developing countries correspond with a localized pattern of indigenous or domestic patent intensity funneled by a lower quality of innovation and related patent intensity overall. Numerous other theoretical ramifications concerning the effect of spatial innovation require further consideration.

6.3   The Empirical Analysis

6.3.1   Methodology

The degree of innovativeness, measured by patenting activity as a proxy for domestic innovation, has for decades been evaluated by examining the proportion of domestic patents for which foreign patent protection is sought, and the number of foreign jurisdictions in which patent protection is sought for a given invention.[163] These two core spatial examinations highlight the relatively low patenting rates by residents of developing countries in Northern patent offices, and particularly in the EPO and USPTO. This ultimately stands for the latter countries' low patent intensity rates by patent clusters.

The following empirical assessment of the three patent clusters per the differing measurements of innovativeness upholds that all series measured included at least one data point for the period 1996–98, and at least one observation for the years 2011–13. After examining each series separately, for each cluster, the log transformation was found to be the best for all.[164] Data were log transformed before imputation, and exponentiated after imputation. The following lists for each series the averages for each country (after imputation), and their log transformation averaged over the years 2003–9 (denoted m_y, Lgm_y respectively). It also specifies the cluster for each country. For quick comparison of these values, the figures below depict the box plots of the average of the ten log series for years 2003–9.

6.3.2   Findings

6.3.2.1   Higher Indigenous Patenting Rates in Lower Patent Clusters
The first measurement of localized rate of innovativeness is, as explained, the relatively higher domestic or indigenous patenting rates by developing countries' residing inventors – that is, in comparison with both relatively

---

[163] Lee Branstetter, Guangwei Li, and Francisco Veloso, supra note 51, at 151.
[164] The minimal value of both series "High-technology exports as a percentage of manufactured exports" and "High-technology exports" was equal to zero, therefore for those two series the value 0.5 was added before applying the transformation.

lower rates of developing countries' residing inventors patenting abroad, as well as in comparison with non-residing inventors patenting in these countries' national or regional patent offices. The data that were used for the following analyses included four series: Resident[165] and abroad[166] count by applicant origin, Resident and non-resident count by filing office,[167] Total count by applicant origin,[168] and Total count by filing office. From those four series, two ratio series were defined and initially analyzed: Resident count by applicants origin/Total count by applicant origin (labeled series 1) and Non-resident count by filing office/Total count by filing office (labeled series 2).

In order to gain further insight into the results obtained from the analyses of these two ratio series, the four series were further separately analyzed, being the numerators and denominators of the latter two ratio series. The purpose herein was to understand which of the four series bear significant relation to the two ratio series. Similar to previous analyses, the additional numerators and denominators analyses have been only for countries for which there was at least one observation between the years 1996 and 1998 and between 2011 and 2014.[169] The first line below corresponds to the original scale and the second to the square root transformed data, per the leaders, followers, and marginalized patent clusters, respectively.

Series 1: *Resident Count by Applicants Origin/Total Count by Applicant Origin*

All countries:

	N	mean	sd	median	min	max
m_y	73	0.57	0.27	0.58	0.06	1
Sq_m_y	73	0.73	0.19	0.76	0.25	1

[165] See, WIPO IP Statistics Data Center – Glossary ("The term 'resident' is used for filings made by applicants at their home office. The home office can be a national office and/or a regional office. The resident figures by origin may thus correspond to the sum of filings made at a national and a regional office").

[166] Ibid., ("The terms 'non-resident' and 'abroad' both relate to filings in a foreign office" while using the "term 'abroad' for statistics by origin").

[167] Ibid., ("we use the term 'non-resident' for statistics by office").

[168] Ibid., ("figures by origin show who seeks protection. Statistics by both office and origin show actual flow of IP rights between countries").

[169] Also, similarly to previous statistical analyses, the best Box-Cox transformation was estimated for selected years, for each series, and separately for each cluster. The conclusion based on that analysis was that the log transformation is optimal for the four original series, but the square root is optimal for the two ratio series. The imputation was done on the transformed series and retransformed to the original scale. The analyses on the series were then performed on the transformed data.

Cluster 1:

	N	mean	sd	median	trimmed	mad	min	max	range	skew
					Mean						
m_y	21	0.37	0.17	0.34	0.35	0.15	0.12	0.75	0.64	0.71
Sq_m_y	21	0.59	0.13	0.58	0.59	0.13	0.34	0.87	0.53	0.31

Cluster 2:

	n	mean	sd	median	trimmed	Mad	min	max	range	skew
					mean						
m_y	31	0.69	0.19	0.67	0.71	0.22	0.25	0.95	0.70	-0.40
Sq_m_y	31	0.82	0.12	0.82	0.84	0.15	0.50	0.98	0.48	-0.68

Cluster 3:

	n	mean	sd	median	trimmed	mad	min	max	range	skew
					mean						
m_y	21	0.60	0.33	0.66	0.61	0.40	0.06	1	0.93	-0.24
Sq_m_y	21	0.73	0.25	0.76	0.75	0.29	0.25	1	0.75	-0.48

Series 2:   **Non-resident Count by Filing Office/Total Count by Filing Office**

All countries

	n	mean	sd	median	min	max
m_y	70	0.46	0.33	0.40	0.01	0.99
Sq_m_y	70	0.62	0.27	0.62	0.12	1.00

Cluster 1:

	N	mean	sd	median	trimmed	mad	min	max	range	skew
					mean						
m_y	21	0.36	0.31	0.20	0.32	0.16	0.08	0.99	0.91	0.89
Sq_m_y	21	0.55	0.24	0.45	0.53	0.19	0.28	0.99	0.71	0.66

Cluster 2:

	N	mean	sd	median	trimmed mean	mad	min	max	range	skew
m_y	31	0.51	0.32	0.43	0.51	0.49	0.05	0.96	0.91	0.00
Sq_m_y	31	0.66	0.26	0.64	0.67	0.37	0.22	0.98	0.76	-0.24

Cluster 3:

	n	mean	sd	median	trimmed mean	mad	min	max	range	skew
m_y	18	0.49	0.37	0.47	0.49	0.55	0.01	0.99	0.98	0.05
Sq_m_y	18	0.62	0.32	0.69	0.63	0.42	0.12	1.00	0.88	-0.24

As the regression model in Appendix G: Indigenous Innovativeness Rates by Patent Cluster confirms, there are significant differences between clusters and between years, and also significant differences in the differences among clusters along time. On the average (of countries and time), the leaders cluster (labeled 1) has lower values in comparison with the followers and marginalized patent clusters. These findings confirm that developing countries indeed are more localized in their patenting activity while being less integrated into the Northern patent offices' innovativeness apparatus, as explained.

The following is an analysis of the numerator and denominators of the previous two series. This added analysis is intended to detail the above findings.

As seen from Graphs 6.1–6.2 per series 1 (Resident count by applicants origin/Total count by applicant origin), there is an increase in time of both numerator (per Graph 6.1) and denominator (per Graph 6.2) for leaders and followers clusters (labeled 1 and 2), but the rate is higher in the denominator, and therefore the ratio decreases with time. In the marginalized cluster (labeled 3), there are almost no changes in the numerator, but an increase with time in the denominator. Therefore, there is also a decrease along time of the ratio. These conclusions are supported by the regression models.[170]

---

[170] The F test for the numerator of changes along time in cluster 3 is $F(15,1050)=0.62$, $P=0.86$, while for the leaders and followers clusters (labeled 1 and 2) the F tests are $F(15,1050)=5.79$, $P<.0001$, $(15,1050)=11.95$ $P<.0001$, respectively. For the denominators all three F tests show a significant increase $P<.0001$.

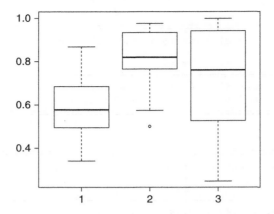

Figure 6.1 Average Square Root Values Averaged over the Years of Resident Count by Applicants Origin/Total Count by Applicant Origin (2003–9)

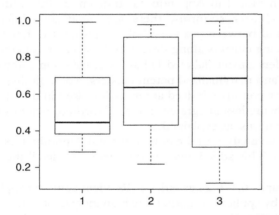

Figure 6.2 Average Square Root Values Averaged over the Years of Non-resident Count by Filing Office/Total Count by Filing Office (2003–9)

In series 2 (Non-resident count by applicants origin/Total count by applicant origin) as Figure 6.3 indicates, there were no significant differences between clusters on the average along time. A relatively large difference between the leaders and followers clusters (labeled 1 and 2) was observed on the average, but it does not reach statistical significance.

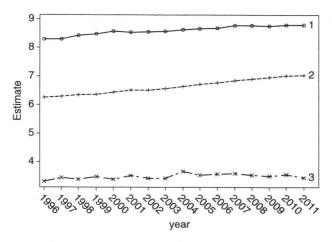

Graph 6.1 Numerator of Series 1 (log resident count by applicants origin)

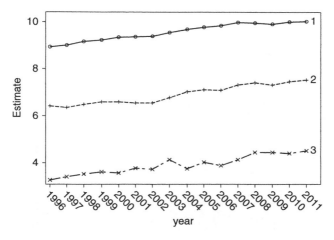

Graph 6.2 Denominator of Series 1 (log total count by applicant origin)

Examining the changes along time, Figure 6.4 and Graph 6.3 show that starting in 2005 there were no significant differences among the three patent clusters. Therefore, while on the long average there is a borderline significance, the differences are essentially only in the earlier nine years.

Type 3 Tests of Fixed Effects

Effect	Num DF	Den DF	F Value	Pr > F
Year	15	1005	7.88	<.0001
r_cluster_km	2	67	2.21	0.1182
r_cluster_km*year	30	1005	2.18	0.0003

Figure 6.3 Series 2 Type 3 Tests of Fixed Effects

Differences of Least Squares Means

Effect	r_cluster_km	_r_cluster_km	Adjustment	Adj P
r_cluster_km	1	2	Bonferroni	0.1189
r_cluster_km	1	3	Bonferroni	0.8254
r_cluster_km	2	3	Bonferroni	1.0000

Figure 6.4 Series 2 Differences of Least Squares Means

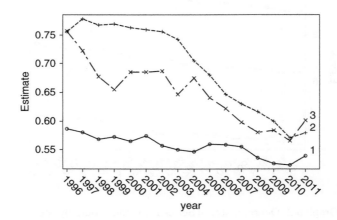

Graph 6.3 Series 2 Square Root of Mean Patent Clusters over Time (1996–2011)

Tests of Effect Slices

Effect	year	Num DF	Den DF	F Value	Pr > F
r_cluster_km*year	1996	2	1005	4.38	0.0127
r_cluster_km*year	1997	2	1005	5.31	0.0051
r_cluster_km*year	1998	2	1005	4.80	0.0084
r_cluster_km*year	1999	2	1005	4.55	0.0108
r_cluster_km*year	2000	2	1005	4.55	0.0108
r_cluster_km*year	2001	2	1005	3.94	0.0198
r_cluster_km*year	2002	2	1005	4.59	0.0104
r_cluster_km*year	2003	2	1005	4.32	0.0136
r_cluster_km*year	2004	2	1005	2.85	0.0583
r_cluster_km*year	2005	2	1005	1.45	0.2344
r_cluster_km*year	2006	2	1005	0.81	0.4469
r_cluster_km*year	2007	2	1005	0.49	0.6106
r_cluster_km*year	2008	2	1005	0.52	0.5920
r_cluster_km*year	2009	2	1005	0.46	0.6333
r_cluster_km*year	2010	2	1005	0.19	0.8248

Graphs 6.4–6.5 offer an analysis of the numerator and denominator of series 2, respectively.

As seen from Graphs 6.4–6.5, there is a decrease with time in the numerator and essentially no changes in time in the denominators for the followers and marginalized clusters (labeled 2 and 3). In cluster 1, there are almost no changes both in the numerator and the denominator. Therefore, the changes along time in the ratio for followers and marginalized clusters (labeled 2 and 3) are due to the decrease along time of the denominator. These conclusions are supported by the mixed regression models.[171]

With almost no changes in these variables over time, it is asserted that the leaders patent cluster is possibly more internationalized per its patenting intensity as opposed to the two lower patent clusters which offer a higher degree of domestic or indigenous patenting rates. Although it is unclear how knowledge spillovers are distinct from the correlation of variables at the geographic level, it is nevertheless clear that indigenous or domestic agglomeration of patent intensity significantly differ across patent clusters.

6.3.2.2 Technology Export and Growth-Related Discrepancies

Indigenous patenting rates may further interrelate with numerous growth-related indicators, including the rates of high-technology exports.

---

[171] For the denominators all three F tests show a non-significant change (P=0.60, 0.84, 0.19, respectively) for the three clusters. The F test on the numerator of changes along time for the leaders cluster (labeled 1) is $F(15,1005)=0.94$, P=0.52, while for the followers and marginalized clusters (labeled 2 and 3) the F tests are $F(15,1005)=1.91$, P.019, $F(15,1050)=1.92$ P=0.019, respectively.

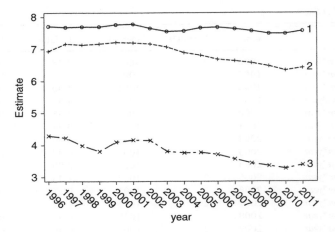

Graph 6.4 Numerator of Series 2 (log non-resident count by applicants origin)

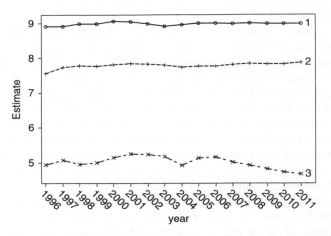

Graph 6.5 Denominator of Series 2 (log total count by applicant origin)

These indicators may account for discrepancies over the degree of spatiality across patent clusters. The findings herein relate to eight such indicators: GERD as a percentage of GDP, GERD per capita, GDP per capita, GNI per capita, GNI, GDP, high-technology exports as a percentage of manufactured export, and high-technology exports. These

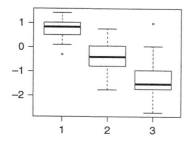

Figure 6.5 Mean Log GERD as a Percentage of GDP

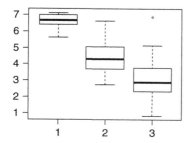

Figure 6.6 Mean Log GERD Per Capita

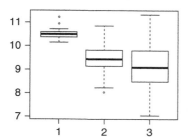

Figure 6.7 Mean Log GDP Per Capita

findings are described in Figures 6.5–6.12 respectively, as well as the relevant regression models in Appendix G.

As confirmed by the regression models of Appendix G, for the mean log of GERD as a percentage of GDP in Figure 6.5, significant differences between all three clusters were found, with the lowest mean value for the marginalized cluster (labeled 3) and the highest for the leaders cluster (labeled 1). There is no difference among the three patent clusters in

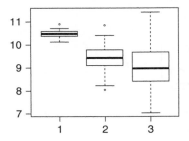

Figure 6.8 Mean Log GNI Per Capita

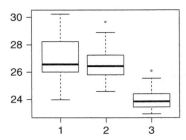

Figure 6.9 Mean Log GNI

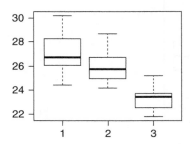

Figure 6.10 Mean Log GDP

their change pattern along time. For the mean log of GERD per capita, see Figure 6.6, where there were significant differences between all three clusters, with the lowest mean value for the marginalized cluster (labeled 3) and the highest for the leaders cluster (labeled 1). For the mean log of GDP per capita per Figure 6.7 above, there were significant differences between the leaders and followers clusters (labeled 1 and 2), as well as the leaders

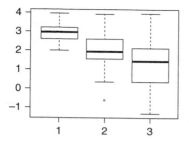

Figure 6.11 Mean Log High-Technology Exports as Percent of Manufactured Exports

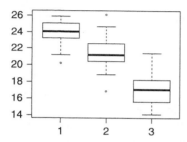

Figure 6.12 Mean Log High-Technology Exports

and marginalized clusters (labeled 1 and 3). Yet, there is no significant difference between the mean values of the followers and the marginalized clusters (labeled 2 and 3). For the mean log of GNI per capita per Figure 6.8, there are significant differences between the leaders and followers clusters (labeled 1 and 2) and between the leaders and marginalized clusters (labeled 1 and 3). Yet, similarly to the findings concerning the mean log of GDP per capita, there was no significant difference between mean values of the followers and marginalized clusters (labeled 2 and 3). For the mean log GNI per Figure 6.9, there were significant differences between the leaders and followers clusters (labeled 1 and 3) and between the followers and marginalized clusters (labeled 2 and 3). Yet, significant differences were not found between the followers and leaders clusters (labeled 2 and 1). For the mean log of GDP per Figure 6.10, there are significant differences between all three clusters with the lowest mean value for the marginalized cluster (labeled 3), and the highest is for the leaders cluster (labeled 1). For the mean log of high-technology exports as a percentage of manufactured exports, significant differences are found between the leaders and marginalized clusters (labeled 1 and 3) and between the leaders and followers clusters (labeled 1 and 2), and none between the followers and marginalized clusters (labeled 2 and 3). Lastly, for the mean log of high-technology exports, significant differences between all three clusters are found, with the lowest mean value being for the marginalized cluster (labeled 3), and the highest for the leaders cluster (labeled 1).

Per the followers and marginalized patent clusters, the medium- and low-income growth rates, respectfully measured, plausibly correspond with lower localization of knowledge creation through local innovation industries when funneled by high internationalized patenting rates. These growth rates instead correspond with higher indigenous or domestic patenting rates in the followers and marginalized clusters. High indigenous patenting rates per the latter patent clusters further relate to low technology export rates by the developing countries of these two clusters.

6.4   Theoretical Ramifications

Maryann Feldman rightly reminds us that economic geographers still lack a precise understanding of the way in which spillovers occur and are realized at the geographic level.[172] In other words, it remains difficult to separate spillovers from the correlation of variables at the geographic level. Economic activity may be co-located, but the pattern of causality is difficult to discern.[173] Economic geographers and regional innovation system theorists thus still need to account for plentiful challenging developmental aspects of regional innovation and patenting across the North–South divide or variations thereof. As a result, the empirics above focused on a more modest assertion concerning indigenous or domestic patenting, negating knowledge spillovers in the lower patent clusters instead of affirming such spatial patterns in the more patent-intense leaders cluster. Even so, three theoretical ramifications relating to spatial patterns of innovation and patent intensity appear to require further elaboration.

Firstly, in the realm of knowledge spillover literature in economic geography, one type of spillover remains unaccounted for from the perspective of innovation theory. This aspect, first postulated in 1969, is known as the Jacobs Spillovers. In her book *The Economy of Cities*, Jane Jacobs depicts this type of knowledge spillover as related to the diversity of industries in an urban area. This contrasts with the above-mentioned MAR spillovers, which focus on firms in a common industry.[174] Drawing on the example

---

[172] Maryann P. Feldman, supra note 1, at 8.   [173] Ibid.

[174] See Jane Jacobs, *The Economy of Cities*, Random House, (1969), at 123–25. See also, Edward Glaeser, Hedi Kallal, José Scheinkman, and Andrei Shleifer, supra note 10, at 1130–31.

of Detroit's shipbuilding industry, Jacobs explains how that industry was the critical antecedent leading to the development of the auto industry in Detroit.[175] Jacobs argues that an industrially varied urban setting stimulates innovation, as it includes people with diverse backgrounds and interests, thereby facilitating knowledge exchange between individuals with diverse perspectives. Such exchanges can lead to the development of new ideas, products, and processes. Jacobs adds that the rate of innovation is greater in cities with competitive market structures. Local monopolies are therefore seen to stifle innovation, whereas competitive local environments foster the introduction of new methods and products.[176]

To date, the connection between the economic concept of Jacobs Spillovers in terms of technological diversity and innovation and patent intensity remains under-theorized. One intriguing correlation with the patent originality measure was first proposed by Trajtenberg, Jaffe, and Henderson.[177] In the context of patented industries, the authors show that the scope of patent originality refers to the breadth of the technology fields on which a patent relies. The authors operationalize this concept of knowledge diversification and emphasize its importance for innovation. Inventions relying on a large number of diverse knowledge sources should lead to original results, namely to patents belonging to a wide array of technology fields. Patent originality has been applied in a wide range of studies relating to aspects such as the creation of venture-backed start-ups,[178] the duration and outcome of the patent examination procedure at the EPO,[179] and the value of post-merger patents compared to pre-merger ones.[180] However, the authors do not attach any specific significance to geographic knowledge spillovers or to potential urban spillovers in keeping with Jacobs' theory. Nevertheless, it seems reasonable to suggest that diverse innovative environments may trigger diversity-based patent originality.

---

[175] Jacobs foretells how several of Detroit's pioneers in the automobile industry had their roots in the boat engine industry. See ibid., at 123–125.

[176] Ibid., at 60, 65.

[177] Adam B. Jaffe, Manuel Trajtenberg, and Rebecca Henderson, University versus Corporate Patents: A Window on the Basicness of Inventions, *Economics of Innovation and New Technology*, 5(1): 19 (1997).

[178] Paul Gompers, Josh Lerner, and David Scharfstein, Entrepreneurial Spawning: Public Corporations and the Genesis of New Ventures, 1986 to 1999, *The Journal of Finance*, 60(2), 577 (2005).

[179] Dietmar Harhoff and Stefan Wagner, The Duration of Patent Examination at the European Patent Office, *Management Science*, vol. 55(12) 1969 (2009).

[180] Jessica C. Stahl, "Mergers and Sequential Innovation: Evidence from Patent Citations." Finance and Economics Discussion Series, No. 2010–12 Division of Research and Statistics and Monetary Affairs Federal Reserve Board, Washington, DC. (2010).

A second theoretical ramification acknowledges that the rate of patenting may be greater in denser locations for reasons other than knowledge spillovers. For example, it may be harder to keep information secret in urban areas, with the result that firms resort to patents. Wesley Cohen, Richard Nelson, and John Walsh examined this hypothesis in the well-known Carnegie Mellon Survey (CMS).[181] The CMS study was based on a 1994 study of R&D in 1,478 US manufacturing firms. The results of the CMS show that, in the case of the United States, manufacturing firms typically protect the profits from their innovations by diverse mechanisms, including patents, secrecy, and first-to-market advantages. The majority of manufacturing firms stated that they rely on secrecy and first-to-market benefits more than on patents.

Although the CMS does not consider the location of the firms in its sample, its findings nonetheless suggest that firms may be forced to rely on patenting to a greater degree in dense areas, where it is harder and more expensive to uphold secrecy. Accordingly, it is possible that increased difficulty in maintaining secrecy, rather than knowledge spillovers, account for the positive correlation between patents per capita and metropolitan density. This inquiry has yet to be made for emerging economies and other developing countries, in the context of these countries' lower patent propensity rates relative to advanced economies such as that examined by the CMS study.

A third theoretical ramification concerning the lower degree of innovativeness and patent quality in developing countries relates to these countries' relatively high adoption rates of utility models[182] – that is, when both patents and utility models are counted as innovation output.[183] The TRIPS Agreement establishes minimum substantive standards for each of the major intellectual property regimes, but it makes no explicit mention of second-tier or utility model protection. The TRIPS Agreement thus leaves WTO member states free to formulate or reject second-tier protection, such as utility model regimes, as member states see fit.[184]

---

[181] See Wesley M. Cohen, Richard R. Nelson, and John P. Walsh, Protecting Their Intellectual Assets: Appropriability Conditions and Why US Manufacturing Firms Patent (OR NOT), Working Paper 7552, National Bureau of Economic Research (2000).

[182] Of the 59 countries and regional patent offices registering utility models, 44 are developing countries. See WIPO, Protecting Innovations by Utility Models, at: www.wipo.int/sme/en/ip_business/utility_models/utility_models.htm. For more details on the definition and special features of a utility model, see ibid.

[183] See Pilar Beneito, Choosing among Alternative Technological Strategies: An Empirical Analysis of Formal Sources of Innovation, Research Policy, 32, 693 (2003) (showing that firms exhibiting a higher/lower propensity to patent also show a higher/lower propensity to register utility models).

[184] See, e.g., Uma Suthersanen, Utility Models and Innovation in Developing Countries, UNCTAD-ICTSD Project on IPRs and Sustainable Development, Issue Paper No. 13 (February 2006), at 3. Uma Suthersanen adds that whilst utility model protection is not

In theory, utility models and patents are said to account for incremental and significant innovativeness respectively.[185] In developing countries, however, utility models often branch off patent application by residing inventors in developing countries.[186] Thus, because the utility model registration is issued speedily, inventors tend to file for patents and utility models simultaneously, and subject to much demand elasticity. The impact of underdevelopment on patenting quality, and possibly the propensity to patent at large, is also reflected in the inclination by Southern inventors to prefer the archetypal less-innovative utility models to the higher newness standard of patents.

One aspect of this phenomenon has already been explained. US firms focus their patent filings in China on invention patents, filing few design patents and almost completely ignoring utility model patents. By contrast, Chinese inventors file more utility model and design patents than invention patents, although filings were more balanced across all three types of patents than they were for foreign inventors.[187] However, these geographical discrepancies did not lead to broader understanding of the utility models–patents exchangeability, given the uneven spatial patterns of innovativeness.

Conclusion

The empirical performance of innovation in geographic locations as one source of sustained growth is still under-theorized and mostly focuses on advanced economies. Although it is unclear how knowledge spillovers are precisely distinct from the correlation of variables at the geographic level, it is plausible that indigenous or domestic agglomeration of patent intensity significantly differs across patent clusters. Different patent clusters may ultimately necessitate differing patent policies and considerations, infrequently present across the development divide.

This chapter cautiously offers an analytical framework based on two measurements of spatial innovation: patent quality, and the degree of newness across patent clusters. The first feature of indigenous or domestic

specifically referred to under the TRIPS Agreement, it is arguable that by reference to Article 2(1), TRIPS Agreement, the relevant provisions of the Paris Convention provisions (including Article 1(2)) are extended to all WTO countries. Ibid.

[185] WIPO, Protecting Innovations by Utility Models, supra note 182 ("Utility models are considered particularly suited for SMEs that make "minor" improvements to, and adaptations of, existing products"); Pilar Beneito, The Innovative Performance of In-house and Contracted R&D in terms of Patents and Utility Models, *Research Policy* 35 (2006) 502 (2006), at 505.

[186] See, e.g., Uma Suthersanen, supra note 184, at 2.

[187] See U.S Int'l Trade Comm'n (USITC), supra note 65, figs 4.1 and 4.2), at sec. 4–4.

narration of patent intensity in developing countries is arguably funneled by a lower quality of innovation, as proxied through patenting activity. Numerous working examples across the developing world, and especially China, shed light on this relation.

The second factor, and the focal point of the empirical part herein, is the degree of innovativeness estimated by numerous indicators. At a start are the ratios of domestic patents for which foreign patent protection is sought, and the number of foreign jurisdictions in which overseas patent protection is wanted. This dual geographic locational examination highlights the relatively low oversea patenting rates by residents of developing countries and of Northern inventors in patent offices of developing countries, as a share of patenting by inventors from developing countries. With almost no changes in these variables over time, it is asserted that the leaders patent cluster is possibly more internationalized per its relatively high patenting intensity, as opposed to the two lower patent clusters, which offer a higher degree of domestic or indigenous patenting rates.

Moreover, the degree of innovativeness correlated with indigenous patenting rates relates to eight growth-related indicators, including the rates of high-technology exports. These indicators may further account for growth-related discrepancies over the degree of spatiality across patent clusters. Eight such indicators are accounted for: GERD as a percentage of GDP, GERD per capita, GDP per capita, GNI per capita, GNI, GDP, high-technology exports as percent of manufactured export, and high-technology exports. As seen, significant differences between all three clusters were found, with the lowest mean value for the marginalized cluster and the highest for the leaders cluster. Per the followers and marginalized patent clusters, the medium- and low-income growth rates measured, respectively, plausibly correspond with lower localization of knowledge creation through local innovation industries when funneled by high internationalized patenting rates. These growth rates instead correspond with higher indigenous or domestic patenting rates in the followers and marginalized clusters. High indigenous patenting rates per the latter patent clusters further relate to low technology export rates by the developing countries of these two clusters.

Appendix G   Indigenous Innovativeness Rates by Patent Cluster

G.1   Series 1 Regression Model

Regressions examining resident count by applicants origin/total count by applicant origin changes along time follow these analyses, and will add to the current inference, which only pertains to the averages over time. The regressions were done for each of the two series on their square root transformation for years 1996–2011, per the leaders, followers, and marginalized patent clusters, labeled 1–3 respectively.

Type 3 Tests of Fixed Effects

Effect	Num DF	Den DF	F Value	Pr > F
year	15	1050	33.15	<.0001
r_cluster_km	2	70	23.60	<.0001
r_cluster_km*year	30	1050	1.53	0.0356

There are significant differences between clusters and between years, and also significant differences in the differences among clusters along time. On the average (of countries and time) cluster 1 has lower values. The differences are changing with time, getting smaller between clusters 1 and 3, and larger between the followers and clusters 2 and 3.

Differences of Least Squares Means

Effect	r_cluster_km	_r_cluster_km	Adjustment	Adj P
r_cluster_km	1	2	Bonferroni	<.0001
r_cluster_km	1	3	Bonferroni	0.0251
r_cluster_km	2	3	Bonferroni	0.4105

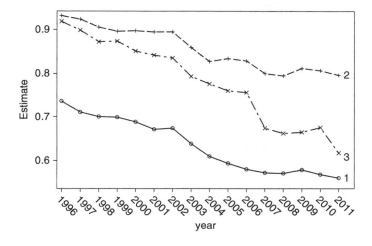

G.2   Regression Models (Eight Series)

Following the Box-Cox analysis, the log transformed variables were fitted in all models. The data are for the years 1996–2011.

G.2.1   Log GERD as a Percentage of GDP

The Mixed Procedure

Type 3 Tests of Fixed Effects

Effect	Num DF	Den DF	F Value	Pr > F
Year	15	990	3.46	<.0001
r_cluster_km	2	66	63.29	<.0001
r_cluster_km*year	30	990	0.96	0.5303

Least Squares Means

Effect	r_cluster_km	Estimate	Standard Error	DF	t Value	Pr > |t|
r_cluster_km	1	0.6987	0.1003	66	6.97	<.0001
r_cluster_km	2	−0.4729	0.1185	66	−3.99	0.0002
r_cluster_km	3	−1.3569	0.1714	66	−7.92	<.0001

Differences of Least Squares Means

Effect	r_cluster_km	_r_cluster_km	Estimate	Standard Error	DF	t Value	Pr > |t|
r_cluster_km	1	2	1.1715	0.1553	66	7.54	<.0001
r_cluster_km	1	3	2.0555	0.1986	66	10.35	<.0001
r_cluster_km	2	3	0.8840	0.2084	66	4.24	<.0001

Differences of Least Squares Means

Effect	r_cluster_km	_r_cluster_km	Adjustment	Adj P
r_cluster_km	1	2	Bonferroni	<.0001
r_cluster_km	1	3	Bonferroni	<.0001
r_cluster_km	2	3	Bonferroni	0.0002

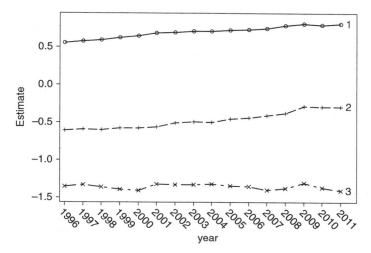

There are significant differences between all three clusters, with the lowest mean value for cluster 3 and the highest for cluster 1. There is no difference among clusters in their change pattern along time.

G.2.2   Log GERD Per Capita

The Mixed Procedure

Type 3 Tests of Fixed Effects

Effect	Num DF	Den DF	F Value	Pr > F
year	15	990	36.36	<.0001
r_cluster_km	2	66	91.58	<.0001
r_cluster_km*year	30	990	2.64	<.0001

Least Squares Means

Effect	r_cluster_km	Estimate	Standard Error	DF	t Value	Pr > |t|
r_cluster_km	1	6.4793	0.1065	66	60.85	<.0001
r_cluster_km	2	4.2711	0.2054	66	20.79	<.0001
r_cluster_km	3	3.0304	0.2960	66	10.24	<.0001

Differences of Least Squares Means

Effect	r_cluster_km	_r_cluster_km	Estimate	Standard Error	DF	t Value	Pr > |t|
r_cluster_km	1	2	2.2082	0.2314	66	9.54	<.0001
r_cluster_km	1	3	3.4489	0.3145	66	10.97	<.0001
r_cluster_km	2	3	1.2407	0.3603	66	3.44	0.0010

Differences of Least Squares Means

Effect	r_cluster_km	_r_cluster_km	Adjustment	Adj P
r_cluster_km	1	2	Bonferroni	<.0001
r_cluster_km	1	3	Bonferroni	<.0001
r_cluster_km	2	3	Bonferroni	0.0030

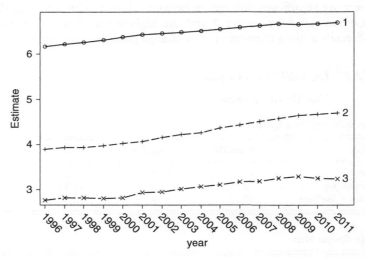

There are significant differences between all three clusters, with the lowest mean value for cluster 3 and the highest for 1. There are significant differences among the clusters in their change pattern, as we can see in the plot above.

G.2.3 Log GDP Per Capita (in PPP Expressed in US$ Constant Price of 2005)

The Mixed Procedure

Type 3 Tests of Fixed Effects

Effect	Num DF	Den DF	F Value	Pr > F
year	15	1140	387.41	<.0001
r_cluster_km	2	76	58.06	<.0001
r_cluster_km*year	30	1140	9.42	<.0001

Least Squares Means

Effect	r_cluster_km	Estimate	Standard Error	DF	t Value	Pr > |t|
r_cluster_km	1	10.3993	0.05055	76	205.72	<.0001
r_cluster_km	2	9.2951	0.1112	76	83.59	<.0001
r_cluster_km	3	8.9178	0.2155	76	41.38	<.0001

Differences of Least Squares Means

Effect	r_cluster_km	_r_cluster_km	Estimate	Standard Error	DF	t Value	Pr > |t|
r_cluster_km	1	2	1.1042	0.1222	76	9.04	<.0001
r_cluster_km	1	3	1.4816	0.2214	76	6.69	<.0001
r_cluster_km	2	3	0.3773	0.2425	76	1.56	0.1239

Differences of Least Squares Means

Effect	r_cluster_km	_r_cluster_km	Adjustment	Adj P
r_cluster_km	2	3	Bonferroni	0.3716

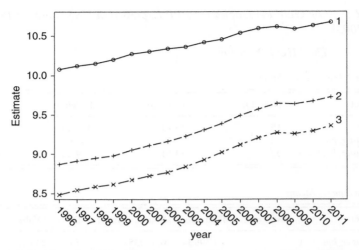

There are significant differences between clusters 1 and 2, and 1 and 3, but no significant difference between clusters 2 and 3 mean values. The lowest mean value was for cluster 3, and the highest for cluster 1. There are significant differences among the clusters in their change pattern along time.

G.2.4 Log GNI Per Capita (in PPP Expressed in US$ Constant Price of 2005)

The Mixed Procedure

Type 3 Tests of Fixed Effects

Effect	Num DF	Den DF	F Value	Pr > F
Year	15	1110	327.94	<.0001
r_cluster_km	2	74	59.51	<.0001
r_cluster_km*year	30	1110	9.41	<.0001

Least Squares Means

Effect	r_cluster_km	Estimate	Standard Error	DF	t Value	Pr > |t|
r_cluster_km	1	10.3800	0.04564	74	227.42	<.0001
r_cluster_km	2	9.2738	0.1105	74	83.89	<.0001
r_cluster_km	3	8.8561	0.2317	74	38.22	<.0001

Differences of Least Squares Means

Effect	r_cluster_km	_r_cluster_km	Estimate	Standard Error	DF	t Value	Pr > |t|
r_cluster_km	1	2	1.1062	0.1196	74	9.25	<.0001
r_cluster_km	1	3	1.5239	0.2362	74	6.45	<.0001
r_cluster_km	2	3	0.4177	0.2567	74	1.63	0.1080

Differences of Least Squares Means

Effect	r_cluster_km	_r_cluster_km	Adjustment	Adj P
r_cluster_km	1	2	Bonferroni	<.0001
r_cluster_km	1	3	Bonferroni	<.0001
r_cluster_km	2	3	Bonferroni	0.3239

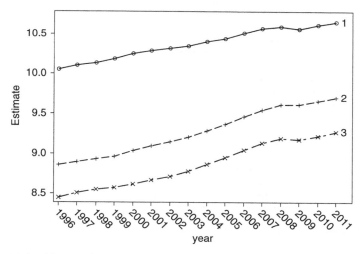

There are significant differences between clusters 1 and 2, and between 1 and 3, but no significant difference between clusters 2 and 3 mean values. The lowest mean value was for cluster 3, and the highest for 1. There are significant differences among the clusters in their change pattern along time.

G.2.5   Log GNI (in PPP Expressed in US$ Constant Price of 2005)

The Mixed Procedure

Type 3 Tests of Fixed Effects

Effect	Num DF	Den DF	F Value	Pr > F
Year	15	1110	448.29	<.0001
r_cluster_km	2	74	62.56	<.0001
r_cluster_km*year	30	1110	7.27	<.0001

Least Squares Means

Effect	r_cluster_km	Estimate	Standard Error	DF	t Value	Pr > |t|
r_cluster_km	1	26.8918	0.3204	74	83.93	<.0001
r_cluster_km	2	26.4266	0.2139	74	123.56	<.0001
r_cluster_km	3	23.8307	0.1704	74	139.86	<.0001

Differences of Least Squares Means

Effect	r_cluster_km	_r_cluster_km	Estimate	Standard Error	DF	t Value	Pr > |t|
r_cluster_km	1	2	0.4652	0.3852	74	1.21	0.2310
r_cluster_km	1	3	3.0610	0.3629	74	8.43	<.0001
r_cluster_km	2	3	2.5958	0.2734	74	9.49	<.0001

Differences of Least Squares Means

Effect	r_cluster_km	_r_cluster_km	Adjustment	Adj P
r_cluster_km	1	2	Bonferroni	0.6931
r_cluster_km	1	3	Bonferroni	<.0001
r_cluster_km	2	3	Bonferroni	<.0001

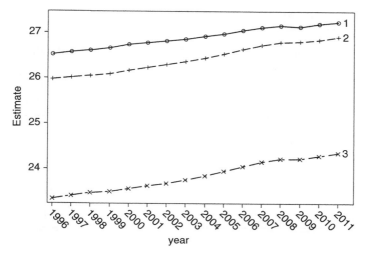

There are significant differences between clusters 1 and 3 and between 2 and 3, and none between 2 and 1. The lowest mean value was for cluster 3. There are significant differences among the clusters in their change pattern along time.

G.2.6   Log GDP (in PPP Expressed in US$ Constant Price of 2005)

The Mixed Procedure

Type 3 Tests of Fixed Effects

Effect	Num DF	Den DF	F Value	Pr > F
year	15	1140	271.41	<.0001
r_cluster_km	2	76	85.03	<.0001
r_cluster_km*year	30	1140	6.96	<.0001

Least Squares Means

Effect	r_cluster_km	Estimate	Standard Error	DF	t Value	Pr > |t|
r_cluster_km	1	26.9252	0.3159	76	85.24	<.0001
r_cluster_km	2	25.6608	0.2209	76	116.18	<.0001
r_cluster_km	3	22.9347	0.1708	76	134.27	<.0001

Differences of Least Squares Means

Effect	r_cluster_km	_r_cluster_km	Estimate	Standard Error	DF	t Value	Pr > |t|
r_cluster_km	1	2	1.2643	0.3854	76	3.28	0.0016
r_cluster_km	1	3	3.9904	0.3591	76	11.11	<.0001
r_cluster_km	2	3	2.7261	0.2792	76	9.76	<.0001

Differences of Least Squares Means

Effect	r_cluster_km	_r_cluster_km	Adjustment	Adj P
r_cluster_km	1	2	Bonferroni	0.0047
r_cluster_km	1	3	Bonferroni	<.0001
r_cluster_km	2	3	Bonferroni	<.0001

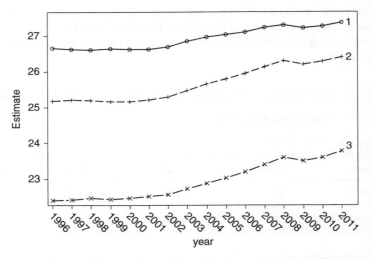

There are significant differences between all three clusters, with the lowest mean value for cluster 3 and the highest for 1. There are also significant differences among the clusters in their change pattern along time.

G.2.7   Log High-Technology Exports as Percent of Manufactured Exports
The Mixed Procedure

Type 3 Tests of Fixed Effects

Effect	Num DF	Den DF	F Value	Pr > F
year	15	1035	1.44	0.1229
r_cluster_km	2	69	23.77	<.0001
r_cluster_km*year	30	1035	3.59	<.0001

Least Squares Means

Effect	r_cluster_km	Estimate	Standard Error	DF	t Value	Pr > |t|
r_cluster_km	1	3.0090	0.1017	69	29.59	<.0001
r_cluster_km	2	2.0537	0.1553	69	13.22	<.0001
r_cluster_km	3	1.6835	0.2098	69	8.02	<.0001

Differences of Least Squares Means

Effect	r_cluster_km	_r_cluster_km	Estimate	Standard Error	DF	t Value	Pr > |t|
r_cluster_km	1	2	0.9552	0.1857	69	5.15	<.0001
r_cluster_km	1	3	1.3255	0.2331	69	5.69	<.0001
r_cluster_km	2	3	0.3703	0.2610	69	1.42	0.1606

Differences of Least Squares Means

Effect	r_cluster_km	_r_cluster_km	Adjustment	Adj P
r_cluster_km	1	2	Bonferroni	<.0001
r_cluster_km	1	3	Bonferroni	<.0001
r_cluster_km	2	3	Bonferroni	0.4818

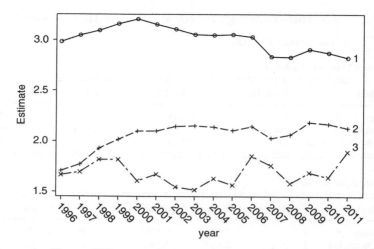

There are significant differences between clusters 1 and 3 and between 1 and 2, and none between 2 and 3. The highest mean value was for cluster 1. There are also significant differences among the clusters in their change pattern along time.

G.2.8   Log High-Technology Exports

The Mixed Procedure

Type 3 Tests of Fixed Effects

Effect	Num DF	Den DF	F Value	Pr > F
year	15	1050	17.54	<.0001
r_cluster_km	2	70	73.81	<.0001
r_cluster_km*year	30	1050	5.31	<.0001

Least Squares Means

Effect	r_cluster_km	Estimate	Standard Error	DF	t Value	Pr > |t|
r_cluster_km	1	23.8380	0.3420	70	69.71	<.0001
r_cluster_km	2	21.0422	0.3527	70	59.66	<.0001
r_cluster_km	3	16.8592	0.4628	70	36.43	<.0001

Differences of Least Squares Means

Effect	r_cluster_km	_r_cluster_km	Estimate	Standard Error	DF	t Value	Pr > |t|
r_cluster_km	1	2	2.7958	0.4913	70	5.69	<.0001
r_cluster_km	1	3	6.9788	0.5755	70	12.13	<.0001
r_cluster_km	2	3	4.1830	0.5819	70	7.19	<.0001

Differences of Least Squares Means

Effect	r_cluster_km	_r_cluster_km	Adjustment	Adj P
r_cluster_km	1	2	Bonferroni	<.0001
r_cluster_km	1	3	Bonferroni	<.0001
r_cluster_km	2	3	Bonferroni	<.0001

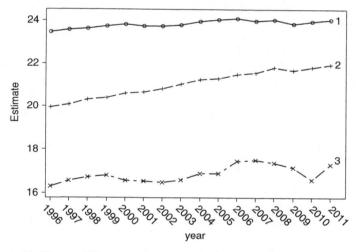

There are significant differences between all three clusters, with the lowest mean value for cluster 3, and the highest for 1. There are also significant differences among the clusters in their change pattern along all time-points.

General Conclusion

Insufficient attention has been devoted to patent harmonization efforts that address the actual properties of the international patent system.[1] The resulting gridlock has been most noticeably evident in WIPO, where the work program of the Standing Committee on Patents (SCP) has been at an impasse for numerous years.[2] Our discussion here took as its starting point the need to identify a more credible approach to patent harmonization,[3] and the country-group coalitions surrounding its efforts. In so doing, we have explained both how and why countries across the development divide differ in terms of patent intensity as a proxy for innovation-based economic growth.

Our discussion has elaborated core differences between the three patent clusters based on comparisons drawn from other World Bank and IMF country-group classifications. Such differences include the income level, geographic region, and type of an economy. We then offered a profile of the three country groups based on aspects such as the type of institutions performing and financing Gross Domestic expenditure on Research and Development (GERD), GERD by type of R&D, human capital and human resources indicators, and spatial growth-related indicators.

As is apparent from a cursory review of the intellectual property, trade, and development indices in use throughout the UN and the WTO, no single innovation-based growth apparatus has prevailed to date. On the

---

[1] See, e.g., Ruth L. Okediji, Public Welfare and the International Patent System, in *Patent Law in Global Perspective* (Ruth L. Okediji and Margo A. Bagley, eds.) Oxford University Press (2014), at 4.

[2] See Kaitlin Mara, Standing Committee on the Law of Patents to Reconvene After Two Year Hiatus, *Intellectual Property Watch* (June 19, 2008); Rachel Marusuk Hermann, WIPO Patent Committee Moves Quickly Through Agenda; Heavy Lifting to Come, *Intellectual Property Watch* (Feb. 26, 2013); Rachel Marusuk Hermann, WIPO Patent Law Committee cinches Agreement on Future Work, Intellectual Property Watch (Mar. 1, 2013); 2006 Patent Cooperation Treaty Conference: Transcript of Proceedings, 32 *William Mitchell Law Review* 1603 (2006), at 1645–46.

[3] Cf: ibid., at 3.

one hand, development settings, and most notably WIPO's Development Agenda, clearly require WIPO to promote norm-setting in adherence with different levels of development and the needs of developing countries. On the other hand, WIPO's norm-setting preferences, much like those of the WTO, are still widely perceived to favor developed countries due to equal-country paternalism. The 1996 diplomatic conference aimed at updating WIPO treaties to the digital age, as well as the more recent Substantive Patent Law Treaty promoting patent harmonization, have served only to underscore the evident complications in WIPO's norm-setting approach.

UN officials, national governments, nongovernmental organizations, and researchers in the field must henceforth be guided by more distinct, country-specific understandings regarding all aspects of patenting and innovation-related norms. These understandings will address dimensions such as pharmaceutical patents, plant genetics, and software protection. The methodology to be employed in classifying countries as patent prone or patent averse as a proxy for their relative domestic innovation demands granular, empirical, and conceptual scrutiny, and this is the central task we sought to address.

Our examination of 79 innovating countries worldwide over the period 1996–2013 raised three empirical findings. The first core finding identifies three domestic innovation-related convergence clubs (referred to as patent clusters) with markedly different levels of patent propensity and GERD intensity rates (referred to as patent intensity). The clustering analysis reveals two substantial gaps between the innovating countries and the global innovation-based economy: the first is found between the middle group of "followers" and the stronger "leaders," while the second occurs between the weaker "marginalized" group and the "followers" club.

This first finding is accompanied by numerous additional insights. Firstly, the relationship between each cluster and six other archetypal patent activity intensity indicators identified as relevant to patent activity has important policy-oriented implications. As we have seen, the only significant relationship between the groupings of the three clusters and the indicators reviewed was found regarding the economy category. These are the advanced economies, emerging economies, and other developing countries (excluding emerging economies) as defined by the IMF. All of the other relationships reviewed were found to be statistically insignificant. The World Bank's income groups and geographic regions were found to be irrelevant in terms of the clusters. Similarly, the percentage of patents granted, out of patents submitted to the USPTO or EPO, was found to play an insignificant role in predicting any of the cluster patenting patterns. Finally, neither the percentage of patent applications submitted solely to PCT (and not UE) out of the USPTO, EPO, or PCT, nor the relationship between clusters and family sizes, were found to be significant policy levers.

The first finding also shows that the leaders cluster includes 21 out of the 35 OECD countries. The remainder of OECD countries belong to the followers (e.g. Australia, Norway, and Spain), and two even belong to the marginalized cluster (Cyprus and Iceland). Thus, this analysis effectively divides the OEDC countries into two approximate halves bridging an internal OECD patenting divide that has yet to be explained. Similar findings apply to other country groups as shown, and notably the EU, whose member states can be found in each of the three patenting clusters.

The analysis upholds a second finding concerning convergence between the three clusters. The results show that, with the exception of Azerbaijan, in the two other cases (all years and only early years) the clustering results remain identical. These results are at odds with the notion that the North–South divide, or variations thereof, is gradually diminishing with the enactment of TRIPS.

In continuation, an intra-club convergence examination followed. As shown, all three clusters witnessed an increase in their GERD over time. However, this is matched by a decrease in the ratio of patents to GERD, defined here as patent propensity, in the leaders cluster, although no significant convergence occurred in the other two clusters. Similar results were shown for the changes over time by the analogous economy category country-group classification. In conclusion, the gap between the lower followers and marginalized clusters on the one hand, and the leaders cluster on the other, is slowly and steadily decreasing following a homogeneous marginal return of decrease in comparison with the former clusters over their slow and steady upward convergence in their propensity to patent.

Lastly, the GERD-related findings further serve to demonstrate how through global investment in R&D, and possibly FDI through the intervention of MNCs and foreign governments, the marginalized and followers clusters and their respective economy groups have seen a slow but stable upward convergence toward the remaining leaders cluster or advanced economy country group. These findings correspond with recent analyses focused almost exclusively on endogenous growth theory income-related indications. They suggest that, unlike orthodox neoclassical models, regional convergence rates are also generally much slower.

The book's novel clustering analysis has numerous theoretical ramifications, particularly in terms of the need for further explanation of the remaining intricacies in accounting for shifts and reversals in rates of regional convergence. Such discrepancies arise from the fact that there is still little evidence accounting for the slowness or nonexistence of inner club convergence – especially in advanced economies, but also in emerging ones. In political terms, it remains unclear to what extent mismatched countries can join new coalitions, given the conflicting interests many of them may retain within their overall WTO bargaining position.

For now, on the broad policy level, legislation to facilitate patenting entices firms either to patent a higher percentage of their innovations or even to invest more in innovation. Yet even this rather basic proposition requires further consolidation.

Such pro-patenting policies indeed carry complex implications that lie beyond the scope of this book. Policies such as those intended to reduce the cost of a patent application may instead increase patent propensity rates in some sectors that currently have low rates, while having little effect on the firms or sectors where a majority of innovations are already patented. These changes may otherwise lead to an empirically uncorroborated reduction in patent quality overall in advanced and emerging economies alike.

As seen through the nonlinearity in the two patent intensity divides, although TRIPS require the harmonization of IPRs protection, such harmonization is not clearly necessary, empirically based, or otherwise sufficient for the South. As a result, the latter's lower patent propensity rates continue to contradict prevailing "one-size-fits-all" innovation and patent-related policy.

After reviewing how countries differ in terms of their patent intensity, we offered core growth-related explanations for these differences. Starting with an institutional characterization of the three patent clusters, we saw that UN-level organs are only loosely concerned with the role institutions take in promoting patenting activity. To date, innovation-based economic growth theory has emphasized how R&D, and particularly internationalized R&D, should be promoted by MNCs worldwide. Such R&D activity is also strongly consistent with a higher yield of patenting activity as measured by comparable national patent propensity rates. Rooted in dependency theories of development whereby developing countries were flatly perceived to be dependent on developed ones, the TRIPS Agreement implicitly pledges for a leading "freer trade" role for the business sector in directly fostering domestic innovative activity directed by a higher yield of patenting activity. The reason for this is that TRIPS corresponded, and largely continues to do so, with the World Bank and UNCTAD's labeling of technology transfer as a reactive form of innovation-based economic growth for developing countries. Thus, rather than promoting domestic innovation by the promotion of local technological capacity, innovation was to be received and, at best, adapted. Thereafter, the business sector was meant to foster technologically based trade.

However, a careful look suggests that MNCs and the business sector in general have not met these high expectations. Contrary to the UN's internationalized R&D view, they have not played a dominant role in promoting an internationalized form of innovation in the developing world.

Against this backdrop, there are possible statistical connections between the government and the business sector (domestic and foreign) based on patent intensity measurement. Our analysis of this aspect proposed two R&D-related indicators: the financing of GERD and the performance of GERD by three types of innovating sectors, namely government, the business sector, and private investment by multinational enterprises.

Our analysis criticized the current business sector pattern seen in developing and developed countries and offered two key findings. Firstly, in accounting for the relatively lower patent intensity rates in the lower patent clusters in comparison with the leaders cluster, we showed that, relatively, the business sector in the former finances and performs much less GERD-related innovative activity by comparison to public sector institutions. This hypothesis may substantiate UNCTAD's 2005 World Investment Report's key findings. Accordingly, the share of middle-income and emerging economies in global business R&D spending (with an emphasis on advanced economies) is lower than the share in total R&D spending. Moreover, these findings implicitly correspond with WIPO's 2011 report on innovation. As the WIPO report shows, government, rather than universities, are often the main R&D actors in low- and middle-income economies, while industry often contributes little to scientific research.

To conclude, the relatively lower patent intensity witnessed in the two lower clusters seemingly relates to both a sub-optimal process of "second best" government political pulling of innovation activity. The latter is directed in tandem by a deficient form of IPR regulatory framework promoted by the WTO apparatus, and TRIPS in particular. As a generalization, the lower patent clusters illustrate how the business sector is sub-optimally related to the increase in patent intensity rates as a proxy for domestic innovation.

The analysis also estimates the central role of the government public sector in financing and performing GERD-related innovative activity in the lower patent clusters in comparison with the leaders patent cluster.

Governments are repeatedly and unreservedly assumed to be benign institutions that are merely, or mostly, driven by their desire to exploit social welfare (even if their limited competence in execution is frequently recognized). This supposition plainly differs from research on neopatrimonialism and from rent seeking that emphasizes the function of states – particularly in developing countries – as entities that follow their individual monetary and political interests and may demonstrate predatory behavior.

Patent intensity also differs across the three patent clusters in terms of R&D type. Little attention has been given to fluctuations in scientific research by comparison to other R&D-related measurements, such as human resources, institution funding and performance, or internationalized R&D spillovers. Numerous insights may already suggest why Southern economies generally show higher basic R&D as a percentage of GERD rates than advanced countries. In absolute terms, R&D expenditures on basic and applied research as well as on experimental development are clearly higher in advanced economies by comparison to middle-income or emerging economies. The comparison between the two economy types becomes more nuanced when patent activity intensity factors are accounted for in terms of their percentage of GDP and GERD. From a patent-policy standpoint, these measurements may go a long way to explaining why different countries and country groups differ in patent intensity as measured by multiple types of R&D rates.

We did not find any significant differences between our three patent clusters in terms of basic and applied research. The only significant difference is between the leaders and marginalized clusters, relating to experimental development as a percentage of GERD – the more patentable and commercial type of non-basic R&D. This finding, limited though it is, nevertheless supports the conclusion that advanced economies, or more narrowly countries belonging to the leaders patent cluster, are more significantly characterized by added commercial and thus patentable types of R&D activity. The latter finding may ultimately explain why advanced countries closely associated with the leaders patent cluster are characterized by higher patent intensity rates in general. It may also offer insights for emerging economies closely associated with the two lower patent clusters regarding their archetypal industry–academia gap challenge, with its ramifications for innovation and patent policy.

The next set of characteristics of the three patent clusters were human capital-related indicators. Most studies at the firm level confirm that firms with a higher proportion of scientists and engineers may show superior innovation-based economic growth, which in turn may also increase these firms' patenting rates. Regrettably, human capital literature consistently focuses on developed or advanced countries, while largely overlooking developing countries, including the emerging economies.

Our analysis presented a country-level examination using six yearly series. These are: GERD per FTE researcher, GERD per HC researcher per country, government expenditure per tertiary student as percentage of GDP per capita (in percentage), scientific and technical journal articles per national researcher HC, scientific and technical journal articles per national researcher FTE, and lastly, the number of researchers (HC) per labor force per country. The findings show that, with the exception of government expenditure per tertiary student as a percentage of GDP per capita, for which there were no differences among the patent clusters, all the other variables showed higher mean values for the leaders cluster than for the followers patent cluster.

For three of the six series, the leaders cluster also showed higher mean values compared with the marginalized cluster. These were GERD per HC researcher, articles per HC researcher, and HC researcher per labor force. For GERD per FTE researcher, there were no significant changes over time. For government expenditure per tertiary student as a percentage of GDP per capita, the same changes over time were seen for all three patent clusters. For all the other series, changes over time differed between the three clusters. The significant differences between clusters were also tested for each year. A consistently significant difference among clusters was found for each of the years for the three series GERD per FTE researcher, GERD per HC researcher, and HC researcher per labor force. For the two articles' series, the difference became insignificant toward the end of the series. These findings highlight the differences between the three patent clusters in terms of patent intensity by relation to human capital measurements.

Lastly, our analysis examined core empirical performance of patent intensity by geographic locations. Although it is unclear how knowledge spillovers are precisely distinct from the correlation of variables at the geographic level, it is plausible that indigenous or domestic agglomeration of patent intensity significantly differs across patenting clusters. Different patent clusters may ultimately necessitate differing patent policies and considerations, infrequently present across the development divide.

Our analysis cautiously offers a conceptual framework based on two measurements of spatial innovation: patent quality, and the degree of newness across patent clusters. The first feature of indigenous or domestic narration of patent intensity in developing countries is arguably funneled by a lower quality of innovation as proxied through patenting activity. Numerous working examples across the developing world, and particularly China, shed light on this relation.

The second factor, and the focal point of the empirical part of our discussion, is the degree of innovativeness as estimated by numerous indicators. We begin with the ratio of domestic patents for which foreign patent protection is sought and the number of foreign jurisdictions in which oversea patent protection is wanted. This dual geographic locational examination highlights the relatively low overseas patenting rates by residents of developing countries and of Northern inventors in patent offices of developing countries as a share of patenting by inventors from developing countries. With almost no changes in these variables over time, we can suggest that the leaders patent cluster is more internationalized in terms of its relatively high patenting intensity than the two lower patent clusters, which show a higher degree of domestic or indigenous patenting rates.

Moreover, the degree of innovativeness correlated with indigenous patenting rates relates to eight growth-related indicators, including the rates of high-technology exports. These indicators may further account for growth-related discrepancies over the degree of spatiality across patent clusters. Eight such indicators are accounted for: GERD as a percentage of GDP, GERD per capita, GDP per capita, GNI per capita, GNI, GDP, high-technology export rates as percent of manufactured export, and high-technology export rates. As we saw, significant differences exist between all three clusters, with the lowest mean value for the marginalized cluster and the highest for the leaders cluster. Regarding the followers and marginalized patent clusters, the medium- and low-income growth rates shown, respectively, may plausibly correspond with lower localization of knowledge creation through local innovation industries when funneled by high internationalized patenting rates. These growth rates instead correspond with higher indigenous or domestic patenting rates in the followers and marginalized clusters. High indigenous patenting rates, as per the latter patent clusters, further relate to low technology export rates by the developing countries of these two clusters.

This book has aimed to make a contribution to our understanding of how and why countries differ in their patent intensity. It is hope that the insights it offers may, in the years to come, provide the foundation for a novel theoretical framework for enhancing innovation-based economic growth.

References

2006 Patent Cooperation Treaty Conference: Transcript of Proceedings, 32 *William Mitchell Law Review*, 1603 (2006)

Abbott, Frederick M., Intellectual Property Provisions of Bilateral and Regional Trade Agreements in Light of US. Federal Law 1 (International Centre for Trade and Sustainable Development, Issue Paper No. 12, Feb. 2006), at http://www.unctad.org/en/docs/iteipc20064_en.pdf

Abbott, Frederick M., The Cycle of Action and Reaction: Developments and Trends in Intellectual Property and Health, in *Negotiating Health: Intellectual Property and Access to Medicines* 31–33 (Pedro Roffe, Geoff Tansey and David Vivas-Eugui, eds.), Earthscan (2006)

Abbott, Frederick M., The WTO TRIPS Agreement and Global Economic Development, in *Public Policy and Global Technological Integration* 39 (Springer) (Frederick M. Abbott and David J. Gerber, eds. (1997)

Abbott, Frederick, Carlos Correa and Peter Drahos, Emerging Markets and the World Patent Order: The Forces of Change, in *Emerging Markets and the World Patent Order*, Edward Elgar (Frederick. Abbott, Carlos Correa and Peter Drahos, eds.) 3 (2013)

Abramovitz, Moses, Rapid Growth Potential and Its Realization: The Experience of Capitalist Economics in the Postwar Period, in 1 *Economic Growth and Resources*, 191 (Edmond Malinvaud, ed., 1979)

Abreau, Maria and Vadim Grinevich, The Nature of Academic Entrepreneurship in the UK: Widening the Focus on Entrepreneurial Activities, *Research Policy*, 42, 408 (2013)

Acemoglu, Daron, and Fabrizio Zilibotti, Productivity Differences, *Quarterly Journal of Economics*, 116, 563–606 (2001)

Acemoglu, Daron, Philippe Aghion, and Fabrizio Zilibotti, Distance to Frontier, Selection and Economic Growth, unpublished, MIT (2002)

Acs, Zoltan J. and David B. Audretsch, Innovation in Large and Small Firms: An Empirical Analysis, *The American Economic Review*, 678 (1988)

Acs, Zoltan J., David B. Audretsch, Maryann P. Feldman, Real Effects of Academic Research: Comment, *American Economic Review*, 82, 363 (1992)

Adams, James D., Fundamental Stocks of Knowledge and Productivity Growth, *Journal of Political Economy*, 98(4) 673 (1990)

African group, http://www.wto.org/english/tratop_e/trips_e/trips_groups_e.htm

Aghion, Philippe and Peter Howitt, A Model of Growth Through Creative Destruction, 60(2) *Econometrica*, 323 (1992)

Aghion, Philippe, Peter Howitt and David Mayer-Foulkes, *The Effect of Financial Development on Convergence: Theory and Evidence*, The Quarterly Journal of Economics, MIT Press, vol. 120(1) 173 (2005 January)

Agreement between the United Nations and the World Intellectual Property Organization (Dec. 17, 1974)

Agreement on Trade-Related Aspects of Intellectual Property Rights in the Marrakesh Agreement Establishing the World Trade Organization, Annex 1C, 1869 U.N.T.S. 299 (Apr. 15, 1994), at http://www.wto.org/eng lish/docs_e/legal_e/27-trips.pdf

Alexiadis, Stilianos, *Convergence Clubs and Spatial Externalities, Advances in Spatial Science* (Springer-Verlag 2013)

Alfaro, Laura, Areendam Chanda, Sebnem Kalemli-Ozcan and Selin Sayek, *How does Foreign Direct Investment Promote Economic Growth? Exploring the Effects of Financial Markets on Linkages*, NBER Working Papers 12522, National Bureau of Economic Research, Inc. (2006)

Alfrancam, Oscar and Wallace E. Huffman, Aggregate Private R&D Investments in Agriculture: The Role of Incentives, Public Policies, and Institutions, 52 *Economic Development and Cultural Change* 1 (2003)

Allison, John R. and Starling David Hunter, On the Feasibility of Improving Patent Quality One Technology at a Time: The Case of Business Methods, *Berkeley Technology Law Journal*, 21(2) 729 (2015)

Almeida, Paul and Bruce Kogut, Localization of Knowledge and the Mobility of Engineers in Regional Networks, *Management Science*, 45, 905 (1999)

Altenburg, Tilman, Building Inclusive Innovation in Developing Countries: Challenges for IS research, in *Handbook of Innovation Systems and Developing Countries* (Bengt-Åke Lundvall K. J. Joseph, Cristina Chaminade and Jan Vang, eds.) (2009)

Amann, Edmund and John Cantwell, *Innovative Firms in Emerging Markets Countries*, Oxford: Oxford University Press (2012)

Amsden, Alice and Ted Tschang, A New Approach to Assess the Technological Complexity of Different Categories of R&D (with examples from Singapore), *Research Policy*, 32(4), 553 (2003)

Amsden, Alice H., *The Rise of "The Rest": Challenges to the West from Late-Industrializing Economies*, Oxford University Press (2001)

Anand, Sudhir and Amartya Sen, Human Development and Economic Sustainability, *World Development*, Elsevier, vol. 28(12) 2029 (2000)

Andrés, Javier and Ángel de la Fuente & Rafael Doménech, *Human Capital in Growth Regressions: How much Difference Does Data Quality Make?*, CEPR discussion paper, no. 2466, London: Centre for Economic Policy Research (2000)

Andries, Door Petra, Julie Delanote, Sarah Demeulemeester, Machteld Hoskens, Nima Moshgbar, Kristof Van Criekingen and Laura Verheyden, (2009), O&O-Activiteiten van de Vlaamse bedrijven, in (Koenraad Debackere and Reinhilde Veugelers (eds.), *Vlaams Indicatorenboek Wetenschap, Technologie en Innovatie 2009* (Vlaamse Overheid, Brussel), 53 (2009)

Antonucci, Tommaso and Mario Pianta, Employment Effects of Product and Process Innovation in Europe, *International Review of Applied Economics*, Vol. 16 (3), 295 (2002)

Applied Management Sciences, Inc., *NSF workshop on Industrial S&T data needs for the 1990s: final report, National Science Foundation* report 1 December, (Washington, DC) (1989)

Archibugi, Daniele and Alberto Coco, The Globalization of Technology and the European Innovation System, in Manfred M. Fischer and Josef Fröhlich (eds.) *Knowledge, Complexity and Innovation Systems* 58 (2001)

Archibugi, Daniele and Alberto Coco, The Globalization of Technology and the European Innovation System, IEEE Working Paper DT09/2001. No (2001)

Archibugi, Daniele and Mario Pianta, Measuring Technological Change through Patents and Innovation Surveys, *Technovation*, 16, 451 (1996)

Lutz G., Arnold, Basic and Applied Research, *Finanzarchiv*, vol. 54, 169 (1997)

Arnold, Lutz G., Growth, Welfare, and Trade in an Integrated Model of Human Capital Accumulation and R&D, *Journal of Macroeconomics*, 20(1) 81 (1998)

Arrow, Kenneth, Economic Welfare and the Allocation of Resources for Invention, in *The Rate and Direction of Inventive Activity* (Princeton University Press, New Jersey, 1962) (Richard R. Nelson, ed.)

Arundel, Anthony and Aldo Geuna, Does Proximity Matter for Knowledge Transfer from Public Institutes and Universities to Firms?, Working Paper 73, SPRU (2001)

Arundel, Anthony and Isabelle Kabla, What Percentage of Innovations Are Patented? Empirical Estimates for European Firms, *Research Policy* 27(2), 127 (1988)

Asheim, Bjørn T. and Arne Isaksen, Regional Innovation Systems: The integration of local "sticky" and global "ubiquitous" knowledge, *Journal of Technology Transfer*, 27, 77 (2002)

Asheim, Bjørn T. and Meric S. Gertler, The Geography of Innovation – Regional Innovation Systems, in *The Oxford Handbook of Innovation*, Oxford University Press, 291 (Jess Fagerberg, David C. Mowery, Richard R. Nelson, eds.) (2006)

Asheim, Bjørn T., *Lars Coenen and Martin Svensson-Henning, Nordic SMEs and Regional Innovation Systems*, Oslo: Nordisk Industrifond (2003)

Asian developing members, http://www.wto.org/english/tratop_e/trips_e/trips_groups_e.htm

Attaran, Amir, An Immeasurable Crisis? A Criticism of the Millennium Development Goals and Why They Cannot Be Measured, *PLOS Medicine* 2 (10):318 (October 2005)

Atuahene-Gima, Kwaku, An Exploratory Analysis of the Impact of Market Orientation on New Product Performance: A Contingency Approach, *Journal of Product Innovation Management*, 12, 275 (1995)

Audretsch, David and Maryann P. Feldman, Knowledge Spillovers and the Geography of Innovation and Production, Discussion Paper 953, London: Centre for Economic Policy Research (1994)

Audretsch, David B. and Maryann P. Feldman, R&D Spillovers and the Geography of Innovation and Production, *American Economic Review*, vol. 86, 630 (1996)

Audretsch, David B., Erik E. Lehmann and Susanne Warning, University Spillovers: Strategic Location and New Firm Performance, Discussion Paper 3837, CORE (2003)

Averch, Harvey Allen, *A Strategic Analysis of Science and Technology Policy* (Johns Hopkins Press, Baltimore, 1985)

Baden, Sally and Anne Marie Goetz, Who Needs [Sex] When You Can Have [Gender]? Conflicting Discourses on Gender at Beijing, *Feminist Review*, 56 (Summer): 3 (1997)

Bailén, Jose Maria, Basic Research, Product Innovation, and Growth, Economics Working Papers 88, Department of Economics and Business, Universitat Pompeu Fabra (1994)

Baily, Martin and Hans Gersbach, Efficiency in Manufacturing and the Need for Global Competition, Brookings Papers on Economic Activity, Microeconomics, 307 (1995)

Balcão Reis, Ana and Tiago Neves Sequeira, Human Capital and Overinvestment in R&D, *The Scandinavian Journal of Economics*, Vol. 109(3) 573 (2007)

Baldwin, John R. and Moreno Da Pont, *Innovation in Canadian Manufacturing Enterprises*, Ottawa: Statistics Canada, Micro Economic Analysis Division (1996)

Balkin, Jack, What Is Access to Knowledge? (April 21, 2006), Balkanization, at: http://balkin.blogspot.co.il/2006/04/what-is-access-to-knowledge.html (on A2K)

Baloch, Irfan, Acting Director, Development Agenda Coordination Division (DACD), WIPO, interview in Geneva, Switzerland on 16 October 2014 (file with author)

Barasa, Laura, Peter Kimuyu, Patrick Vermeulen, Joris Knoben, Bethuel Kinyanjui, Institutions, Resources and innovation in Developing Countries: A Firm Level Approach, Creating Knowledge for Society, Nijmegen (December 2014)

BarNir, Anat, Starting Technologically Innovative Ventures: Reasons, *Human Capital, and Gender, Management Decision*, 50(3), 399 (2012)

Barro, Robert J. and Xavier Sala-i-Martin, Convergence across States and Regions, *Brookings Papers on Economic Activity*, 2:107, 58 (1991)

Barro, Robert J. and Xavier Sala-i-Martin, Convergence, *Journal of Political Economy*, 100, 223 (1992)

Barro, Robert J., Economic Growth in a Cross-Section of Countries, *Quarterly Journal of Economics*, vol. 106, 407 (1991)

Bartelsman, Eric J., *Federally Sponsored R&D and Productivity Growth*, Federal Reserve Economics Discussion Paper No. 121. Federal Reserve Board of Governors, Washington, DC (1990)

Barton, John H. and Keith E. Maskus, Economic Perspective on a Multilateral Agreement on Open Access to Basic Science and Technology 349, in *Economic Development and Multilateral Trade Cooperation* (Bernard Hoekman and Simon J. Evenett, eds.) (2006)

Baskaran, Angathevar, From Science to Commerce: The Evolution of Space Development Policy and Technology Accumulation in India, *Technology in Society*, 27(2), 155–179 (2005)

Bassanini, Andrea, Stefano Scarpetta and Philip Hemmings, Economic Growth: The Role of Policies and Institutions, Panel Data Evidence from OECD Countries, OECD working paper, STI 2001/9 (2001)

Basu, Susanto and David N. Weil, Appropriate Technology and Growth, *Quarterly Journal of Economics*, 113, 1025–54 (1998)

Baumol, William J. and Edward N. Wolff, Productivity Growth, Convergence, and Welfare: Reply, *The American Economic Review*, Vol. 78(5) 1155 (1988)

Baumol, William J., Education for Innovation, in *Innovation Policy and the Economy* (Adam Jaffe, Josh Lerner and Scott Stern, eds.), vol. 5 (MIT Press, 2005)

Bayh–Dole Act or Patent and Trademark Law Amendments Act (Pub. L. 96–517, December 12, 1980)

Becker, Gary S. and Barry R. Chiswick, Education and the Distribution of Earnings, *American Economic Review*, vol. 53, 358 (1966)

Becker, Gary S., Kevin M. Murphy, Robert Tamura, Human Capital, Fertility, and Economic Growth, *Journal of Political Economy*, vol. 98, no. 5, pages S12–37 (October 1990), reprinted in Human Capital: A Theoretical and Empirical Analysis with Special Reference to Education (3rd Edition) (1994)

Becker, Gary Stanley, *Human Capital: A Theoretical and Empirical Analysis with Special Reference to Education*, New York: Columbia University Press (1964)

Becker, Gary Stanley, Investment in Human Capital: A Theoretical Analysis, *Journal of Political Economy*, vol. 70, 9 (1975)

Beijing Treaty on Audiovisual Performances (June 24, 2012)

Bell, Martin and Keith Pavitt, Technological Accumulation and Industrial Growth: Contrasts between Developed and Developing Countries, *Industrial Corporate Change*, 2(2) 157 (1993)

Belussi, Fiorenza and Katia Caldari, At the Origin of the Industrial District: Alfred Marshall and the Cambridge School, *Cambridge Journal of Economics*, 33, 335 (2009)

Ben-David, Dan, Convergence Clubs and Subsistence Economies, *Journal of Development Economics*, vol. 55(1) 155 (1988)

Beneito, Pilar, Choosing among Alternative Technological Strategies: An Empirical Analysis of Formal Sources of Innovation, Research Policy, 32, 693 (2003)

Beneito, Pilar, The Innovative Performance of In-house and Contracted R&D in Terms of Patents and Utility Models, *Research Policy* 35 (2006) 502 (2006)

Benhabib, Jess and Mark M. Spiegel, The Role of Human Capital in Economic Development: Evidence from Aggregate Cross-Country Data, *Journal of Monetary Economics*, vol. 34, 143 (1994)

Benoliel, Daniel and Bruno M. Salama, Toward an Intellectual Property Bargaining Theory: The Post-WTO Era. 32 University of Pennsylvania *Journal of International Law*, 265 (2010)

Benoliel, Daniel, Patent Convergence Club Among Nations, *Marquette Intellectual Property Law Review*, vol. 18(2) 297 (2014)

Benoliel, Daniel, The Impact of Institutions on Patent Propensity Across Countries, *Boston University International Law Journal*, vol. 33(1) 129 (2015)

Benoliel, Daniel, The International Patent Propensity Divide, *North Carolina Journal of Law and Technology*, vol. 15(1) 49 (2013)

Bergek, Anna and Maria Bruzelius, Patents with Inventors from Different Countries: Exploring Some Methodological Issues through a Case Study, presented at the DRUID conference, Copenhagen, 27–29 June (2005)

Bessen, James and Eric Maskin, Sequential Innovation, Patents, and Imitation, Department of Economics, Massachusetts Institute of Technology working paper no. 00–01 (2000)

Bessen, James and Michael J. Meurer, *Patent Failure: How Judges, Lawyers and Bureaucrats Put Innovators at Risk*, Princeton University Press (2008)

Bessen, James and Robert M. Hunt, An Empirical Look at Software Patents, *Journal of Economics & Management Strategy, Wiley Blackwell*, vol. 16(1) 157 (2007)

Bhagwati, Jagdish N. (ed.), *The New International Economic Order: The North-South Debate*, MIT Press (1977)

Bianchi, Tito, With and Without Co-operation: Two Alternative Strategies in the Food Processing Industry in the Italian South, Entrepreneurship & Regional Development, nr. 13, 117 (2001)

Bilbao-Osorio, Beñat and Andrés Rodríguez-Pose, From R&D to Innovation and Economic Growth in the EU, *Growth and Change*, 35(4), 434 (2004)

Birdsall, Nancy and Changyong Rhee, *Does R&D Contribute to Economic Growth in Developing Countries?*, *Mimeo*, The World Bank, Washington, DC (1993)

Bitton, Miriam, Patenting Abstractions, *North Carolina Journal of Law and Technology*, vol. 15(2) 153 (2014)

Black, Duncan and J. Vernon Henderson, A Theory of Urban Growth, *Journal of Political Economy*, vol. 107(2) 252 (1999)

Blackburn, Keith, Victor T.Y. Hung and Alberto F. Pozzolo, Research, Development and Human Capital Accumulation, *Journal of Macroeconomics*, 22(2) 189 (2000)

Blakeney, Michael, The International Protection of Industrial Property: From the Paris Convention to the TRIPS Agreement, WIPO National Seminar on Intellectual Property, 2003, WIPO/IP/CAI/1/03/2

Blaug, Mark, *The Economics of Education and the Education of an Economist*, Edward Elgar Publishing, no. 48 (1987)

Boards of appeal of the European Patent Office decision T 388/04, OJ 1/2007, 16

Bogliacino, Francesco, Mariacristina Piva and Marco Vivarelli, R&D and Employment: An Application of the LSDVC Estimator using European Data, *Economics Letters*, 116(1) 56 (2012)

Bonin, Bernard y Claude Desranleau, *Innovation Industrielle et Analyse économique* (Montréal: HEC, 1987)

Borges Barbosa, Denis, Patents and the emerging markets of Latin America-Brazil, In Emerging Markets and the World Patent Order, Edward Elgar (Frederick Abbott, Carlos Correa and Peter Drahos, eds.) 135 (2013)

Borts, George H., Jerome L. Stein, *Economic Growth in a Free Market*, Columbia University Press (1964)

Bottazzi, Laura and Giovanni Peri, The International Dynamics of R&D and Innovation in the Short and in the Long Run, NBER Working Paper No. 11524 (July 2005) at: http://www.nber.org/papers/w11524.pdf?new_window=1

334   References

Botto, Mercedes, *Research and International Trade Policy Negotiations: Knowledge and Power in Latin America*, Routledge/IDRC (2009)

Bowles, Sam and Herb Gintis, *Schooling in Capitalist America: Educational Reform and the Contradictions of Economic Life* (Routledge and Kegan Paul, 1976)

Bradley, Steve W., Jeffery S. McMullen, Kendall Artz and Edward M. Simiyu, Capital Is Not Enough: Innovation in Developing Economies, *Journal of Management Studies*, 49(4), 684 (2012)

Braithwaite, John and Peter Drahos, *Global Business Regulation*, Cambridge University (2000)

Branstetter, Lee, Guangwei Li and Francisco Veloso, The Rise of International Coinvention, in *The Changing Frontier: Rethinking Science and Innovation Policy* (Adam B. Jaffe and Benjamin F. Jones, eds.), University of Chicago Press (2015)

Brenner, Thomas and Tom Broekel, *Methodological Issues in Measuring Innovation Performance of Spatial Units*, Papers in Evolutionary Economic Geography, No. 04-2009, Urban & Regional Research Centre Utrecht, Utrecht University (2009)

Breschi, Stefano and Francesco Lissoni, *Knowledge Spillovers and Local Innovation Systems: A Critical Survey*, Industrial and Corporate Change, Oxford University Press, vol. 10(4) 975 (2001)

BRICS, The BRICS Report: A Study of Brazil, Russia, India, China, and South Africa with Special Focus on Synergies and Complementarities (September 2012)

Brouwer, Erik and Alfred Kleinknecht Innovative Output, and a Firm's Propensity to Patent. *An Exploration of CIS Microdata, Research Policy* 28(6) 615 (1999)

Brouwer, Erik, Alfred Kleinknecht and Jeroen O. N. Reijnen, Employment Growth and Innovation at the Firm Level, Journal of Evolutionary Economics 3, 153 (1993)

Brouwer, Erik, Hana Budil-Nadvornikova and Alfred Kleinknecht, Are Urban Agglomerations a Better Breeding Place for Product Innovations? An Analysis of New Product Announcements, Regional Studies, 33, 541 (1999)

Bucci, Alberto, Monopoly Power in Human Capital-Based Growth, *Acta Oeconomica*, Vol. 55(2) 121 (2005)

Buerger, Matthias, Tom Broekel and Alex Coad, Regional Dynamics of Innovation: Investigating the Coevolution of Patents, Research and Development (R&D), and Employment, *Regional Studies*, vol. 46(5) 565 (2012)

Bureau of the Census, *Evaluation of Proposed Changes to the Survey of Industrial Research and Development*, National Science Foundation report, June (Washington, DC) (1993)

Bush, Vannevar, *Science: The Endless Frontier, A Report to the President by Vannevar Bush, Director of the Office of Scientific Research and Development*, July 1945 (United States Government Printing Office, Washington: 1945)

Buy Brazil Act 12.349/10 of 15 December 2010 (Brazil)   .

Caillaud, Bernard and Anne Duchêne, *Patent Office in Innovation Policy: Nobody's Perfect*, Paris School of Economics, Working Paper, 2009–39 (2009)

Calvert, Jane and Ben R. Martin, Changing Conceptions of Basic Research?, Background Document for the Workshop on Policy Relevance and Measurement of Basic Research Oslo 29–30 October 2001, SPRU – Science and Technology Policy Research (September 2001)

Canova, Fabio and Albert Marcet, The Poor Stay Poor: Non-Convergence Across Countries and Regions, Discussion Paper 1265, London Centre for Economic Policy Research (1995)

Canova, Fabio, Testing for Convergence Clubs in Income Per Capita: A Predictive Density Approach, *International Economic Review*, Vol. 45(1) (2004)

Cao, Cong, Richard P. Suttmeier and Denis F. Simon, China's 15-Year Science and Technology Plan, 59 *Physics Today* 38 (2006)

Carden, Fred, *Knowledge to Policy: Making the Most of Development Research*, Sage Publications Ltd. (2009)

Cardoso, Fernando Henrique and Enzo Faletto, *Dependency and Development in Latin America 149–71*, Marjory Mattingly Uriquidi, trans., University of California Press (1979)

Carillo, Maria Rosaria and Erasmo Papagni, *Social Rewards in Science and Economic Growth*, MPRA Paper 2776, University Library of Munich, Germany (2007)

Carlino, Gerald A., Knowledge Spillovers: Cities' Role in the New Economy, *Business Review*, Q4 17 (2001)

Cass, David, Optimum Growth in an Aggregative Model of Capital Accumulation, *The Review of Economic Studies*, 32, 233 (1965)

Castellacci, Fulvio, Convergence and Divergence among Technology Clubs, DRUID Working Paper No. 06–21 1 (2006)

Chang, Ha-Joon, *Globalization, Economic Development and the Role of the State* (Zed Books) (2003)

Chang, Ha-Joon, *Kicking Away the Ladder: Policies and Institutions for Economic Development in Historical Perspective*, London: Anthem Press (2003)

Chang, Ha-Joon, Understanding the Relationship between Institutions and Economic Development: Some Key Theoretical Issues 1 (U.N. World Institute for Development Economics Research, Discussion Paper No. 2006/05, 2006), available at http://www.wider.unu.edu/publications/working-papers/discussion-papers/2006/en_GB/dp2006-05/

Charrad, Malika, Nadia Ghazzali, Véronique Boiteau and Azam Niknafs, NbClust: An R Package for Determining the Relevant Number of Clusters in a Data Set, *Journal of Statistical Software*, Vol. 61(6) (2014)

Chen, Yongmin and Thitima Puttitanun, Intellectual Property Rights and Innovation in Developing Countries, 78 *Journal of Development Economics* 474 (2005)

Chenery, Hollis and Moshe Syrquin, Typical Patterns of Transformation, in (Hollis Chenery, Sherman Robinson and Moshe Syrquin, eds.), *Industrialization and Growth: A Comparative Study*, Oxford: Oxford University Press (1986)

Chon, Margaret, Denis Borges Barbosa and Andrés Moncayo von Hase, Slouching Toward Development in International Intellectual Property, *Michigan State Law Review*, 71 (2007)

Chowdhury, Anis and Iyanatul Islam, The Newly Industrializing Economies of East Asia 4 (1997)

Cimoli, Mario, João Carlos Ferraz and Annalisa Primi, Science and technology policies in open economies: the case of Latin America and the Caribbean, Paper presented at the first meeting of ministers and high authorities on science and technology, Lima, Peru, 1112 November (2004)

Cimoli, Mario, Networks, market structures and economic shocks: the structural changes of innovation systems in Latin America, Paper presented at the seminar on Redes productivas e institucionales en America Latina, Buenos Aires, 9–12 April (2001)

Cincera, Michele, Firms' Productivity Growth and R&D Spillovers: An Analysis of Alternative Technological Proximity Measures, Economics of Innovation and New Technology, *Taylor and Francis Journals*, vol. 14(8), 657 (2005)

Coad, Alexander and Rekha Rao, The Employment Effects of Innovations in High-Tech Industries, Papers on Economics & Evolution, #0705, Max Planck Institute of Economics, Jena (2007)

Coase, Ronald, The Nature of the Firm, in Economica NS 4, 386–405 (1937)

Cockburn, Iain M., Rebecca M. Henderson, Publicly Funded Science and the Productivity of the Pharmaceutical Industry, in *Innovation Policy and the Economy*, vol. 1, 1 (Adam B. Jaffe, Josh Lerner and Scott Stern, eds.) (MIT Press, Cambridge, MA) (2001)

Coe, David T. and Elhanan Helpman, International R&D Spillovers, *European Economic Review* 39: 859–887 (1995)

Coe, David T., Elhanan Helpman and Alexander W. Hoffmaister, North-South R&D Spillovers, *The Economic Journal*, Vol. 107(440) 134 (1997)

Cohen, Daniel and Marcelo Soto, *Growth and Human Capital: Good Data, Good Results*, CEPR discussion paper, no. 3025, London: Centre for Economic Policy Research (2001)

Cohen, Linda R. and Roger G. Noll, *The Technology Pork Barrel*. The Brookings Institution Press, Washington, DC (1997)

Cohen, Wesley M. and Daniel A. Levinthal, Absorptive Capacity: A New Perspective on Learning and Innovation, *Administrative Science Quarterly* 35(1), 128 (1990)

Cohen, Wesley M., Akira Goto, Akiya Nagata, Richard R. Nelson, J. Walsh, R&D Spillovers, Patents and the Incentives to Innovate in Japan and the United States, *Research Policy*, 31, 1349 (2002)

Cohen, Wesley M., Richard R. Nelson and John P. Walsh, Links and Impacts: The Influence of Public Research on Industrial R&D, *Management Science*, 48(1), 1 (2002)

Cohen, Wesley M., Richard R. Nelson and John Walsh, Appropriability Conditions and Why Firms Patent and Why They Do Not in the American Manufacturing Sector, Paper presented to the Conference on New S and T

Indicators for the Knowledge Based Economy, OECD, Paris, June 19–21 (1996)

Cohen, Wesley M., Richard R. Nelson, John P. Walsh, Protecting Their Intellectual Assets: Appropriability Conditions and Why US Manufacturing Firms Patent (or Not), NBER Working Paper No. 7552 (February 2000)

Coleman, James S., Social Capital in the Creation of Human Capital, *American Journal of Sociology* (1988), at S95–S120

Collins, Eileen L., *Estimating Basic and Applied Research and Development in Industry: A Preliminary Review of Survey Procedures*, National Science Foundation report (Washington, DC) (1990)

Commission for Africa, Our Common Interest: Report of the Commission for Africa (2005)

Convention Establishing the World Intellectual Property Organization, July 14, 1967, 21 UST. 1749, 828 U.N.T.S. 3

Convention on the Grant of European Patents (European Patent Convention) of 5 October 1973

Cooke, Philip N., P. Boekholt, Franz Tödtling, *The Governance of Innovation in Europe*, London: Pinter (2000))

Cooke, Philip, Regional Innovation Systems, Clusters, and the Knowledge Economy, *Industrial and Corporate Change*, 10(4), 945 (2001)

Cooter, Robert D. and Hans-Bernd Schaefer, *Solomon's Knot: How Law Can End the Poverty of Nations*, Princeton University Press (2009)

Cornwall, John, Modern Capitalism: Its Growth and Transformation (1977)

Correa, Carlos M., *Intellectual Property Rights, the WTO and Developing Countries: The TRIPS Agreement and Policy Options* (Zed books) (2000)

Court, Julius and John Young, Bridging Research and Policy in International Development, in *Global Knowledge Networks and International Development*, Routledge 18 (Diane Stone and Simon Maxwell, eds.) (2005)

Cozzi, Guido and Silvia Galli, Privatization of Knowledge: Did the US Get It Right? MPRA Paper 29710 (2011)

Cozzi, Guido and Silvia Galli, Science-Based R&D in Schumpeterian Growth, *Scottish Journal of Political Economy*, Vol. 56(4) 474 (2009)

Cozzi, Guido and Silvia Galli, Upstream Innovation Protection: Common Law Evolution and the Dynamics of Wage Inequality, MPRA Paper 31902 (2011)

Crescenzi, Riccardo and Andrés Rodríguez-Pose, *Innovation and Regional Growth in the European Union* (Springer Berlin Heidelberg, 2011)

Crespi, Gustavo and Pluvia Zuñiga, Innovation and Productivity: Evidence from Six Latin American Countries, *World Development*, 40(2), 273 (2011)

Crespi, Gustavo and Ezequiel Tacsir, *Effects of Innovation on Employment in Latin America, Inter-American Development Bank Institutions for Development (IFD) Technical Note*, Inter-American Development Bank, Washington, DC (2012)

Crewe, Emma and John Young, Bridging Research and Policy: Context, Links and Evidence (2002)

Crosby, Mark, Patents, Innovation and Growth, *The Economic Record*, 76 (234), 255 (2000)

Cuneo, Philippe and Jacques Mairesse, Productivity and R&D at the Firm Level in French Manufacturing, in Zvi Griliches (ed.), *R&D, Patents and Productivity*, Chicago: University of Chicago Press, 339 (1984)

Cyert, Richard M. and James G. March, *A Behavioral Theory of the Firm*, Prentice-Hall (1963)

Cypher, James M. and James L. Dietz, The Process of Economic Development (2009)

Czarnitzki, Dirk and Susanne Thorwarth, Productivity Effects of Basic Research in Low-tech and High-tech Industries, *Research Policy*, 41(9) 1555 (2012)

Dai, Yixin, David Popp and Stuart Bretschneider, *Journal of Policy Analysis and Management*, Vol. 24(3) 579, (2005)

Dam, Kenneth W., The Economic Underpinnings of Patent Law, 23 *Journal of Legal Studies*, 247 (1994)

Danguy, Jérôme, Gaétan de Rassenfosse and Bruno van Pottelsberghe de la Potterie, The R&D-Patent Relationship: An Industry Perspective, ECARES working paper 2010–038 (September 2010)

Dasgupta, Partha, *The Economic Theory of Technology Policy: An Introduction, in Paul Stoneman Partha Dasgupta, Economic Policy and Technological Performance*, Cambridge University Press, Cambridge (1987)

Dasgupta, Partha and Eric Maskin, The Simple Economics of Research Portfolios, *The Economic Journal*, vol. 97, 581 (1987)

David, Matthew and Debora Halbert, IP and Development 89, in *Sage Handbook on Intellectual Property*, SAGE Publications Ltd (Matthew David and Debora Halbert, eds.) (2014)

David, Paul A. and Bronwyn H. Hall, Property and the Pursuit of Knowledge: IPR Issues Affecting Scientific Research, *Research Policy*, 35(6), 767–771 (2006)

David, Paul A., Bronwyn H. Hall, and Andrew A. Toole, Is Public R&D a Complement or Substitute for Private R&D? A Review of the Econometric Evidence, *Research Policy*, vol. 29 (4–5) 497 (2000)

David, Paul, Clio and the Economics of QWERTY, *American Economic Review*, vol. 75 332 (1985)

de Ferranti, David, Guillermo Perry, Indermit Gill, William Maloney, Jose Luís Guasch, Carolina Sanchez-Paramo and Norbert Schady, *Closing the Gap in Education and Technology* (Washington, DC: World Bank) (2003)

de la Mothe, John and Gilles Paquet (eds.), *Local and Regional Systems of Innovation*, Amsterdam: Kluwer Academics Publishers (1998)

de Moraes Aviani, Daniela, Data from National Service for the Protection of Cultivars, available at: www.sbmp.org.br/6congresso/wp-content/uploads/2011/08/1.-Daniela-Aviani-Panorama-Atual-no-Brasil.pdf

de Vibe, Maja, Ingeborg Hovland and John Young, Bridging Research and Policy: An Annotated Bibliography, ODI Working Paper No 174, ODI, London (2002)

de Melo-Martín, Inmaculada, Patenting and the Gender Gap: Should Women Be Encouraged to Patent More? *Science and Engineering Ethics*, 19, 491 (2013)

Dearborn, DeWitt C., Rose W. Kneznek and Robert N. Anthony, *Spending for industrial research, 1951–52*, Harvard University Graduate School of Business report (Cambridge, MA) (1953)

Declaration of the Fourth BRICS Summit: BRICS Partnership for Stability, Security and Prosperity, 29 March 2012, at: http://www.itamaraty.gov.br/en/press-releases/9428-fourth-brics-summit-new-delhi-29-march-2012-brics-par tnership-for-global-stability-security-and-prosperity-delhi-declaration

Dedrick, Jason and Kenneth L Kraemer, *Asia's Computer Challenge*, Oxford University Press (1998)

Deere Birkbeck, Carolyn and Santiago Roca, An External Review of WIPO Technical Assistance in the Area of Cooperation for Development (31 August 2011)

Deere, Carolyn, Developing Countries in the Global IP System, in *The Implementation Game: The TRIPS Agreement and the Global Politics of Intellectual Property Reform in Developing Countries*, Oxford University Press (2009)

Deere, Carolyn, *The Implementation Game: The TRIPS Agreement and the Global Politics of Intellectual Property Reform in Developing Countries*, Oxford University Press (2009)

Deloitte, Development Trends and Practical Aspects of the Russian Pharmaceutical Industry – 2014 (2014)

Dennison, Edward, *Accounting for Growth*, Harvard University Press (1985)

Diamond v. Diehr, 450 US 175 (1981)

Ding, Waverly W., Fiona Murray and Toby E. Stuart, Gender Differences in Patenting in the Academic Life Sciences, *Science*, 313 (5787), 665 (2006)

Directive 98/44/EC of the European Parliament and of the Council of 6 July 1998 on the legal protection of biotechnological inventions

Doloreux, David, Innovative Networks in Core Manufacturing Firms: Evidence from the Metropolitan Area of Ottawa, *European Planning Studies*, vol. 12(2) (2004) 178

Doloreux, David, Regional Innovation Systems in the Periphery: The Case of the Beauce in Québec (Canada), *International Journal of Innovation Management*, 7(1) 67 (2003)

Doloreux, David, What We Should Know about Regional Systems of Innovation, *Technology in Society*, 24 (2002)

Dosi, Giovanni, Institutional Factors and Market Mechanisms in the Innovative Process, SERC, University of Sussex, mimeo (1979)

Dosi, Giovanni, Institutions and Markets in a Dynamic World, *The Manchester School* 56(2), 119–146 (1988)

Dosi, Giovanni, The Nature of the Innovation Process, Chapter 10 in *Technical Change and Economic Theory* (Giovanni Dosi, Christopher Freeman, Richard Nelson, Gerarld Silverberg and Luc Soete, eds.), LEM Book Series (1988)

Dowrick, Steve and Duc-Tho Nguyen, OECD Comparative Economic Growth: Catch Up and Convergence, *American Economic Review*, 79 (1989) 1010

Dowrick, Steve and Norman Gemmell, Industrialization, Catching Up and Economic Growth: A Comparative Study Across the World's Capitalist Countries, *The Economic Journal*, 101, 263 (1991)

Drahos, Peter and John Braithwaite, *Information Feudalism: Who Owns the Knowledge Economy*, Earthscan Publications Ltd., London (2003)

Drahos, Peter, An Alternative Framework for the Global Regulation of Intellectual Property Rights, 21 *Austrian Journal of Development Studies* 1 (2005)

Drahos, Peter, Developing Countries and International Intellectual Property Standards-Setting, 5 *Journal of World Intellectual Property*, 765 (2002)

Dreyfuss, Rochelle C., Intellectual Property Lawmaking, Global Governance and Emerging Economies, 53 In *Patent Law in Global Perspective* (Ruth L. Okediji and Margo A. Bagley, eds.), Oxford University Press (2014)

Dreyfuss, Rochelle C., The Role of India, China, Brazil and the Emerging Economies in Establishing Access Norms for Intellectual Property and Intellectual Property Lawmaking, IILJ Working Paper 2009/5

Drug Price Competition and Patent Term Restoration Act of 1984, Public Law 98–417, codified in pertinent part at 35 USC § 271(e)

Duch, Jordi, Xiao Han T. Zeng, Marta Sales-Pardo, Filippo Radicchi, Shayna Otis, Teresa K. Woodruff and Luís A. Nunes Amaral, The Possible Role of Resource Requirements and Academic Career-Choice Risk on Gender Differences in Publication Rate and Impact, *PLoS ONE*, 7(12): e51332 (2012)

Dudley, John M., Defending Basic Research, *Nature Photonics*, 7, 338 (2013)

Duguet, Emmanuel, and Isabelle Kabla, Appropriation Strategy and the Motivations to Use the Patent System: An Econometric Analysis at the Firm Level, Ann. INSEE (2010)

Duguet, Emmanuel, and Isabelle Kabla, Appropriation Strategy and the Motivations to Use the Patent System: An Econometric Analysis at the Firm Level in French Manufacturing, *Annales D'Économie et de Statistique*, 49/50, 289–327 (1998)

Dunning, John H. and Sarianna M. Lundan, The Internationalization of Corporate R&D: A Review of the Evidence and Some Policy Implications for Home Countries, *Review of Policy Research*, Vol. 26, Numbers 1–2, 13 (2009)

Durlauf, Steven Neil and Paul A. Johnson, Multiple Regimes and Cross-Country Growth Behavior, *Journal of Applied Econometrics*, vol. 10, 365 (1995)

Eaton, Jonathan and Samuel Kortum, Josh Lerner, International Patenting and the European Patent Office: A Quantitative Assessment, Patents, Innovation and Economic Performance: OECD Conference Proceedings (2004)

Eberhardt, Markus, Christian Helmers and Zhihong Yu, Is the Dragon Learning to Fly? An Analysis of the Chinese Patent Explosion, CSAE Working Paper no. 2011–15, Centre for the Studies of African Economies (2011)

Eckstein, Zvi and Jonathan Eaton, *Cities and Growth: Theory and Evidence from France and Japan*, Working paper, Economics Department, Tel-Aviv University, September (1994)

Ederer, Peer, Lisbon Council Policy Brief, Innovation at Work: The European Human Capital Index (2006)

Edwards, Michael, R&D in Emerging Markets: A new Approach for a New Era, McKinsey, February 2010, at: http://www.mckinsey.com

Einhaäupl, Karl Max, *What does "Basic Research" mean in Today's Research Environment?, Keynote address to the OECD workshop on "Basic Research: Policy Relevant Definitions and Measurement,"* 28–30 October, Oslo, Norway (2001)

Eisenberg, Rebecca, Intellectual Property Rights and the Dissemination of Research Tools in Molecular Biology: Summary of a Workshop Held at the National Academy of Sciences, February 15–16 (1996)

Eisenstadt, Shmuel N., *Traditional Patrimonialism and Modern Neo-Patrimonialism,* London Sage (1973)

Elms, Deborah K., The Trans-Pacific Partnership Trade Negotiations: Some Outstanding Issues for the Final Stretch, *Asian Journal of the WTO and International Health Law and Policy,* vol. 8(2) 379 (2013)

Elzinga, Aant, Research, Bureaucracy and the Drift of Epistemic Criteria, in *The University Research System: The Public Policies of the Home of Scientists* (Björn Wittrock and Aant Elzinga, eds.) Stockholm: Almqvist and Wiksell International (1985)

Encaoua, David, Dominique Guellec and Catalina Martínez, Patent Systems for Encouraging Innovation: Lessons from Economic Analysis, *Research Policy,* 35(9) 1423 (2006)

Engelbrecht, Hans-Jürgen, International R&D Spillovers, Human Capital and Productivity in OECD Economies: An Empirical Investigation, *European Economic Review,* 41(8): 1479–1488 (1997)

Epstein, Richard A., *Steady the Course: Property Rights in Genetic Material* (University of Chicago Law School, John M. Olin Law and Economics Working Paper No. 152 [2d set.], June 2002)

Esteban, Joan-María and Debraj Ray, On the Measurement of Polarization, *Econometrica,* 62(4):819 (1994)

Ettlie, John E. and Albert H. Rubenstein, Firm Size and Product Innovation, *Journal of Product Innovation Management,* 4, 89 (1987)

European Commission (EC), 2013 EU Survey on R&D Investment Business Trends, Monitoring Industrial Research: The 2006 EU Survey on R&D Investment Business Trends (Alexander Tübke and René van Bavel, eds.) (2013)

European Commission (EC), Benchmarking policy measures for gender equality in science (2008)

European Commission (EC), *Communication from the Commission to the European Parliament, the Council, the European Economic and Social Committee and the Committee of the Regions, Europe 2020 Flagship Initiative Innovation Union, COM* (2010) 546 final, European Commission, Brussels (2010)

European Commission (EC), Innovation Union Competitiveness report 2011 (2011)

European Commission (EC), She Figures 2015: Gender in Research and Innovation, Brussels: European Commission (2015)

European Commission (EC), *Waste of Talents: Turning Private Struggles into a Public Issue; Women and Science in the ENWISE Countries, A Report to the European Commission from the ENWISE Expert Group on Women Scientists in Central and Eastern European Countries and in the Baltic States,* Luxembourg (2003)

European Patent Office (EPO), EPO Economic and Scientific Advisory Board (ESAB), Recommendations for Improving the Patent System: 2012 Statement (February 2013)

European Patent Office (EPO), Patent Quality Workshop Report, EPO-Economic and Scientific Advisory Board (ESAB) (May 2012), at: www.epo.org/about-us/office/esab/workshops.html

European Union (EU) members, www.wto.org/english/tratop_e/trips_e/trips_groups_e.htm (1 April 2017)

Eurostat—Statistics explained, Glossary: R&D intensity, at: http://ec.europa.eu/eurostat/statistics-explained/index.php/Main_Page

Evangelista, Rinaldo and Maria Savona, Innovation, Employment and Skills in Services: Firm and Sectoral Evidence, Structural Change and Economic Dynamics, 14, 449 (2003)

Everitt, Brian S., Sabine Landau, Morven Leese, Daniel Stahl, *Cluster analysis* (5th edition) London: Arnold (2011)

Fagerberg, Jan, Bart Verspagen and Marjolein Canieëls, Technology, Growth and Unemployment Across European Regions, *Regional Studies*, 31, 457 (1997)

Falvey, Rod, David Greenaway, and Zhihong Yu, Extending the Melitz Model to Asymmetric Countries (University of Nottingham Research Paper Series, Research Paper 2006/07)

Falvey, Rod, Neil Foster and David Greenaway, Intellectual Property Rights and Economic Growth, *Review of Development Economics*, 10(4): 700 (2006)

Falvey, Rod, Neil Foster and David Greenaway, Trade, Imitative Ability and Intellectual Property Rights, *Review World Economics*, 145(3), 373 (2009)

Farrell, Joseph and Robert P. Merges, Implementing Reform of the Patent System: Incentives to Challenge and Defend Patents: Why Litigation Won't Reliably Fix Patent Office Errors and Why Administrative Patent Review Might Help, 19 *Berkeley Technology and Law Journal*, 943 (2004)

Feldman, Maryann P., and David B. Audretsch, Innovation in Cities: Science Based Diversity, Specialization and Localized Competition, *European Economic Review*, 409 (1999)

Feldman, Maryann P. and Frank R. Lichtenberg, The Impact and Organization of Publicly-Funded Research and Development in the European Community, NBER Working Paper 6040) (1997)

Feldman, Maryann P., and David B. Audretsch, R&D Spillovers and the Geography of Innovation and Production, *American Economic Review*, vol. 86 (3) 630 (1996)

Feldman, Maryann P., *The Geography of Innovation*, Kluwer: Boston, MA (1994)

Feldman, Maryann P., The New Economics of Innovation, Spillovers and Agglomeration: A Review of Empirical Studies, *Economics of Innovation and New Technology*, 8, 5 (1999)

Feldmann, Horst, Technological Unemployment in Industrial Countries, *Journal of Evolutionary Economics*, 23(5) 1099 (2013)

Ferretti, Marco, Adele Parmentola, *The Creation of Local Innovation Systems in Emerging Countries: The Role of Governments*, Firms and Universities, Springer (2015)

Findlay, Ronald, Relative Backwardness, Direct Foreign Investment and the Transfer of Technology: A Simple Dynamic Model, *Quarterly of Journal of Economics*, 92(1) 1 (1978)

Fine, Ben and Ellen Leopold, *The World of Consumption*, Routledge (1993)

Fine, Ben and Pauline Rose, Education and the Post-Washington Consensus, in *Development Policy in the Twenty-First Century: Beyond the Post-Washington Consensus* (Ben Fine, Contas Lapavitas and Jonathan Pincus, eds.), Routledge (2001)

Fine, Ben, Endogenous Growth Theory: A Critical Assessment, 24 *Cambridge Journal of Economics*, 245 (2000)

Fleisher, Belton M., Yifan Hu, Haizheng Li, Seonghoon Kim, Economic Transition, Higher Education and Worker Productivity in China, *Journal of Development Economics*, 94(1), 86 (2011)

Foray, Dominique, Knowledge Policy for Development, in *OECD, Innovation and the Development Agenda*, Published by OECD and the International Development Research Centre (IDRC), Canada (Kraemer-Mbula Erika and Wamae Watu, eds.) (2010)

Frantzen, Dirk, R&D, Human Capital and International Technology Spillovers: A Cross-Country Analysis, *The Scandinavian Journal of Economics*, Vol. 102(1) 57 (2000)

Freeman, Christopher, *Technology Policy and Economic Performance*, London: Pinter (1987)

Freeman, Christopher, *The Economics of Industrial Innovation*, Penguin: Harmondsworth (1974)

Freeman, Christopher, The National System of Innovation in Historical Perspective, *Cambridge Journal of Economics*, 19, 5 (1995)

Freitas, Isabel Maria Bodas and Nick von Tunzelmann, Mapping Public Support for Innovation: A Comparison of Policy Alignment in the UK and France, *Research Policy*, Vol. 37(9) 1446 (2008)

Freitas, Isabel Maria Bodas and Nick von Tunzelmann, Alignment of Innovation Policy Objectives: A Demand Side Perspective, DRUID Working Paper No. 13–02 (2008)

Frietsch, Rainer, Inna Haller, Melanie Funken-Vrohlings, and Hariolf Grupp, Gender-specific Patterns in Patenting and Publishing, *Research Policy*, 38 590 (2009)

Fu, Xiaolan and Yundan Gong, Indigenous and Foreign Innovation Efforts and Drivers of Technological Upgrading: Evidence from China, *World Development*, Vol. 39(7) 1213 (2011)

Fu, Xiaolan, *China's Path to Innovation*, Cambridge University Press (2015)

Funk Bros. Seed Co. v. Kalo Inoculant Co., 333 US 127 (1948)

Furman, Jeffrey L., Michael E. Porter and Scott Stern, The Determinants of National Innovative Capacity, 31 *Research Policy*, 899 (2002)

Gallini, Nancy, Patent Policy and Costly Imitation, *Rand Journal of Economics*, Vol. 23, 52 (1992)

Gallini, Nancy, The Economics of Patents: Lessons from Recent US Patent Reform, *Journal of Economic Perspectives*, 16(2) 131 (2002)

Galor, Oded, From Stagnation to Growth: Unified Growth Theory, in *Handbook of Economic Growth* (Philippe Aghion and Steven N. Durlauf, eds.), North-Holland, Amsterdam, 171 (2005)

Garcia, Rosanna and Roger Calantone, A Critical Look at Technological Innovation Typology and Innovativeness Terminology: A Literature Review, *The Journal of Product Innovation Management*, 19(2) 110 (2002)

Genolini, Christophe, Xavier Alacoque, Mariane Sentenac, Catherine Arnaud, kml and kml3d: R Packages to Cluster Longitudinal Data, *Journal of Statistical Software*, 65(4), 1 (2015)

Gereffi, Gary and Karina Fernandez-Stark, Global Value Chain Analysis: A Primer, Center on Globalization, Governance & Competitiveness (CGGC) Duke University (May 31, 2011)

Gereffi, Gary, John Humphrey, and Timothy Sturgeon, The Governance of Global Value Chains, *Review of International Political Economy*, vol. 12(1) (2005)

Gerhart, Peter M., The Two Constitutional Visions of the World Trade Organization, 24 *University of Pennsylvania Journal of International Economic Law*, 1, (2003)

Gersbach, Hans and Maik T. Schneider, On the Global Supply of Basic Research, CER-ETH Economics working paper series 13/175, CER-ETH – Center of Economic Research (CER-ETH) at ETH Zurich (2013)

Gersbach, Hans, Gerhard Sorger and Christian Amon, Hierarchical Growth: Basic and Applied Research, CER-ETH Working Papers 118, CER-ETH – Center of Economic Research at ETH Zurich (2009)

Gersbach, Hans, Ulrich Schetter and Maik Schneider, How Much Science? The 5 Ws (and 1 H) of Investing in Basic Research, CEPR Discussion Papers 10482, C.E.P.R. Discussion Papers (2015)

Gerschenkon, Alexander, Economic Backwardness in Historical Perspective, in *The Progress of Underdeveloped Areas* (Bert F. Hoselitz, ed.). Chicago: University of Chicago Press (1971)

Gerschenkron, Alexander, *Economic Backwardness in Historical Perspective: A Book of Essays*, Cambridge, MA: Belknap Press of Harvard University Press (1962)

Gertler, Meric S. and David A. Wolfe, *Dynamics of the Regional Innovation System in Ontario, in Local and Regional Systems of Innovation*. Amsterdam: Kluwer Academic Publishers (J. de la Mothe and Gilles Paquet, eds.) (1998)

Gervais, Daniel J., Information Technology and International Trade: Intellectual Property, Trade & Development: The State of Play, 74 *Fordham Law Review*, 505 (2005)

Gervais, Daniel, Country Club, Empiricism, Blogs and Innovation: The Future of International Intellectual Property Norm Making in the Wake of ACTA 323, in *Trade Governance in the Digital Age* (Mira Burri and Thomas Cottier, eds.), Cambridge University Press (2012)

Gervais, Daniel, TRIPS and Development, 95, in *Sage Handbook on Intellectual Property*, SAGE Publications Ltd. (Matthew David and Debora Halbert, eds.) (2014)

Gerybadze, Alexander and Guido Reger, Globalization of R&D: Recent Changes in the Management of Innovation in Transnational Corporations, *Research Policy*, Vol. 28, No. 2–3 (special issue) 251 (1999)

Gibbons, Michael T., *Engineering by the Numbers*, ASEE (2011)

Gibson, Christopher S., Globalization and the Technology Standards Game: Balancing Concerns of Protectionism and Intellectual Property in International Standards, 22 *Berkeley Technology Law Journal*, 1403, (2007)

Gicheva, Dora and Albert N. Link, Leveraging Entrepreneurship through Private Investments: Does Gender Matter? *Small Business Economics*, 40, 199 (2013)

Gilbert, Richard and Carl Shapiro, Optimal Patent Length and Breadth, *Rand Journal of Economics*, Vol. 21, 106 (1990)

Gilson, Ronald J. and Curtis J. Milhaupt, Economically Benevolent Dictators: Lessons for Developing Democracies, *American Journal of Comparative Law*, Vol. 59(1), 227–288 (2011)

Gimeno, Javier, Timothy B. Folta, Arnold C. Cooper, Carolyn Y. Woo, Survival of the fittest? Entrepreneurial Human Capital and the Persistence of Underperforming Firms, Administrative Science Quarterly, nr. 42, 750 (1997))

Ginarte, Juan Carlos and Walter Park, Determinants of Patent Rights: Cross-National Study, *Research Policy*, 26 (1997)

Glaeser, Edward, Hedi Kallal, José Scheinkman and Andrei Shleifer, Growth in Cities, *The Journal of Political Economy*, vol. 100(6) Centennial Issue 1126 (1992)

Global Value Chains, *SMEs and Clusters in Latin America* (Carlo Pietrobelli and Roberta Rabellotti, eds.), Cambridge Ma.: Harvard University Press (2007)

Godin, Benoît, Measuring science: is there "basic research" without statistics? Project on the History and Sociology of S&T Indicators, Paper No. 3, Montreal: Observatoire des Sciences et des Technologies INRS/CIRST

Godin, Benoît, On the Origins of Bibliometrics, *Scientometrics*, 68 (1) 109–133 (2006)

Godin, Benoît, The Linear Model of Innovation: The Historical Construction of an Analytical Framework, 31 *Science, Technology and Human Values*, 639 (2006)

Goedhuys, Micheline, Learning, Product Innovation, and Firm Heterogeneity in Developing Countries: Evidence from Tanzania, *Industrial and Corporate Change*, 16(2) 269 (2007)

Goedhuys, Micheline, Norbert Janz and Pierre Mohnen, *Knowledge-Based Productivity in "Low-Tech" Industries: Evidence from Firms in Developing Countries*, UNU-MERIT Working Paper 2008–007, Maastricht: UNU-MERIT (2008)

Goldman Sachs Group, The New Geography of Global Innovation, Global Markets Institute report, 20 September 2010 (2010)

Gompers, Paul, Josh Lerner, and David Scharfstein, Entrepreneurial Spawning: Public Corporations and the Genesis of New Ventures, 1986 to 1999, *The Journal of Finance*, 60(2), 577 (2005)

Görg, Holger and David Greenaway, Much ado about Nothing? Do Domestic Firms Really Benefit from Foreign Direct Investment?, *World Bank Research Observer*, vol. 19, 171 (2004)

Gottschalk v. Benson, 409 US 63 (1972)

Gould, David M. and Roy J. Ruffin, What Determines Economic Growth? *Federal Bank of Dallas Economic Review*, 2:25, 40 (1993)

Govaere, Inge and Paul Demaret, The TRIPS Agreement: A Response to Global Regulatory Competition or an Exercise in Global Regulatory Coercion?, in *Regulatory Competition and Economic Integration: Comparative Perspectives*, Oxford University Press, 364, 368–69 (Daniel C. Esty and Damien Geradin, eds., 2001)

Government of China, Outline of the National Intellectual Property Strategy (June 2008)

Govindarajan, Vijay and Chris Trimble, *Reverse Innovation: Create Far from Home, Win Everywhere* (Harvard Business Review Press, 2012), in Strategy & Innovation, May 2012, Vol. 10(2)

Govindarajan, Vijay, The Case for "Reverse Innovation" Now (October 26, 2009)

Grabel, Ilene, International Private Capital Flows and Developing Countries, in *Rethinking Development Economics*, Chapter 15 (Ha-Joon Chang, Ed.) (2003)

Gradstein, Mark and Moshe Justman, Human Capital, Social Capital, and Public Schooling, *European Economic Review*, nr. 44, 879 (2000)

Graevenitz, Georg, Stefan Wagner and Dietmar Harhoff, Incidence and Growth of Patent Thickets: The Impact of Technological Opportunities and Complexity, *Journal of Industrial Economics*, 61(3), 521 (2013)

Grant, Robert M., Toward a Knowledge-based Theory of the Firm, *Strategic Management Journal*, Special Issue, nr. 17, 109 (1996)

Green, Francis, David Ashton, Donna James and Johny Sung, The Role of the State in Skill Formation: Evidence from the Republic of Korea, Singapore, and Taiwan, *Oxford Review of Economic Policy*, 15, 1 (1999)

Greenan, Nathalie and Dominique Guellec, Technological innovation and employment reallocation, *Labour*, 14(4), 547 (2000)

Greenhalgh, Christine and Mark Rogers, *Innovation, Intellectual Property and Economic Growth*, Princeton University Press (2010)

Greenhalgh, Christine, Mark Longland, Derek Bosworth, Technological Activity and Employment in a Panel of UK Firms, *Scottish Journal of Political Economy*, 48, 260 (2001)

Greer, Douglas F., The Case against Patent Systems in Less-Developed Countries, 8 *Journal of International Law and Economics*, 223 (1973)

Griffith, Rachel, Elena Huergo, Jacques Mairesse, Bettina Peters, Innovation and Productivity Across Four European Countries, NBER Working Paper 12722 (2006)

Griliches, Zvi and Frank R. Lichtenberg, R&D and Productivity Growth at the Industry Level: Is There Still a Relationship? in *R&D, Patents and Productivity* (Zvi Griliches, ed.) Univ. of Chicago Press (1984)

Griliches, Zvi and Jacques Mairesse, Productivity and R&D at the Firm Level, in Zvi Griliches (ed.), *R&D, Patents and Productivity*, Chicago: University of Chicago Press, 339 (1984)

Griliches, Zvi and Jacques Mairesse, R&D and Productivity Growth: Comparing Japanese and US Manufacturing Firms, in *Productivity Growth in Japan and United States* (Charles R. Hulten, ed.) University of Chicago Press (1991)

Griliches, Zvi, Issues in Assessing the Contribution of R&D to Productivity, *Bell Journal of Economics*, vol. 10 92 (1979)

Griliches, Zvi, Productivity Puzzles and R&D: Another Non-explanation, *Journal of Economic Perspectives*, vol. 2, 9 (1988)

Griliches, Zvi, Productivity, Productivity, R&D and Basic Research at the Firm Level in the 1970s, *American Economic Review*, 76(1) 141 (1986)

Griliches, Zvi, R&D and Productivity: Econometric Results and Measurement Issues, in (Paul Stoneman, ed.), *The Handbook of the Economics of Innovation and Technological Change*, Blackwell, Oxford (1995)

Griliches, Zvi, *R&D and Productivity: The Econometric Evidence*, University of Chicago Press (1998)

Griliches, Zvi, R&D and the Productivity Slowdown, *The American Economic Review*, 70(2), 343–348 (1980)

Griliches, Zvi, Patent Statistics as Economic Indicators: A Survey, *Journal of Economic Literature*, 28, 1661–1707 (1990)

Grimpea, Christoph and Wolfgang Sofka, Search Patterns and Absorptive Capacity: Low and High Technology Sectors in European Countries, *Research Policy*, 38(3) 495 (2009)

Grimwade, Nigel, International Trade: New Patterns of Trade, *Production and Investment*, 312 (1989)

Groizard, José L., Technology Trade, *Journal of Development Studies*, 1526 (2009)

Groizard, Jose, Technology Trade, *The Journal of Development Studies*, Taylor and Francis Journals, vol. 45(9) 1526 (2009)

Grossman, Gene M. and Elhanan Helpman, *Innovation and Growth in the Global Economy*, Cambridge: MIT Press (1991)

Grossman, Gene M. and Elhanan Helpman, Quality Ladders in the Theory of Growth, *Review of Economic Studies*, LVIII 43 (1991)

Groups in the WTO (updated 2 March 2013), at http://www.wto.org/english/tr atop_e/dda_e/negotiating_groups_e.pdf

Grundmann, Helge E., Foreign Patent Monopolies in Developing Countries: An Empirical Analysis, 12 *Journal of Development Studies*, 186 (1976)

Grupp, Hariolf and Ulrich Schmoch, Patent Statistics in the Age of Globalization: New Legal Procedures, New Analytical Methods, New Economic Interpretation, *Research Policy*, 28, 377 (1999)

Guellec, Dominique and Bruno van Pottelsberghe de la Potterie, Applications, Grants and the Value of Patent, *Economic Letters*, 69(1), 109 (2000)

Guellec, Dominique and Bruno Van Pottelsberghe de la Potterie, From R&D to Productivity Growth: Do the Institutional Settings and the Source of Funds of R&D Matter?, *Oxford Bulletin of Economics and Statistics*, Department of Economics, University of Oxford, vol. 66(3), 353–378 (2004)

Guellec, Dominique and Bruno van Pottelsberghe de la Potterie, The Value of Patents and Filing Strategies: Countries and Technology Areas Patterns, *Economics of Innovation and New Technology*, 11(2), 133 (2002)

Guellec, Dominique and Bruno van Pottelsberghe de la Potterie, The Impact of Public R&D Expenditure on Business R&D, OECD Science, Technology and Industry Working Papers 2000/4, OECD Publishing (2000)

Guellec, Dominique and Bruno Van Pottelsberghe de la Potterie, The Impact of Public R&D Expenditure on Business R&D, *Economics of Innovation and New Technology*, Vol. 12(3) (2003)

Guellec, Dominique and Bruno van Pottelsberghe de la Potterie, *The Economics of the European Patent System*, Oxford University Press (2007)

Gummett, Philip, The Evolution of Science and Technology Policy: A UK Perspective, *Science and Public Policy*, 1 (1991)

Guo, Shoukang, Some Remarks on the Third Revision Draft of the Chinese Patent Law 713, in *Patents and Technological Progress in a Globalized World* (Wolrad Prinz zu Waldeck und Pyrmont, Martin J. Adelman, Robert Brauneis, Josef Drexl, Ralph Nack) (Springer, 2009)

Haines, Andy and Andrew Cassels, Can The Millennium Development Goals Be Attained?, *British Medical Journal*, Vol. 329, No. 7462 (14 August 2004) 394

Hall, Bronwyn H. and Dietmar Harhoff, Implementing Reform of The Patent System: Post-Grant Reviews in the US Patent System – Design Choices and Expected Impact, 19 *Berkeley Technology Law Journal*, 989 (2004)

Hall, Bronwyn H. and Jacques Mairesse, Exploring the Relationship between R&D and Productivity in French Manufacturing Firms. *Journal of Econometrics*, 65, 263 (1995)

Hall, Bronwyn H. and Raffaele Oriani, Does the Market Value R&D Investment by European Firms? Evidence from a Panel of Manufacturing Firms in France, Germany and Italy, *International Journal of Industrial Organization*, 24, 971 (2006)

Hall, Bronwyn H. and Rosemarie Ham Ziedonis, The Patent Paradox Revisited: An Empirical Study of Patenting in the US Semiconductor Industry, 1979–1995, *Rand Journal of Economics*, Vol. 32(1) 101 (2001)

Hall, Bronwyn H., Adam B. Jaffe and Manuel Trajtenberg, The NBER Patent Citations Data File: Lessons, Insights and Methodological Tools, NBER Working Paper No. 849 (2001)

Hall, Bronwyn H., Industrial Research During the 1980s: Did the Rate of Return Fall? *Brookings Papers on Economic Activity Microeconomics*, (2):289–344 (1993)

Hall, Bronwyn H., *The Internationalization of R&D*, UC Berkeley and University of Maastricht (March 2010)

Hall, Bronwyn H., The Stock Market Valuation of R&D Investment During the 1980s, *American Economic Review*, 83(2), 259 (1993)

Hall, Bronwyn H., Jacques Mairesse and Pierre Mohnen, Measuring the Returns to R&D, Working Paper 15622, National Bureau of Economic Research (2009)

Hamilton, Colleen and John Whalley, Coalitions in the Uruguay Round, *Weltwirtschaftliches Archiv*, 125 (3) (1989)

Hanel, Petr, Skills Required for Innovation: A Review of the Literature, *Note de Recherche* (2008)

Hansen, Povl A. and Göran Serin, Will Low Technology Products Disappear?: The Hidden Innovation Processes in Low Technology Industries, *Technological Forecasting and Social Change*, 55(2) 179 (1997)

Hanushek, Eric A. and Dennis Kimko, Schooling Labor Force Quality, and the Growth of Nations, *American Economic Review*, vol. 90 1184 (2000)

Harhoff, Dietmar and Stefan Wagner, The Duration of Patent Examination at the European Patent Office, *Management Science*, Vol. 55(12) 1969 (2009)

Harhoff, Dietmar, Frederic M. Scherer and Katrin Vopel, Citations, Family Size, Opposition and the Value of Patent Rights, *Research Policy*, 32(8) 1343 (2003)

Harhoff, Dietmar, R&D and Productivity in German Manufacturing Firms, *Economics of Innovation and New Technology*, 6:22 (1998)

Harris, Donald P., Carrying a Good Joke Too Far: TRIPS and Treaties of Adhesion, 27 *University of Pennsylvania Journal of International Law*, 681 (2006)

Harrison, Rupert, Jordi Jaumandreu, Jacques Mairesse and Bettina Peters, Does Innovation Stimulate Employment? A Firm-level Analysis Using Comparable Micro-data From Four European Countries. NBER Working Paper No. 14216 (2008)

Harvey, David, *A Brief History of Neoliberalism*, 2–4 (2005)

Hassan, Emmanuel, Ohid Yaqub and Stephanie Diepeveen, *Intellectual Property and Developing Countries: A Review of the Literature*, Rand Europe (2010)

Hausmann, Ricardo, Dani Rodrik and Andrés Velasco, Getting the Diagnosis Right: A New Approach to Economic Reform, 43 *Finance and Development*, 12 (2006)

Hausmann, Ricardo, Dani Rodrik and Andrés Velasco, Growth Diagnostics, in *The Washington Consensus Reconsidered: Toward a New Global Governance* (Narcís Serra and Joseph Stiglitz, eds.), Oxford University Press (2008)

Heinrich, Ernst Hirschel, Horst Prem and *Gero Madelung: Aeronautical Research in Germany – from Lilienthal until Today*, Springer Verlag (2004)

Helfer, Laurence R., Regime Shifting: The TRIPS Agreement and New Dynamics of International Intellectual Property Lawmaking, 29 *Yale Journal of International Law*, 1 (2004)

Heller, Michael A. and Rebecca S. Eisenberg, Can Patents Deter Innovation? The Anticommons in Biomedical Research, 280 *Science*, 698 (1998)

Henderson, J. Vernon, Ari Kuncoro, Matthew Turner, Industrial Development in Cities, *The Journal of Political Economy*, 103, 1067 (1995)

Herzberg, Frederick, Motivation-hygiene Theory, in *Organization Theory*, Penguin, Harmondsworth (Derek S. Pugh, ed.) (1966)

Hicks, Diana, Published papers, Tacit Competencies and Corporate Management of the Public/Private Character of Knowledge, *Industrial and Corporate Change*, 4: 401 (1995)

Hicks, Dianna, T. Ishizuka and S. Sweet, Japanese Corporations, Scientific Research and Globalization, 23 *Research Policy*, 4 (1994)

Hinz, Thomas and Monika Jungbauer-Gans, Starting a Business after Unemployment: Characteristics and Chances of Success (Empirical Evidence from a Regional German Labor Market), *Entrepreneurship and Regional Development*, nr. 11, 317 (1999)

Hoare, Anthony G., Linkage Flows, Locational Evaluation, and Industrial Geography: A Case Study of Greater London, *Environment and Planning A*, 7:41–58 (1975)

Hoare, Anthony G., Review of Paul Krugman's "Geography and Trade," *Regional Studies*, 26, 679 (1992)

Hoover, Edgar Malone, *Location Theory and the Shoe and Leather Industries*, Cambridge (MA): Harvard University Press (1936)

Howitt, Peter and David Mayer-Foulkes, R&D, Implementation and Stagnation: A Schumpeterian Theory of Convergence Clubs, *Journal of Money, Credit and Banking*, 37(1), 147 (2005)

Hu, Albert G. and Gary H. Jefferson, A Great Wall of Patents: What Is Behind China's Recent Patent Explosion?, *Journal of Development Economics*, 90(1) 57 (2009)

Huang, Can, Estimates of the Value of Patent Rights in China, UNU-MERIT Working Paper no. 004, Maastricht Economic and Social Research Institute on Innovation and Technology (2012)

Huang, Kuo-Feng and Tsung-Chi Cheng, Determinants of Firms' Patenting or not Patenting Behaviors, *Journal of Engineering and Technology Management*, Vol. 36, 52 (2015)

Humphrey, John and Hubert Schmitz, *Developing Country Firms in the World Economy: Governance and Upgrading in Global Value Chains*, INEF Report, No. 61, Duisburg: University of Duisburg (2002)

Hunt, Jennifer, Jean-Philippe Garant, Hannah Herman and David J. Munroe, Why Don't Women Patent?, IZA Discussion Paper No. 6886 (September 2012)

Hunt, Jennifer, Jean-Philippe Garant, Hannah Herman and David J. Munroe, Why are Women Underrepresented amongst Patentees? *Research Policy*, 42, 831 (2013)

Hvide, Hans K. and Benjamin F. Jones, University Innovation and the Professor's Privilege, NBER Working Paper No. 22057 (March 2016)

Idris, Kamil, *Intellectual Property: A Power Tool for Economic Growth* 1 (2d ed. 2003)

Imam, Ali, How Patent Protection Helps Developing Countries, 33 *American Intellectual Property Law Association Quarterly Journal*, 377 (2005)

In re Fisher, 421 F.3d 1365, 76 USPQ2d 1225 (Fed. Cir. 2005)

India Brand Equity Foundation (IBEF), Pharmaceutical Companies in India, at: http://www.ibef.org/industry/pharmaceutical-india/showcase

India-Brazil-South Africa (IBSA) Dialogue Forum: New Delhi Agenda for Cooperation, 4–5 March 2004

Industrial Property Law, of June 27, 1991 (Mexico)

International Monetary Fund (IMF), Balance of Payments Manual (fifth edition, 1993)

International Monetary Fund (IMF), Data and Statistics (2012), at: http://www.imf.org/external/data.htm

International Monetary Fund (IMF), Data and Statistics, at: http://www.imf.org/external/data.htm, (1 April 2017)

International Monetary Fund (IMF), WEO Groups Aggregates Information (1 April 2010)

International Monetary Fund (IMF), IMF Advanced Economies List, World Economic Outlook, (April 2016)

Ismail, Faizel, Reforming the World Trade Organization: Developing Countries in the Doha Round, Geneva: CUTS International and Friedrich Ebert Stifung (FES) (2009)

Ismail, Faizel, The G-20 and NAMA 11: The Role of Developing Countries in the WTO Doha Round, 1 *Indian Journal of International Economic Law*, 80 (2008)

Jacobs, Jane, *The Economy of Cities*, Random House (1969)

Jaffe, Adam and Manuel Trajtenberg, Flows of Knowledge from Universities and Federal Labs: Modeling the Flows of Patent Citations over Time and Across Institutional and Geographic Boundaries, NBER Working Paper 5712) (1996)

Jaffe, Adam B. and Josh Lerner, *Innovation and Its Discontents: How Our Broken Patent System Is Endangering Innovation and Progress, and What to Do About It*, Princeton University Press (2004)

Jaffe, Adam B., Manuel Trajtenberg and Rebecca Henderson, Geographic Localization of Knowledge Spillovers as Evidenced by Patent Citations, *Quarterly Journal of Economics*, 108, 577 (1993)

Jaffe, Adam B., Manuel Trajtenberg and Rebecca Henderson, University versus Corporate Patents: A Window on the Basicness of Inventions, *Economics of Innovation and New Technology*, 5(1): 19 (1997)

Jaffe, Adam B., Real Effects of Academic Research, *American Economic Review*, 79(5), 957 (1989)

Jain, Subhash Chandra, *Emerging Economies and the Transformation of International Business*, Edward Elgar Publishing (2006)

Javorcik, Beata S. and Mariana Spatareanu, Disentangling FDI Spillovers Effects: What Do Firm Perceptions Tell Us?, in *Does Foreign Direct Investment Promote Development? New Methods, Outcomes and Policy Approaches* (Theodore H. Moran, Edward M. Graham and Magnus Blomstrom, eds.) Washington, Institute for International Economics, 45 (2005)

Johnson, Clete D., A Barren Harvest for the Developing World? Presidential "Trade Promotion Authority" and the Unfulfilled Promise of Agriculture Negotiations in the Doha Round, 32 *Georgia Journal of International and Comparative Law*, 437 (2004)

Jones, Charles I. and Paul M. Romer, The New Kaldor Facts: Ideas, Institutions, Population, and Human Capital 8 (Nat'l Bureau Econ. Research, Working Paper No. 15094, 2009)

Jones, Charles, Time Series Tests of Endogenous Growth Models, 110(2) *Quarterly Journal of Economics*, 495 (1995)

Jorge, Katz, M., Patents, the *Paris Convention and Less Developed Countries*, Discussion Paper no. 190 (Yale Univ. Economic Growth Center, Nov. 1973)

Juma, Calestous and Lee Yee-Cheong, United Nations Millennium Project, Innovation: Applying Knowledge in Development (2005)

Kabeer, Naila, *Can the MDGs provide a Pathway to Social Justice?: The Challenge of Intersecting Inequalities*, Institute of Development Studies (2010)

Kabla, Isabelle, The Patent as Indicator of Innovation, *INSEE Studies Economic Statistics*, 1, 56 (1996)

Kadarusman, Yohanes and Khalid Nadvi, Competitiveness and Technological Upgrading in Global Value Chains: Evidence from the Indonesian Electronics and Garment Sectors, *European Planning Studies*, 21(7) 1007 (2013)

Kadarusman, Yohanes, Knowledge Acquisition: Lessons from Local and Global Interaction in the Indonesian Consumer Electronics Sector, *Institutions and Economies*, 4(2) 65 (2012)

Kaeser, Joe, Why a US-European Trade Deal Is a Win-Win, *The Wall Street Journal* (February 2, 2014)

Kahn, Michael, William Blankley and Neo Molotja, Measuring R&D in South Africa and in Selected SADC Countries: Issues in Implementing Frascati Manual Based Surveys, Working Paper prepared for the UIS, Montreal (2008)

Kalaitzidakis, Pantelis, Theofanis Mamuneas, Andreas Savvides and Thanasis Stengos, Measures of Human Capital and Nonlinearities in Economic Growth, *Journal of Economic Growth*, vol. 6, 229 (2001)

Kaldor, Nicholas, A Model of Economic Growth, 67 *Economic Journal*, Dec. 1957, 591 (1957)

Kaldor, Nicholas, The Case for Regional Policies, *Scottish Journal of Political Economy*, November (1970) 337

Kaldor, Nicholas, The Role of Increasing Returns, Technical Progress and Cumulative Causation in the Theory of International Trade and Economic Growth, Economie Appliquee, 34 Reprinted in, *The Essential Kaldor*, (F. Targetti and A. Thirlwall, eds.) 327 (1981)

Kanwar, Sunil and Robert Evenson, Does Intellectual Property Protection Spur Technological Change?, 55 *Oxford Economic Papers*, 235 (2003)

Kao, Chihwa, Min-Hsien Chiang and Bangtian Chen, International R&D Spillovers: An Application of Estimation and Inference in Panel Cointegration, *Oxford Bulleting of Economics and Statistics*, 61(S1): 691–709 (1999)

Kaplinsky, Raphael, The Role of Standards in Global Value Chains and their Impact on Economic and Social Upgrading, Policy Research Paper 5396, World Bank (2010)

Kapur, Devesh, John Lewis and Richard Webb, *The World Bank: Its Half-Century, Vol. I: History* (Washington DC: Brookings Institute, 1997)

Kassambara, Alboukadel and Fabian Mundt, Factoextra: Extract and Visualize the Results of Multivariate Data Analyses, R package version 1.0.3. http://www.sthda.com/english/rpkgs/factoextra (2016)

Katz, J. Sylvan, Geographical Proximity and Scientific Collaboration, *Scientometrics*, 31(1), 31 (1994)

Katz, Michael L. and Carl Shapiro, Systems Competition and Network Effects, *Journal of Economic Perspectives*, vol. 8(2) 93 (1994)

Kay, John and Chris Llewellyn Smith, Science Policy and Public Spending, *Fiscal Studies* 6 (1985) 14–23; P. Dasgupta, The Economic Theory of Technology Policy: An Introduction, in Paul Stoneman and Partha Dasgupta, *Economic Policy and Technological Performance* (Cambridge University Press, Cambridge, 1987)

Keller, Wolfgang and Stephen R. Yeaple, *Multinational Enterprises, International Trade, and Productivity Growth: Firm-level Evidence from the United States*, NBER Working Papers 9504, National Bureau of Economic Research, Inc. (2003)

Keller, Wolfgang, International Technology Diffusion, *Journal of Economic Literature*, 42, 3 (2004)

Keller, Wolfgang, *International Technology Diffusion*, NBER Working Paper Series 8573, Cambridge, Massachusetts (2001)

Kelly, Morgan and Anya Hageman, Marshallian Externalities in Innovation, 1 *Journal of Economic Growth*, Vol. 4(1) 39 (1999)

Kelly, Morgan and Anya Hageman, Marshallian Externalities in Innovation, *Journal of Economic Growth*, 4 (March), 39 (1999)

Kennedy, David, The "Rule of Law," Political Choices, and Development Common Sense, in *The New Law and Economic Development: A Critical Appraisal* 95, 128–150 (David M. Trubek and Alvaro Santos, eds.) Cambridge University Press (2006)

Kenney, Martin and Urs von Burg, Technology Entrepreneurship and Path Dependence: Industrial Clustering in Silicon Valley and Route 128, *Industrial and Corporate Change*, nr. 8, 67 (1999)

Kher, Rajeev, *India in the World Patent Order, in Emerging Markets and the World Patent Order*, Edward Elgar (Frederick. Abbott, Carlos Correa and Peter Drahos, eds.) (2013) 183

Kilkenny, Maureen, Laura Nalbarte and Terry Besser, Reciprocated Community Support and Small Town Small Business Success, Entrepreneurship and Regional Development, nr. 11, 231 (1999)

Kim, Linsu and Richard R. Nelson (eds.), *Technology, Learning, and Innovation: Experiences of Newly Industrializing Economies*, Cambridge University Press (2000)

Kim, Linsu and Richard. R. Nelson, *Introduction, in Technology, Learning and Innovation: Experience of Newly Industrializing Economies* 1 (Linsu Kim and Richard R. Nelson, eds.), Cambridge University Press (2000)

Kitch, Edmund W., The Patent Policy of Developing Countries, 13 *University of California Los Angeles Pacific Basin Law Journal*, 166 (1994)

Kleinknecht, Alfred, Kees van Montfort and Erik Brouwer, The Non-trivial Choice between Innovation Indicators, *Economics of Innovation and New Technology*, 11, 109 (2002)

Kleinschmidt, Elko J. and Robert G. Cooper, The Impact of Product Innovativeness on Performance, *Journal of Product Innovation Management*, 8, 240 (1991)

Klemperer, Paul, How Broad Should the Scope of Patent Protection Be?, *Rand Journal of Economics*, Vol. 21, 113 (1990)

Klette, Tor Jakob and Svein Erik Førre, Innovation and Job Creation in a Small open, Economy: Evidence from Norwegian Manufacturing Plants 1982–92, *Economics of Innovation and New Technology*, 5, 247 (1998)

Klinische Versuche I and Klinische Versuche II (Clinical Trials I and II), ACIP Issues Paper, 4 (2004)

Knodel, John E. and Gavin W. Jones, Post-Cairo Population Policy: Does Promoting Girl's Schooling Miss the Mark?, *Population and Development Review*, 22(4): 683 (1996)

Konings, Joep, Gross Job Flows and Wage Determination in the UK: Evidence from Firm-Level Data. PhD thesis, LSE, London (1994)

Koopmans, Reinout and Ana R. Lamo, Cross-Sectional Firm Dynamics: Theory and Empirical Results from the Chemical Sector, Working paper, Economics Department, LSE, London, April (1994)

Koopmans, Tjalling, On the Concept of Optimal Economic Growth, in (Study Week on the) Econometric Approach to Development Planning, Cowles Foundation Discussion Paper no. 163, (1965), chapter 4, 225 (1965)

Kortum, Samuel, Research, Patenting and Technological Change, *Econometrica*, Vol. 65:6, 1389 (1997)

Krikorian, Gaëlle and Amy Kapczynski, *Access to Knowledge in the Age of Intellectual Property* (eds.) (2010)

Krishnarao Prahalad, Coimbatore, *The Fortune at the Bottom of the Pyramid: Eradicating Poverty through Profits* (Philadelphia, PA: Wharton School Publishing, 2005)

Kristensen, Thorkil, *Development in Rich and Poor Countries*, New York: Praeger (1982)

Krueger, Alan B., Mikael Lindahl, Education for Growth: Why and for Whom, NBER working paper, no. 7591(2000)

Krugman, Paul and A. Venables, Integration and the Competitiveness of Peripheral Industry 56, in *Unity with diversity in the European Community* (Christopher Bliss and Jorge Braga de Macedo, eds.), Cambridge University Press (1990)

Krugman, Paul R., *Geography and Trade*. Cambridge, MA: MIT Press (1991)

Krugman, Paul, A Model of Innovation, Technology Transfer, and the World Distribution of Income, 87 *Journal of Political Economy*, 253 (1979)

Krugman, Paul, A Model of Technology Transfer, and the World Distribution of Income, 87 *Journal of Policy Economics*, 253 (1979)

Krugman, Paul, *Geography and Trade*, Leuven University Press (1991)

Paul Krugman, On the Relationship between Trade Theory and Location Theory, *Review of International Economics*, 1, 110 (1993)

Krugman, Paul, History and Industrial Location: The Case of the Manufacturing Belt, American Economic Review (Papers and Proceedings) 81 (1991)

Krugman, Paul, Increasing Returns and Economic Geography, *Journal of Political Economy*, 99, 483 (1991)

Krugman, Paul, The "New Theory" of International Trade and the Multinational Enterprise, in *The Multinational Corporation in the 1980s* (Charles Kindleberger, ed.) 57, MIT Press (1983)

Krugman, Paul, The Current Case for Industrial Policy 160, in *Protectionism and World Welfare* (Dominick Salvatore, ed.), Cambridge University Press (1993)

Krugman, Paul, The Lessons of Massachusetts for EMU, in *Adjustment and Growth in the European Monetary Union* (Francisco Torres and Francesco Giavazzi, eds.) 241, Cambridge University Press (1993)

Lachenmaier, Stefan and Horst Rottmann, Effects of Innovation on Employment: A Dynamic Panel Analysis, *International Journal of Industrial Organization*, 29(2) 210 (2011)

Lall, Sanjaya and Morris Teubal, Market Stimulating Technology Policies in Developing Countries: A Framework with Examples from East Asia, *World Development*, Vol. 26(8) 1369 (1998)

Lall, Sanjaya and Morris Teubal, *Market-stimulating Technology Policies in Developing Countries: A Framework with Examples from East Asia*, World Development, Elsevier, vol. 26(8) 1369 (1998)

Landes, David S., *The Wealth and Poverty of Nations: Why Some Are So Rich and Some So Poor*, W.W. Norton (1998)

Landes, William M. and Richard A. Posner, *The Economic Structure of Intellectual Property Law*, Harvard University Press (2003)

Lanjouw, Jean O., and Mark Schankerman, Research Productivity and Patent Quality: Measurement with Multiple Indicators, CEPR Discussion Papers in its series CEPR Discussion Papers with number 3623 (October 2002)

Lanjouw, Jean O., Ariel Pakes and Jonathan Putnam, How to Count Patents and Value Intellectual Property: The Uses of Patent Renewal and Application Data, *Journal of Industrial Economics*, 46(4) 404 (1998)

Larch, Martin, *Regional Cross-Section Growth Dynamics in the European Community*, Working paper, European Institute, LSE, London, June (1994)

Latouche, Daniel, Do Regions Make a Difference? The Case of Science and Technology Policies in Quebec, in *Regional Innovation Systems: The Role of Governances in a Globalized World* (Hans-Joachim Braczyk, Philip Cooke and Martin Heidenreich, eds.), London: UCL Press (1998)

Lazaridis, George and Bruno Van Pottelsberghe, The Rigor of EPO's Patentability Criteria: An Insight into the "Induced Withdrawals, CEB Working Paper N° 07/007 (April 2007)

Lecomte, Henri Bernard Solignac, Building Capacity to Trade: A Road Map for Development Partners: Insights from Africa and the Caribbean 7 (European Centre for Dev. Pol'y Mgmt Discussion Paper 33, 2001), available at http://www.ecdpm.org

Lederman, Daniel and William F. Maloney, R&D and Development, Policy Research Working Paper, 3024 (2003)

Lee, Keun, *Schumpeterian Analysis of Economic Catch-up: Knowledge, Path-creation and the Middle Income Trap*, Cambridge University Press (2013)

Léger, Andreanne and Sushmita Swaminathan, Innovation Theories: Relevance and Implications for Developing Country Innovation, German Institute for Economic Research (DIW) Discussion paper 743 (November 2007)

Legislative Affairs Office of the State Council PRC, "12th five-Year Plan (2011–2015) for National Economic and Social Development of P.R China" (17 March 2011)

Lei, Zhen, Zhen Sun and Brian Wright, Patent Subsidy and Patent Filing in China (2012)

Lemley, Mark A. and B. Sampat, Examiner Characteristics and the Patent Grant Rate, mimeo: http://www.nber.org/~confer/2008/si2008/IPPI/lemley.pdf (2008)

Lemley, Mark A., Douglas Lichtman and Bhaven N. Sampat, What to Do About Bad Patents, *Regulation*, Vol. 28(4) 10 (Winter 2005–2006)

Lerner, Josh, Patent Protection and Innovation Over 150 Years, NBER Working Paper No. 8977 (2002)

Levin, Richard C., Alvin K. Klevorick, Richard Nelson and Sidney Winter, Appropriating the Returns from Industrial Research and Development, *Brookings Papers on Economic Activity*, 3 (1987)

Levinthal, Daniel and James G. March, A Model of Adaptive Organizational search, *Journal of Economic Behavior and Organization*, 2, 307–33 (1981)

Levitt, Theodore, Innovative Imitation, *Harvard Business Review* (Sept.-Oct. 1966)

Levy, Charles S., Implementing TRIPS–A Test of Political Will, 31 *Law and Policy in International Business*, 789 (2000)

Lewis, Meredith K., Expanding the P-4 Trade Agreement into a Broader Trans-Pacific Partnership: Implications, Risks and Opportunities, *Asian Journal of the WTO and International Health Law and Policy*, vol. 4(2) 401 (2009)

Licht, Georg and Konrad Zoz, *Patents and R&D: An Econometric Investigation Using Applications for German, European, and US Patents by German Companies*, ZEW Discussion Paper 96–19, Zentrum fur Europaische Wirtschaftsforschung, Mannheim (1996)

Lichtenberg, Frank and Donald Siegel, The Impact of R&D Investment on Productivity – New Evidence Using Linked R&D-LRD Data, *Economic Inquiry*, 29, 203 (1991)

Lichtenberg, Frank R., and Bruno van Pottelsberghe de la Potterie, International R&D Spillovers: A Comment. *European Economic Review*, 42(8): 1483 (1998)

Lichtenberg, Frank R., *R&D Investment and International Productivity Differences*, NBER Working Papers 4161, National Bureau of Economic Research, Inc. (1992)

Lichtenberg, Frank R., R&D Investment and International Productivity Differences, in *Economic Growth in the World Economy* (Horst Siebert, ed.), Tubingen: Mohr (1993)

Lichtman, Douglas G. and Mark A. Lemley, Presume Nothing: Rethinking Patent Law's Presumption of Validity 300, in *Competition Policy and Patent Law under Uncertainty: Regulating Innovation* (Geoffrey A. Manne, Joshua D. Wright, eds.), Cambridge University Press (2011)

Lin, Cheng, Ping Lin, Frank M. Song and Chuntao Li, Managerial Incentives, CEO Characteristics and Corporate Innovation in China's Private Sector, *Journal of Comparative Economics*, 39(2), 176 (2011)

Link, Albert N., and John T. Scott, *Public Accountability: Evaluating Technology-Based Institutions*, Kluwer Academic Publishers, Norwell, MA (1998)

Link, Albert N., Basic Research and Productivity Increase in Manufacturing: Additional Evidence, *American Economic Review*, 71(5) 1111 (1981)

Link, Albert N., On the classification of industrial R&D, *Research Policy*, Vol. 25(3) 397 (May 1996)

Loewe, Markus, Jonas Blume, Verena Schönleber, Stella Seibert, Johanna Speer, Christian Voss, *The Impact of Favoritism on the Business Climate: A Study of Wasta in Jordan*, DIE studies 30, Bonn (German Development Institute) (2007)

Long, J. Scott, ed., *From Scarcity to Visibility: Gender Differences in the Careers of Doctoral Scientists and Engineers*, Washington, DC: National Academy Press (2001)

Love, James, KEI Analysis of Wikileaks Leak of TPP IPR Text, from August 30, 2013, available at: www.keionline.org/node/1825 (2013)

Lu, Qiwen, *China's Leap into the Information Age: Innovation and Organization in the Computer Industry*, Oxford University Press (2000)

Lucas, Robert Emerson, On the Machines of Economic Development, *Journal of Monetary Economics*, vol. 22, 3 (1988)

Luintel, Kul B. and Mosahid Khan, Basic, Applied and Experimental Knowledge and Productivity: Further Evidence, *Economics Letters*, 111(1) 71 (2011)

Lundvall, Bengt-Åke (ed.) *National System of Innovation: Towards a Theory of Innovation and Interactive Learning* (Anthem Press, 1993)

Lundvall, Bengt-Åke and Susana Borrás, The Globalizing Learning Economy: Implications for Technology Policy, Final Report under the TSER Programme, EU Commission (1997)

Lundvall, Bengt-Åke, ed., *National Innovation Systems: Toward a Theory of Innovation and Interactive Learning*, Pinter, London (1992)

Lundvall, Bengt-Åke, Innovation as an Interactive Process: From User-Producer Interaction to the National System of Innovation, Chapter 18 in *Technical Change and Economic Theory* (Giovanni Dosi, Christopher Freeman, Richard Nelson, Gerarld Silverberg and Luc Soete, eds.), LEM Book Series (1988)

Lundvall, Bengt-Åke, Product Innovation and User-Producer Interaction, in 31 *Industrial Development Research Series*, 28–29 (1985)

Madey v. Duke, 307 F 3d 1351 (Fed. Cir. 2002)

Maharajh, Rasigan and Erika Kraemer-Mbula, Innovation Strategies in Developing Countries, in *Innovation and the Development Agenda* (Erika Kraemer-Mbula and Watu Wamae, eds.) (2009)

Mairesse, Jacques and Pierre Mohnen, The Importance of R&D for Innovation: A Reassessment Using French Survey Data, *Journal of Technology Transfer*, 30, 183 (2005)

Mairesse, Jacques and Mohamed Sassenou, R&D and Productivity: A Survey of Econometric Studies at the Firm Level, *Science-Technology-Industry Review*, 8, 317 (1991)

Mäkinen, Iiro, The Propensity to Patent: An Empirical Analysis at the Innovation Level, ETLA – The Research Institute of the Finnish Economy (2007) (File with author)

Mankiw, N. Gregory, David Romer and David N. Weil, A Contribution to the Empirics of Economic Growth, *Quarterly Journal of Economics*, vol. 107, 407 (1992)

Mankiw, N. Gregory, *Principles of Economics*, 4th ed. (2007)

Mansfield, Edwin, Patents and Innovation: An Empirical Study, *Management Science*, 32, 173 (1986)

Mansfield, Edwin, Academic Research and Industrial Innovation: An Update of Empirical Findings, *Research Policy*, 26(7–8), 773–776 (1998)

Mansfield, Edwin, Basic Research and Productivity Increase in Manufacturing, *American Economic Review*, 70(5) 863 (1980)

Mara, Kaitlin, Standing Committee on the Law of Patents to Reconvene After Two Year Hiatus, Intellectual Property Watch (June 19, 2008)

March, James G., Herbert A. Simon, *Organizations*, New York: Wiley (1958)

Marrakesh Treaty to Facilitate Access to Published Works for Persons Who Are Blind, Visually Impaired or Otherwise Print Disabled (2013), at: http://www.wipo.int/treaties/en/text.jsp?file_id=301016

Marshall, Alfred, *Principles of Economics*, London: Macmillan, (8th ed.) (1920)

Martin, Fernand, The Economic Impact of Canadian University R&D, *Research Policy*, 27(7) 677 (1998)

Martin, Ron and Peter Sunley, Paul Krugman's Geographical Economics and Its Implications for Regional Development Theory, *Economic Geography*, Vol. 72 (3) 259 (1996)

Martin, Ron and Peter Sunley, Slow Convergence? The New Endogenous Growth Theory and Regional Development, *Economic Geography*, Vol. 74(3) 201 (1988)

Marusuk Hermann, Rachel, WIPO Patent Committee Moves Quickly Through Agenda; Heavy Lifting to Come, Intell. Prop. Watch (Feb. 26, 2013)

Marusuk Hermann, Rachel, WIPO Patent Law Committee cinches Agreement on Future Work, Intell. Prop. Watch (Mar. 1, 2013)

Marx, Karl, *Capital: A Critical Analysis of Capitalist Production*, Foreign Languages Publishing House (first edn. 1867, Moscow) (1961)

Maskell, Peter, Learning in the Village Economy of Denmark: The Role of Institutions and Policy in Sustaining Competitiveness, in Hans-Joachim Braczyk, Philip Cooke and Martin Heidenreich (eds.) *Regional Innovation Systems: The Role of Governances in a Globalized World*, London: UCL Press (1998)

Maskus, Keith E. and Jerome H. Reichman (eds.), *International Public Goods and Transfer of Technology Under a Globalized Intellectual Property Regime* (2005)

Maskus, Keith E. and Jerome H. Reichman, The Globalization of Private Knowledge Goods and the Privatization of Global Public Goods, 7 *Journal of International Economic Law*, 279 (2004)

Maskus, Keith E., and Mohan Penubarti, How Trade-Related Are Intellectual Property Rights?, 39 *J. Int'l Econ,*. 227 (1995)

Maskus, Keith E., Intellectual Property Rights in the Global Economy 11 (2000)

Maskus, Keith E., The Role of Intellectual Property Rights in Encouraging Foreign Direct Investment and Technology Transfer, 9 *Duke Journal of Comparative & International Law*, 109 (1998)

Matthew Waguespack, David, Jóhanna Kristín Birnir and Jeff Schroeder, Technological Development and Political Stability: Patenting in Latin America and the Caribbean, *Research Policy*, 34: 1570 (2005)

Maurseth, Botolf and Bart Verspagen, *Knowledge Spillovers in Europe: A Patent Citation Analysis, paper presented at the CRENOS Conference on Technological Externalities and Spatial Location*, University of Cagliari, 24–25 September (1999)

May, Christopher T., *The Information Society: A Skeptical View*, Cambridge: Polity Press (2002)

May, Christopher, The Pre-History and Establishment of the WIPO (2009), Journal No. 1, 16, at: www.research.lancs.ac.uk/portal/en/publications/the-pre history-and-establishment-of-the-wipo(4db79f65–30d9-42a3-b7ed-da285 b32f77a).html

McGregor, James, China's Drive for "Indigenous Innovation": A Web of Industrial Policies, US Chamber of Commerce, Global Regulatory Cooperation Project (26 July 2010)

McMillan, G. Steven, Gender Differences in Patent Activity: An Examination of the US Biotechnology Industry, 80 *Scientometrics*, 683 (2009)

Meier, Gerald M., and James E. Rauch (eds.), *Leading Issues in Economic Development*, Oxford University Press (1995)

Mellström, Ulf, The Intersection of Gender, Race and Cultural Boundaries, or Why Is Computer Science in Malaysia Dominated by Women?, *Social Studies of Science*, vol. 39(6) 885 (2009)

Mensch, Gerhard, *Stalemate in Technology: Innovations Overcome the Depression* (Ballinger Publishing Company, 1979)

Meyer-Krahmer, Frieder and Guido Reger, New Perspectives on the Innovation Strategies of Multinational Enterprises: Lessons for Technology Policy in Europe, 28 *Research Policy*, Vol. 28, 751 (1999)

Michalisin, Michael D., Robert D. Smith and Douglas M. Kline, In Search of Strategic Assets, *The International Journal of Organizational Analysis*, 5(4) 360 (1997)

Michalopoulos, Constantine, *The Participation of the Developing Countries in the WTO*, Policy Research Working Paper, World Bank, Washington, DC (1999)

Milhaupt, Curtis and Katharina Pistor, *Law and Capitalism – What Corporate Crises Reveal about Legal Systems and Economic Development Around the World*, Chicago, Chicago Press (2008)

Millennium Development Goals (MDG), MDG 8 – Access to Essential Medicines, at: http://iif.un.org/content/mdg-8-access-essential-medicines (2000)

Millennium Project, Investing in Development: A Practical Plan to Achieve the Millennium Development Goals (2005)

Milligan, Glenn W. and Martha C. Cooper, An Examination of Procedures for Determining the Number of Clusters in a Data Set, *Psychometrika*, 50(2), 159 (1985)

Ministry of Foreign Affairs of Japan, Signing Ceremony of the EU for the Anti-Counterfeiting Trade Agreement (ACTA) (Outline) (26 January 2012)

Ministry of Science and Technology of China (MOST), Statistics of Science and Technology at: www.most.org.cn (2010)

Morales, Maria, Research Policy and Endogenous Growth, *Spanish Economic Review*, 6(3): 179 (2004)

Morgan, Peter, Technical Assistance: Correcting the Precedents, 2 *Development Policy Journal*, 1 (2002)

Moris, Francisco, *R&D Investments by US TNCs in Emerging and Developing Markets in the 1990s*, Background paper prepared for UNCTAD (Arlington, VA: US National Science Foundation), mimeo (2005)

Moritz, Steffen, ImputeTS: Time Series Missing Value Imputation, R package version 1.5, at: https://CRAN.R-project.org/package=imputeTS (2016)

Morrison, Andrea, Carlo Pietrobelli and Roberta Rabellotti, Global Value Chains and Technological Capabilities: A Framework to Study Industrial Innovation in Developing Countries, Quaderni SEMeQ n° 03/2006 (2006)

Mosahid Khan and Kul B. Luintel, Basic, Applied and Experimental Knowledge and Productivity: Further Evidence, *Economics Letters*, 111(1), 71–74 (2011)

Mota, Isabel and António Brandão, Modeling Location Decisions—The Role of R&D Activities, European Regional Science Association in its series ERSA conference papers No. ersa05p612 (2005)

Mowery, David and Nathan Rosenberg, The Influence Of Market Demand Upon Innovation: A Critical Review of Some Recent Empirical Studies, 8 *Research policy*, 102 (1979)

Mowery, David C. and Nathan Rosenberg, *Technology and the Pursuit of Economic Growth*, Cambridge University Press (1989)

Mowery, David C., Economic Theory and Government Technology Policy, *Policy Sciences*, 16, 27 (1983)

Mowery, David C., Richard R. Nelson, Bhaven N. Sampat and Arvids A. Ziedonis, *Ivory Tower and Industrial Innovation University-Industry Technology Transfer Before and After the Bayh-Dole Act*, Stanford University Press (2004)

Musungu, Sisule F., *Susan Villanueva, Roxana Blasetti, Utilizing TRIPS Flexibilities for Public Health Protection Through South-South Regional Frameworks* (South Center 2004)

Myrdal, Gunnar, *Economic Theory and Under-Developed Regions*, Taylor & Francis (1957)

Mytelka, Lynn K. and Keith Smith, Innovation Theory and Innovation Policy: Bridging the Gap 12–17 (2001)

Nadiri, Ishaq and Theofanis P. Mamuneas, The Effects of Public Infrastructure and R&D Capital on the Cost Structure and Performance of US Manufacturing Industries, *Review of Economics and Statistics*, Vol. 76, 22–37 (1994)

Nadiri, Ishaq, *Innovations and Technological Spillovers*, NBER Working Paper Series, 4423, Cambridge, MA (1993)

Naldi, Fulvio and Ilaria Vannini Parenti, *Scientific and Technological Performance by Gender*, European Commission (2002)

Narin, Francis and J. Davidson Frame, The Growth of Japanese Science and Technology, *Science*, 245 (1989)

Narin, Francis, Kimberly S. Hamilton and Dominic Olivastro, The Increasing linkage between US Technology and Public Science, *Research Policy*, 26:317 (1997)

Narlikar, Amrita and John Odell, The Strict Distributive Strategy for a Bargaining Coalition: The Like Minded Group in the World Trade Organization, 1998–2001, in *Negotiating Trade: Developing Countries in the WTO and NAFTA* (John Odell, ed.), Cambridge University Press (2006)

Narlikar, Amrita, Bargaining over the Doha Development Agenda: Coalitions in the World Trade Organization, Serie LATN Papers, N° 34 (2005)

Narlikar, Amrita, *International Trade and Developing Countries: Bargaining Coalitions in the GATT and WTO*, London: Routledge, RIPE Studies in Global Political Economy (2003)

Narula, Rajneesh and Antonello Zanfei, Globalization of Innovation: The Role of Multinational Enterprises, Chapter 12 in *The Oxford Handbook of Innovation* (Jan Fagerberg and David Mowery, eds.), Oxford University Press (2005)

Nason, Howard K., George E. Manners and Joseph A. Steger, *Support of Basic Research in Industry*, National Science Foundation report (Washington, DC) (1978)

National Research Council, *Funding a Revolution: Government Support for Computing Research, Report of the NRC Computer Science and Telecommunications Board Committee on Innovations in Computing: Lessons from History.* National Academy Press, Washington DC (1999)

National Science Foundation (NSF), 1956, *Science and Engineering in American Industry: Final Report on a 1953–1954 Survey*, National Science Foundation report (Washington, DC)

National Science Foundation (NSF), *Science and Engineering in American Industry: Final Report on a 1956 Survey*, National Science Foundation report (Washington, DC) (1959)

Navas-Aleman, Lizbeth, The Impact of Operating in Multiple Value Chains for Upgrading: The case of the Brazilian Furniture and Footwear Industries, *World Development*, 39(8), 1386 (2011)

Nelson, Richard R. (ed.), *National Innovation System: A Comparative Analysis*, Oxford University Press (1993)

Nelson, Richard R., and Edmund S. Phelps, Investment in Humans, Technological Diffusion, and Economic Growth, *American Economic Review*, 56(2) 69 (1966)

Nelson, Richard R. and Nathan Rosenberg, Technical innovation and National Systems, in Richard R. Nelson (editor), *National Innovation Systems: A Comparative Analysis*, Oxford University Press (1993)

Nelson, Richard R., *Government and Technical Progress: A Cross-Industry Analysis*, New York (1982)

Nelson, Richard R., Reflections on "The Simple Economics of Basic Scientific Research": Looking Back and Looking Forward, *Industrial and Corporate Change*, Vol. 15(6) 903 (2006)

Nelson, Richard R., The Co-evolution of Technology, Industrial Structure and Supporting Institutions, *Industrial and Corporate Change*, 3(1) 47–63 (1994)

Nelson, Richard R., The Link between Science and Invention: The Case of the Transistor, in *The Rate and Direction of Inventive Activity* (Richard R. Nelson, ed.) (Princeton University Press, New Jersey, 1962)

Nelson, Richard R., The Roles of Universities in the Advance of Industrial Technology, in *Engines of Innovation* (Richard S. Rosenbloom and William J. Spencer, eds.), Cambridge, MA: Harvard Business School Press (1996)

Netanel, Neil, Introduction: The WIPO Development Agenda and Its Development Policy Context 1, in *The Development Agenda: Global Intellectual*

Property and Developing Countries (Neil Netanel, ed.) Oxford University Press (2009)

New Scientist, Silicon Subcontinent: India Is Becoming the Place to Be for Cutting-edge Research (19 February 2005)

New, William, IP-Watch, USPTO Acting Director Discusses Patent Quality, Pendency, Harmonization (3/03/2015), at: http://www.ip-watch.org/2015/03/03/uspto-acting-director-discusses-patent-quality-pendency-harmonisation/

New, William, WIPO Development Agenda Implementation: The Ongoing Fight for Development in IP, Intellectual Property Watch (9.5.2012)

Niefer, Michaela, Patenting Behavior and Employment Growth in German Start-up Firms: A Panel Data Analysis, Discussion Paper No. 05–03, ZEW (2003)

Nieto, María Jesús and Luís Santamaría, The Importance of Diverse Collaborative Networks for the Novelty of Product Innovation, *Technovation*, 27(6–7) 367 (2007)

North, Douglas, *Institutional Change and Economic Performance*, Cambridge University Press (1990)

NovaMedica, The Pharma Letter, Russian Government to Change Rules on Public Procurement of Drugs (28 August 2013), available at: http://novamedica.com/media/theme_news/p/631#sthash.lFR4tPlr.dpuf

Nunnenkamp, Peter and Julius Spatz, Intellectual Property Rights and Foreign Direct Investment: The Role of Industry and Host-Country Characteristics 2 (Kiel Instit. for World Econ., Working Paper No. 1167, June 2003)

Oakey, Ray P. and Sarah. Y. Cooper, High Technology Industry, Agglomeration and the Potential for Peripherally Sited Small Firms, *Regional Studies*, 23, 347 (1989)

Odagiri, Hiroyuki, Akira Goto, Atsushi Sunami, and Richard R. Nelson (eds.), *Intellectual Property Rights, Development and Catch UP*, Oxford University Press (2010)

Oded Galor, Convergence? Inferences from Theoretical Models, *The Economic Journal*, 106, 1056 (1996)

Odell, John (ed.), *Negotiating Trade: Developing Countries in the WTO and NAFTA*, Cambridge University Press (2006)

Odell, John S. and Susan K. Sell, Reframing the Issue: The WTO Coalition on Intellectual Property and Public Health 85, in *Negotiating Trade* (John S. Odell, ed.), Cambridge University Press (2006)

Okediji, Ruth L. (Gana), The Myth of Development, The Progress of Rights: Human Rights to Intellectual Property and Development, 18 *Law and Policy*, 315 (1996)

Okediji, Ruth L., Public Welfare and the International Patent System 1, in *Patent Law in Global Perspective* (Ruth L. Okediji, Margo A. Bagley, eds.), Oxford University Press (2014)

Okediji, Ruth L., Public Welfare and the Role of the WTO: Reconsidering the TRIPS Agreement, 17 *Emory International Law Review*, 819 (2003)

Olwan, Rami M., *Intellectual Property and Development: Theory and Practice*, Springer (2013)

Onyeama, Geoffrey, Deputy Director General, Cooperation for Development, WIPO, interview in Geneva, Switzerland, on October 15, 2014 (file with author)

Opinions on the Provision of Judicial Support and Service, Supreme People's Court, http://www.chinacourt.org/flwk/show.php?file_id=144434 (link to text in Chinese), June 29, 2010

Order of the President of the People's Republic of China No.8, The Decision of the Standing Committee of the National People's Congress on Amending the Patent Law of the People's Republic of China, adopted at the 6th Meeting of the Standing Committee of the Eleventh National People's Congress on December 27, 2008

Organization for Economic Co-operation and Development (OECD) (2011), Science, Technology and Industry Scoreboard (20 September 2011)

Organization for Economic Co-operation and Development (OECD) and Eurostat (2005), *Oslo Manual: Guidelines for Collecting and Interpreting Innovation Data* (Paris: OECD) (Oslo Manual)

Organization for Economic Co-operation and Development (OECD), *Innovative Clusters: Drivers of National Innovation Systems*, Paris: OECD publication (2001)

Organization for Economic Co-operation and Development (OECD), Benchmark Definition of Foreign Direct Investment (third edition, 1995)

Organization for Economic Co-operation and Development (OECD), Compendium of Patent Statistics, Economic Analysis and Statistics Division of the OECD Directorate for Science, Technology and Industry (2004)

Organization for Economic Co-operation and Development (OECD), Compendium of Patent Statistics (2008)

Organization for Economic Co-operation and Development (OECD), Developed and Developing Countries (4 January 2006)

Organization for Economic Co-operation and Development (OECD), Knowledge Networks and Markets (KNM) "Expert Workshop on Patent Practice and Innovation" (May 2012)

Organization for Economic Co-operation and Development (OECD), Mariagrazia Squicciarini, Hélène Dernis and Chiara Criscuolo, Measuring Patent Quality: Indicators of Technological and Economic Value (OECD France) (06 June 2013)

Organization for Economic Co-operation and Development (OECD), OECD Reviews of Innovation Policy: Russian Federation (2011)

Organization for Economic Co-operation and Development (OECD), Patents and Innovation Trends and Policy Challenges: Trends and Policy Challenges (2004)

Organization for Economic Co-operation and Development (OECD), Patent Statistics Manual (2009), at: http://www.oecdbookshop.org/en/browse/title-d etail/?ISB=9789264056442

Organization for Economic Co-operation and Development (OECD), *Proposed Standard Practice for Surveys on Research and Experimental Development* (Paris: OECD) (2002) (Frascati Manual)

Organization for Economic Co-operation and Development (OECD), Research Use of Patented Knowledge: A Review by Chris Dent, Paul Jensen, Sophie Waller and Beth Webster, STI Working Paper 2006/2 (2006)

Organization for Economic Co-operation and Development (OECD), Science, Technology and Industry Scoreboard 2011: Innovation and Growth in Knowledge Economies (2011)

Organization for Economic Co-operation and Development (OECD), *The Measurement of Scientific and Technological Activities: Standard practice for Surveys of Research and Experimental Development* (1994) (Frascati Manual 1993: OECD Publications)

Organization for Economic Co-operation and Development (OECD), Turning Science into Business: Patenting and Licensing at Public Research Organizations, OECD (2003)

Organization for Economic Co-operation and Development (OECD)/OCDE, Background report to the Conference on internationalization of R&D, Brussels (March 2005)

Osano, Hiroshi, Basic Research and Applied R&D in a Model of Endogenous Economic Growth, *Osaka Economic Papers*, 42(1–2) 144 (1992)

Ostry, Sylvia, After Doha: Fearful New World?, Bridges, Aug. 2006, at 3, at http://www.ictsd.org/monthly/bridges/BRIDGES 10–5.pdf (2006)

Ostry, Sylvia, The Uruguay Round North-South Bargain: Implications for Future Negotiations, in *The Political Economy of International Trade Law: Essays in Honor of Robert E. Hudec* (Daniel L. M. Kennedy and James D. Southwick, eds.), Cambridge University Press (2002) 285 (September 2000), at http://www.utoronto.ca/cis/Minnesota.pdf

Owen-Smith, Jason, Massimo Riccaboni, Fabio Pammolli and Walter W. Powell, A Comparison of US and European University-Industry Relations in the Life Sciences, *Management Science*, 48(1), 24–43 (2002)

Parente, Stephen L., and Edward C. Prescott, Monopoly Rights: A Barrier to Riches, *American Economic Review*, 89, 1216 (1999)

Parente, Stephen L. and Edward C. Prescott, Technology Adoption and Growth, *Journal of Political Economy*, 102, 298 (1994)

Park, Walter G., A Theoretical Model of Government Research and Growth, *Journal of Economic Behavior & Organization*, 34(1), 69 (1998)

Park, Walter G., International Patent Protection: 1960–2005, Research Policy, 37, 761 (2008)

Park, Walter G., International R&D Spillovers and OECD Economic Growth, *Economic Inquiry*, Vol. 33, 571 (1995)

Patel, Mayur, New Faces in the Green Room: Developing Country Coalitions and Decision-Making in the WTO, GEG Working Paper Series, 2007/WP33

Patel, Pari and Keith Pavitt, The Nature and Economic Importance of National Innovation Systems, *STI Review*, 14 (1994)

Patel, Parimal, Localized Production of Technology for Global Markets, *Cambridge Journal of Economics*, Vol. 19(1), 141 (1991)

Patel, Parimal and Modesto Vega, Patterns of internationalization of corporate technology: location vs. home country advantages, *Research Policy*, Vol. 28, No. 145–155 (1999)

Pavitt, Keith, Academic Research, Technical Change and Government Policy, in *Companion Encyclopedia of Science in the Twentieth Century* 143 (John Krige and Dominique Pestre, eds.) (2013)

Pavitt, Keith, The Social Shaping of the National Science Base, *Research Policy*, 27 793 (1998)

Pavitt, Keith, What Makes Basic Research Economically Useful?, *Research Policy*, Vol. 20(2), 109 (April 1991)

Payosova, Tetyana, *Russian Trip to the TRIPS: Patent Protection, Innovation Promotion and Public Health, in Emerging Markets and the World Patent Order*, Edward Elgar (Frederick Abbott, Carlos Correa and Peter Drahos, eds.) (2013) 225

Peeters, Carine and Bruno Van Pottelsberghe, *Economics and Management Perspectives on Innovation and Intellectual Property Rights* (Palgrave Macmillan) (2006)

Pelloni, Alessandra, Public Financing of Education and Research in a Model of Endogenous Growth, *Labour*, 11(3) 517 (1997)

Pennings, Johannes M., Kyungmook Lee and Arjen van Witteloostuijn, Human Capital, Social Capital, and Firm Dissolution, Academy of Management Journal, nr. 41, 425 (1998)

Perez, Carlota and L. Luc Soete, Catching-Up in Technology: Entry Barriers and Windows of Opportunity, in *Technical Change and Economic Theory* 458 (Giovanni Dosi, Christopher Freeman, Richard Nelson, Gerarld Silverberg and Luc Soete, eds.), LEM Book Series (1988)

Perroux, François, Economic Space: Theory and Applications, *Quarterly Journal of Economics*, 64, 89–104 (1950)

Perroux, François, Note sur la Notion des "Poles du Croissance", *Economie Appliquee*, 1 and 2, 307–320 (1955)

Pessoa, Argentino, R&D and Economic Growth: How Strong Is the Link?, *Economics Letters*, Vol. 107(2) 152 (May 2010)

Petit, Pascal, Employment and Technological Change, in *Handbook of the Economics of Innovation and Technological Change* (Paul Stoneman, ed.), North Holland, Amsterdam (1995)

Pianta, Mario, Innovation and Employment, in *The Oxford Handbook of Innovation* (Jan Fagerberg and David C. Mowery, eds.), Oxford University Press (2005)

Piore, Michael J., Charles F. Sabel, *The Second Industrial Divide*, Basic Book, New York (1984)

Piva, Mariacristina and Marco Vivarelli, Innovation and Employment: Evidence from Italian Microdata, *Journal of Economics*, 86, 65 (2005)

Piva, Mariacristina and Marco Vivarelli, Technological Change and Employment: Some Micro Evidence from Italy, *Applied Economics Letters*, 11, 373 (2004)

Plechero, Monica and Cristina Chaminade, From New to the Firm to New to the World: Effect of Geographical Proximity and Technological Capabilities on the Degree of Novelty in Emerging Economies, Paper no. 2010/12 (2010)

Polk Wagner, Patent Quality Index (PQI), at: https://www.law.upenn.edu/blogs/polk/pqi/faq.html

Poole, Erik and Jean-Thomas Bernard, Defense Innovation Stock and Total Factor Productivity Growth, *Canadian Journal of Economics*, Vol. 25, 438 (1992)

Porter, Michael E., Clusters and Competition: New Agendas for Companies, Governments, and Institutions 197, in Michael E. Porter, *On competition* (Harvard Business School Press) (1998)

Porter, Michael E., Location, Competition and Economic Development: Local Clusters in a Global Economy, *Economic Development Quarterly*, 14(1) 15 (2000)

Porter, Michael E., The Role of Location in Competition, *Journal of the Economics of Business*, 1(1) 35 (1994)

Porter, Michael E., *The Competitive Advantage of Nations*, London: Macmillan (1990)

Porter, Michael, *The Competitive Advantage of Nations*, New York: Free Press (1990); Richard R. Nelson, *National Innovation Systems: A Comparative Analysis*, Oxford: Oxford University Press (1993)

Prebisch, Raul, International Trade and Payments in an Era of Coexistence: Commercial Policy in the Underdeveloped Countries, 49 *American Economic Review*, 251, (1959)

Pred, Allen R., *The Spatial Dynamics of US Urban Industrial Growth, 1800–1914*, Harvard University Press (1966)

President's Council of Advisors on Science and Technology, University-Private Sector Research Partnerships in the Innovation Ecosystem (2008)

Prieur, Jerome and, Omar R. Serrano, Coalitions of Developing Countries in the WTO: Why Regionalism Matters?, Paper Presented at the WTO Seminar at the Department of Political Science at the Graduate Institute of International Studies in Geneva (May 2006)

Primo Braga, Carlos A. and Carsten Fink, The Relationship Between Intellectual Property Rights and Foreign Direct Investment, 9 *Duke Journal of Comparative and International Law*, 163 (1998)

Pritchett, Lant, Where Has All the Education Gone, World Bank working papers, no. 1581 (1996)

Provisional Committee on Proposals Related to a WIPO Development Agenda (PCDA), WIPO Development Agenda; Preliminary Implementation Report in Respect of 19 Proposals, Feb. 28, 2008, available at http://ip-watch.org/files/WIPO%20comments%20on%20DA%20recs%20-%20part%201.pdf

Psacharopoulos, George and Harry Partinos, *Returns on Investment in Further Update*, World Bank Policy Research Working Paper 2881, Washington, DC (2002)

Psacharopoulos, George, *Returns on Education: An International Comparison*, Amsterdam: Elsevier (1973)

Psacharopoulos, George, Returns on Education: An Updated International Comparison, *Comparative Education*, 17(3) 321 (1981)

Psacharopoulos, George, Returns to Education: A Further International Update and Implications, *Journal of Human Resources*, 20(4) 683 (1985)

Psacharopoulos, George, Returns to Investment in Education: A Global Update, *World Development*, 22(9) 1325 (1994)

Qian, Yi, Do National Patent Laws Stimulate Domestic Innovation in a Global Patenting Environment? A Cross-Country Analysis of Pharmaceutical Patent Protection, 1978–2002, *The Review of Economics and Statistics*, Vol. 89(3) 436 (2007)

Quah, Danny T., *Convergence across Europe*, Working paper, Economics Department, LSE, London, June (1994)

Quah, Danny T., Empirics for Economic Growth and Convergence, LSE Economics Department and CEP – Center for Economic Performance, Discussion Paper No. 253 (July 1995)

Quah, Danny T., One Business Cycle and One Trend from (Many,) Many Disaggregates, *European Economic Review*, 38(3/4):605 (1994)

R Core Team, R: A Language and Environment for Statistical Computing, R Foundation for Statistical Computing, Vienna, Austria, at: https://www.R-project.org/. (2016)

Radoševic, Slavo and Esin Yoruk, SAPPHO Revisited: Factors of Innovation Success in Knowledge-Intensive Enterprises in Central and Eastern Europe, (DRUID Working Paper No. 12–11), available at http://www3.druId.dk/wp/20120011.pdf

Radoševic, Slavo, Domestic Innovation Capacity – Can CEE Governments correct FDI-driven Trends through R&D Policy? 135, in *Closing the EU East-West Productivity Gap: Foreign Direct Investment, Competitiveness, and Public Policy* (David A. Dyker, ed.), Imperial College Press (2006)

Radoševic, Slavo, Patterns of Preservation, Restructuring and Survival: Science and Technology Policy in Russia in the Post-Soviet Era, *Research Policy*, vol. 32 (6) 1105 (2003)

Rai, Arti K., US Executive Patent Policy, Global and Domestic 85, in *Patent Law in Global Perspective*, (Ruth L. Okediji and Margo A. Bagley, eds.) (2014)

Rammer, Christian, Dirk Czarnitzki and Alfred Spielkamp, Innovation Success of Non-R&D-performers: Substituting Technology by Management in SMEs, *Small Business Economics*, 33, 35 (2009)

Ramo, Joshua Cooper, The Beijing Consensus The Foreign Policy Centre (May 2004)

Ramsey, Frank P., A Mathematical Theory of Saving, *Economic Journal*, vol. 38 (1928) 543

Ravix, Jacques-Laurent, Localization, Innovation and Entrepreneurship: An Appraisal of the Analytical Impact of Marshall's Notion of Industrial Atmosphere, *Journal of Innovation Economics and Management*, 2014/2 (n°14), 63 (2014)

Redding, Stephen, The Low-Skill, Low-Quality Trap: Strategic Complementarities between Human Capital and R&D, *The Economic Journal*, Vol. 106, No. 435, 458 (1996)

Reddy, Prasada, *Global Innovation in Emerging Economies*, Routledge (2011)

Reed, Richard and Robert J. DeFillippi, Casual Ambiguity, Barriers to Imitation and Sustainable Competitive Advantage, *Academic Management Review* 15(1) 88 (1990)

Reichman, Jerome H. and Rochelle Cooper Dreyfuss, Harmonization without Consensus: Critical Reflections on Drafting a Substantive Patent Law Treaty, 57 *Duke Law Journal*, 85 (2007)

Reichman, Jerome H., The TRIPS Component of the GATT's Uruguay Round: Comparative Prospects for Intellectual Property Owners in an Integrated World Market, 4 *Fordham Intellectual Property, Media & Entertainment Law Journal*, 171 (1993)

Report of the National Institute of Health (NIH) Working Group on Research Tools (June 4, 1998) at: https://www.mmrrc.org/about/NIH_research_tools_policy/

Ricardo, David, Principles of Political Economy, in *The Works and Correspondence of David Ricardo* (Piero Saffra, ed.), Cambridge University Press, Cambridge, third edn. 1821 (1951)

Rist, Clibert, *The History of Development: From Western Origins to Global Faith* (2002)

Robson, Paul J., Helen M. Haugh and Bernard A. Obeng, Entrepreneurship and Innovation in Ghana: Enterprising Africa, *Small Business Economics*, 32(3), 331 (2009)

Roche Products v. Bolar Pharmaceutical Company, 733 F.2d 858 (Fed. Cir.). cert. denied, 469 US 856 (1984)

Rodríguez-Pose, Andrés, Innovation Prone and Innovation Averse Societies, *Economic Performance in Europe, Growth and Change* Vol. 30 75 (1999)

Rodrik, Dani, Goodbye Washington Consensus, Hello Washington Confusion? A Review of the World Bank's "Economic Growth in the 1990s: Learning from a Decade of Reform", 44(4) *Journal of Economic Literature*, 973 (2006)

Rodrik, Dani, *One Economics, Many Recipes: Globalization, Institutions and Economic Growth*, Princeton University Press (2007)

Rodrik, Dani, The New Development Economics: We Shall Experiment, But How Shall We Learn? 24–28 (Harvard University, John F. Kennedy School of Government Faculty Research Working Papers Series, Paper No. RWP08–055, 2008)

Rodrik, Dani, *The New Global Economy and Developing Countries: Making Openness Work*, MD Policy Essay No. 24., Baltimore, Overseas Development Council (1999)

Rolland, Sonia E., Developing Country Coalitions at the WTO: In Search of Legal Support, *Harvard International Law Journal*, Vo. 48(2) 483 (2007)

Romer, Paul M., Endogenous Technological Change, 98 *Journal of Political Economy*, S71, S72 (1990)

Romer, Paul M., The Origins of Endogenous Growth, 8 *Journal of Economic Perspectives*, 3, 4–10 (1994)

Romer, Paul, *What Determines the Rate of Growth and Technological Change?*, working paper #279, The World Bank (September 1989)

Rose, Pauline, From Washington to Post-Washington Consensus: The Triumph of Human Capital, in *The New Development Economics: Post Washington Consensus Neoliberal Thinking* (Jomo KS and Ben Fine, eds.) 162 (2006)

Rosenberg, Nathan, *Inside the Black Box: Technology and Economics*, Cambridge University Press (1982)

Rosenberg, Nathan, *Perspectives on Technology*, Cambridge University Press (1976)

Rosenberg, Nathan, Why Do Firms Do Basic Research (with Their Own Money)?, *Research Policy*, 19, 165 (1990)

Rostow, Walt Whitman, *The Stages of Economic Growth: A Non-Communist Manifesto*, Cambridge University Press, Cambridge, 3d ed. (1991)

Rousseeuw, Peter J., Silhouettes: A Graphical Aid to the Interpretation and Validation of Cluster Analysis, *Journal of Computational and Applied Mathematics*, Vol. 20, 53 (1987)

Roy, Vincent Van, Daniel Vertesy and Marco Vivarelli, Innovation and Employment in Patenting Firms: Empirical Evidence from Europe, IZA Discussion Paper No. 9147 (June 2015)

Russel, Daniel R., Transatlantic Interests in Asia, United States Department of State (13 January 2014), at: https://2009–2017.state.gov/p/eap/rls/rm/2014/01/219881.htm

Saez, Catherine, Crisis at WIPO Over Development Agenda; Overall Objectives In Question, Intellectual Property Watch (24 May 2014)

Sakakibara, Mariko and Lee Branstetter, Do Stronger Patents Induce More Innovation? Evidence from the 1988 Japanese Patent Law Reforms, NBER working paper 7066 (1999)

Sala-i-Martin, Xavier, The Classical Approach to Convergence Analysis, *The Economic Journal*, 106 (July 1996)

Salomon, Jean-Jacques, Science Policy Studies and the Development of Science Policy, in *Science, Technology and Society: A Cross Disciplinary Perspective* (Ina Spiegel-Rösing and Derek de Solla Price, eds.) London: Sage (1977)

Salter, Amnon J. and Ben R. Martin, The Economic Benefits of Publicly Funded Basic Research: A Critical Review, *Research Policy*, 30(3), 509 (2001)

Samimiand, Ahmad Jafari and Seyede Monireh Alerasoul, R&D and Economic Growth: New Evidence from Some Developing Countries, *Australian Journal of Basic and Applied Sciences*, 3(4):3464 (2009)

Samuelson, Pamela, Benson Revisited: The Case Against Patent Protection for Algorithms and Other Computer Program-Related Inventions, 39 *Emory Law Journal*, 1025, (1990)

Samuelson, Pamela, The US Digital Agenda at WIPO, 37 *Virginia Journal of International Law*, 369 (1997)

Santos, Alvaro, The World Bank's Uses of the "Rule of Law" Promise in Economic Development, in *The New Law and Economic Development: A Critical Appraisal* 253 (David M. Trubek and Alvaro Santos, eds., 2006)

Sauvant, Karl, Jaya Pradhan, Ayesha Chatterjee and Brian Harley (eds.), *The Rise of Indian Multinationals: Perspectives on Indian Outward Foreign Direct Investment*, New York: Palgrave Macmillan (2010)

Savvides, Andreas and Thanasis Stengos, *Human Capital and Economic Growth*, Stanford University Press (2009)

Saxenian, AnnaLee, *Regional Advantage: Culture and Competition in Silicon Valley and Route 128*, Cambridge, Harvard University Press (1994)

Say, Jean-Baptiste, *A Treatise on Political Economy; or, The Production, Distribution and Consumption of Wealth*, Augustus M. Kelley Publishers (first edn. 1803), New York (1964)

Schankerman, Mark and Ariel Pakes, Estimates of the Value of Patent Rights in European Countries During the Post-1950 Period, *The Economic Journal*, 96, 1052 (1986)

Scherer, Frederic M., The Political Economy of Patent Policy Reform in the United States, 7 *Journal of Telecommunication and High Technology Law*, 167, 205 (2009)

Scherer, Frederic M., The Propensity to Patent, *International Journal of Industrial Organization*, 1, 107 (1983)

Scherer, Frederic M., Firm Size, Market Structure, Opportunity, and the Output of Patented Inventions, *American Economic Review*, 55, 319 (1965)

Schlag, Pierre, An Appreciative Comment on Coase's The Problem of Social Cost: A View from the Left, *Wisconsin Law Review*, 919 (1986)

Schnaars, Steven, *Managing Imitation Strategy: How Later Entrants Seize Markets from Pioneers*, New York: Free Press (1994)

Schnepf, Randall D., Erik Dohlman and Christine Bolling, Agriculture in Brazil and Argentina: Developments and Prospects for Major Fields Crops, Agriculture and Trade Report No. WRS013, Economic Research Service, USDA 85 (2001)

Schultz, Theodore William, Capital Formation by Education, *Journal of Political Economy*, vol. 68, 571 (1960)

Schultz, Theodore, *The Economic Value of Education*, Columbia University Press (1963)

Schulz, Theodore William, *Investment in Human Capital: The Role of Education and of Research*, London: Free Press: Collier-Macmillan (1971)

Science Foundation (NSF), 1959, *Science and Engineering in American Industry: Final Report on a 1956 Survey*, National Science Foundation report (Washington, DC)

Science Policy Research Unit, Report on Project SAPPHO (1971)

Segerstrom, Paul, Endogenous Growth Without Scale Effects, *American Economic Review*, 88(5) 1290 (1998)

Segran, Grace, As Innovation Drives Growth in Emerging Markets, Western Economies need to Adapt, at: http://knowledge. insead.edu/innovation-emer ging-markets-110112.cfm?vid=515 (2011)

Sell, Susan, *Private Power, Public Law: The Globalization of Intellectual Property Rights*, Cambridge University Press (2003)

Sen, Amartya, *Commodities and Capabilities*, Amsterdam: North-Holland (1985)

Sen, Amartya, *Development as Freedom*, Alfred A. Knopf (1999)

Sen, Amartya, Well-being, Agency and Freedom: The Dewey Lectures 1984, *Journal of Philosophy*, 82(4) 169 (1985)

Shaffer, Gregory, Can WTO Technical Assistance and Capacity Building Serve Developing Countries?, *Wisconsin International Law Journal*, Vol. 23 643 (2006)

Shaffer, Gregory, *Power, Governance and the WTO: A Comparative Institutional Approach, in Power and Global Governance* 130 (Michael Barnett and Raymond Duvall, eds.), Cambridge University Press (2004)

Shell, Karl, A Model of Inventive Activity and Capital Accumulation, in *Essays on the Theory of Optimal Economic Growth*, MIT Press, Cambridge, Massachusetts (Karl Shell, ed.) (1967)

Shell, Karl, Toward a Theory of Inventive Activity and Capital Accumulation, *American Economic Review*, 56(1/2), 62 (1966)

Sherwood, Robert M., Global Prospects for the Role of Intellectual Property in Technology Transfer, 42 IDEA 27, 30 (1997)

Sherwood, Robert M., Human Creativity for Economic Development: Patents Propel Technology, 33 *Akron Law Review*, 351 (2000)

Sherwood, Robert M., Some Things Cannot Be Legislated, 10 *Cardozo Journal of International and Comparative Law*, 37 (2002)

Shim, Yosung, Ji-won Chung and In-chan Choi, A Comparison Study of Cluster Validity Indices using a Nonhierarchical Clustering Algorithm, Proceedings – International Conference on Computational Intelligence for Modeling, Control and Automation, CIMCA 2005 and International Conference on Intelligent Agents, *Web Technologies and Internet*, Vol. 1, 199 (2005)

Shin, Hochul and Keun Lee, Asymmetric Trade Protection leading not to Productivity but to Export Share Change, The Economics of Transition, *The European Bank for Reconstruction and Development*, vol. 20(4) 745 (2012)

Siegel, Robin, Eric Siegel and Ian C. Macmillan, Characteristics Distinguishing High-growth Ventures, *Journal of Business Venturing*, nr. 8, 169 (1993)

Simmie, James, (ed.) *Innovative cities*, London: Spon Press (2001)

Singer, Hans Wolfgang, The Sussex Manifesto: Science and Technology for Developing Countries during the Second Development Decade, in I.D.S. Reprints no. 101 (Institute of Development Studies 1974) (1970)

Sixty-First World Health Assembly (WHA), WHO Global Strategy and Plan of Action on Public Health, Innovation and Intellectual Property, at 1, WHA61.21, (May 24, 2008), available at http://apps.who.int/medicinedocs/documents/s214 29en/s21429en.pdf

Sjostedt, Gunnar, Negotiating the Uruguay Round of the General Agreement on Tariffs and Trade, in *International Multilateral Negotiation: Approaches to the Management of Complexity* 44 (I. William Zartman, ed.), Jossey-Bass Publishers (1994)

Smith, Keith, *Economic Returns to R&D: Method, Results, and Challenges*, Science Policy Support Group Review Paper No. 3, London (1991)

Smolny, Werner, Innovation, Prices and Employment – A Theoretical Model and an Empirical Application for West German manufacturing firms, *The Journal of Industrial Economics*, XLVI(3), 359 (1998)

Soete, Luc and Parimal Patel, Recherche-Développement, Importations Technologiques et Croissance Economique, *Revue Economique*, Vol. 36, 975 (1985)

Soete, Luc L., Firm Size and Inventive Activity: The Evidence Reconsidered, *European Economic Review*, 12(4) 319 (1979)

Solow, Robert M., A Contribution to the Theory of Economic Growth, 70 *Quarterly Journal of Economics*, vol. 70, 65 (1956)

Sorensen, Anders, R&D, Learning and Phases of Economic Growth, *Journal of Economic Growth*, 4, 429 (1999)

Spiezia, Vincenzo and Marco Vivarelli, Innovation and Employment: A Critical Survey, in *Productivity, Inequality and the Digital Economy: A Transatlantic Perspective* (Nathalie Greenan, Yannick L'Horty and Jacques Mairesse, eds.) 101 MIT Press (2002)

Srholec, Martin, A Multilevel Analysis of Innovation in Developing Countries, UNU-MERIT TIK working paper on Innovation Studies 20080812 (2008)

Sridharan, Eswaran, *The Political Economy of Industrial Promotion: Indian, Brazilian, and Korean Electronics in Comparative Perspective 1969–1994,* Praeger (August 1996)

Stahl, Jessica C., *Mergers and Sequential Innovation: Evidence from Patent Citations,* Finance and Economics Discussion Series, No. 2010–12, Division of Research and Statistics and Monetary Affairs Federal Reserve Board, Washington, DC (2010)

Steed, Guy P. F., Internal Organization, Firm Integration and Locational Change: The Northern Ireland Linen Complex, 1954–64, *Economic Geography,* 47 371 (1971)

Steger, Debra P., The Future of the WTO: The Case of Institutional Reforms, *Journal of International Economic Law,* 12(4) 803 (2009)

Steinberg, Richard H., in the Shadow of Law or Power? Consensus-Based Bargaining and Outcomes in the GATT/WTO, 56(2) *International Organization,* 339 (2002)

Steinberg, Richard H., In the Shadow of Law or Power? Consensus-Based Bargaining and Outcomes in the GATT/WTO, 56(2) *Int'l. Org.* 339 (2002)

Stiglitz, Joseph E. and Andrew Charlton, *Fair Trade for All: How Trade Can Promote Development,* Oxford University Press (2005)

Stiglitz, Joseph E., Chief Economist, World Bank, More Instruments and Broader Goals: Moving Toward the Post-Washington Consensus, address at the 1998 WIDER Annual Lecture 17 (Jan. 7, 1998), at: http://time.dufe.edu.cn/wencong/washingtonconsensus/instrumentsbroadergoals.pdf

Stiglitz, Joseph E., *Making Globalization Work,* W.W. Norton & Co. (2006)

Stiglitz, Joseph E., Social Absorption Capability and Innovation, Stanford University CERP Publication, 292 (1991)

Stokes, Donald, *Pasteur's Quadrant: Basic Science and Technological Innovation, The Brookings Institution,* Washington, DC (1997)

Stokey, Nancy, Human Capital, Product Quality, and Growth, *Quarterly Journal of Economics,* 106(2) 587 (1991)

Stoneman, Paul and John Vickers, The Assessment: The Economics of Technology Policy, *Oxford Review of Economic Policy,* 4 (1988)

Stoneman, Paul, *The Economic Analysis of Technology Policy,* Oxford, University Press (1987)

Storey, David J. and Bruce S. Tether, Public Policy Measures to Support New Technology-based Firms in the European Union, *Research Policy,* 26, 1037 (1998)

Storper, Michael and Anthony J. Venables, Buzz: Face-to-Face Contact and the Urban Economy, *Journal of Economic Geography,* Oxford University Press, vol. 4(4) 351 (2004)

Storper, Michael, *The Regional World: Territorial Development in a Global Economy*, New York: Guildford Press (1997)

Strategic Plan for New and Renewable Energy Sector for the Period 2011–17, Ministry of New and Renewable Energy, The Government of India (February 2011)

Straus, Joseph, Comment, Bargaining Around the TRIPS Agreement: The Case for Ongoing Public-Private Initiatives to Facilitate Worldwide Intellectual Property Transactions, 9 *Duke Journal of Comparative & International Law*, 91 (1998)

Straus, Joseph, The Impact of the New World Order on Economic Development: The Role of the Intellectual Property Rights System, 6 *John Marshall Review of Intellectual Property Law*, 1 (2006)

Suarez-Villa, Luis and Wallace Walrod, Operational Strategy, R&D and Intra-metropolitan Clustering in a Polycentric Structure: The Advanced Electronics Industries of the Los Angeles Basin, *Urban Studies*, 34, 1343 (1997)

Sugimoto, Cassidy R., Chaoqun Ni, Jevin D. West, Vincent Larivière, The Academic Advantage: Gender Disparities in Patenting (May 27, 2015)

Sun, Xiuli, *Firm-level Human Capital and Innovation: Evidence from China*, Partial Doctor of Philosophy thesis in the School of Economics, Georgia Institute of Technology (2015)

Sun, Yifei, Yu Zhou, George C. S. Lin, and Yehua H. Dennis Wei, Subcontracting and Supplier Innovativeness in a Developing Economy: Evidence from China's Information and Communication Technology Industry, *Regional Studies*, 47(10) 1766 (2013)

Sunley, Peter J., Marshallian Industrial Districts: The Case of the Lancashire Cotton Industry in the Inter-war Years, *Transactions of the Institute of British Geographers*, n.s. 17, 306 (1992)

Suthersanen, Uma, Utility Models and Innovation in Developing Countries, UNCTAD-ICTSD Project on IPRs and Sustainable Development, Issue Paper No. 13 (February 2006)

Swan, Trevor W., Economic Growth and Capital Accumulation, *Economic Record*, vil. 32, no. 63, 334 (1956)

Takagi, Yoshiyuki (Yo), *Assistant Director General*, Global Infrastructure, WIPO, interview in Geneva, Switzerland on October 16, 2015 (file with author)

Tassey, Gregory, The Economics of R&D Policy (Quorum Books) 54–55, 226 (1997)

Tellez, Viviana Munoz, The Changing Global Governance of Intellectual Property Enforcement: New Challenges for Developing Countries, in *Intellectual Property and Enforcement* (Xuan Li and Carlos M. Correa, eds.), Edward Elgar (2009)

Tether, Bruce, Who Co-operates for Innovation, and Why: An Empirical Analysis, *Research Policy*, 31 947 (2002)

The Community Innovation Survey 2010, at http://epp.eurostat.ec.europa.eu/por tal/page/portal/microdata/documents/CIS_Survey_form_2010.pdf

The Economist (US Edition), Technology in Emerging Economies (February 9, 2008)

The Global Innovation Index (GII) 2014 (7th edition) (Cornell University, INSEAD and WIPO)

374   References

The Global Innovation Index 2015: Effective Innovation Policies for Development (2015)

The Manual of Patent Examining Procedure (MPEP) 2107 Guidelines for Examination of Applications for Compliance with the Utility Requirement [R-11.2013]

The Observatory of Economic Complexity (OEC), at: http://atlas.media.mit.edu/en/visualize/tree_map/hs92/export/chn/show/all/2014/

The Organization for Economic Co-operation and Development (OECD), Agricultural Policies in OECD Countries: Monitoring and Evaluation 2000: Glossary of Agricultural Policy Terms, OECD (defining – Central and Eastern European Countries (CEECs)) (2001)

The Patents Decree Law of 1995 (Turkey)

The Agreement on Trade-Related Aspects of Intellectual Property Rights (TRIPS), 1994

The Community Patent Convention (CPC), signed on December 15, 1975

Thelen, Christine, Carrots and Sticks: Evaluating the Tools for Securing Successful TRIPs Implementation, XXIV, Temple Journal of Science, Technology & Environmental Law, 519 (2006)

Therneau, Terry, Beth Atkinson and Brian Ripley, rpart: Recursive Partitioning and Regression Trees (2015)

Thomas Taylor, Christopher, Z. A. Silberston and Aubrey Silberston, The Economic Impact of the Patent System: a Study of the British Experience, Cambridge University Press, Cambridge (1973)

Thomas, Lacy Glenn, Implicit Industrial Policy: The Triumph of Britain and the Failure of France in Global Pharmaceuticals, Industrial and Corporate Change, 1(2), 451 (1994)

Thompson, Peter and Melanie Fox-Kean, Patent Citations and the Geography of Knowledge Spillovers: A Reassessment, American Economic Review, vol. 95(1) 450 (2005)

Thompson, Wilbur R., Locational Differences in Inventive Efforts and Their Determinants 253, in Richard R. Nelson (ed.), The Rate and Direction of Inventive Activity: Economic and Social Factors, Princeton University Press: Princeton, NJ (1962)

Thursby, Jerry G. and Marie Thursby, Gender Patterns of Research and Licensing Activity of Science and Engineering Faculty, Journal of Technology Transfer, 30, 343 (2005)

Thursby, Jerry G. and Marie Thursby, Here or There? A Survey of Factors in Multinational R&D Location, Washington, DC: National Academies Press (2006)

Thursby, Marie and Richard Jensen, Proofs and Prototypes for Sale: The Licensing of University Inventions, American Economic Review, American Economic Association, vol. 91(1), 240 (2001)

Tondl, Gabriele, The Changing Pattern of Regional Convergence in Europe, Jahrbuch für Regionalwissenschaft, 19(1) 1 (1999)

Toole, Andrew A., The Impact of Public Basic Research on Industrial Innovation: Evidence from the Pharmaceutical Industry, Research Policy, 41(1) 1 (2012)

Trajtenberg, Manuel, A Penny for your Quotes, *Rand Journal of Economics*, 21(1) 172 (1990)

Tussie, Diana and David Glover, eds., *Developing Countries in World Trade: Policies and Bargaining Strategies*, Boulder, CO: Lynne Rienner (1995)

Tussie, Diana, *The Less Developed Countries and the World Trading System: A Challenge to the GATT*, London: Francis Pinter, (1987)

Tybout, James, Manufacturing Firms in Developing Countries: How Well Do They Do, and Why?, *Journal of Economic Literature*, 38(March): 11 (2000)

United Kingdom Commission on Intellectual Property Rights, *U.K Intellectual Property Rights Report, Integrating Intellectual Property Rights and Development Policy* (London, September 2002)

United Nations (UN) General Assembly, Resolution 55/2 – The United Nations Millennium Declaration (Sept. 18, 2000)

United Nations (UN) General Assembly, Resolution, Declaration for the Establishment of a New International Economic Order, U.N. Doc. A/RES/S-6/3201 (1 May 1974)

United Nations Conference on Trade and Development (UNCTAD) and ICTSD (International Centre for Trade and Sustainable Development), Intellectual Property Rights: Implications for Development, Intellectual Property Rights and Sustainable Development Series Policy Discussion Paper, ICTSD, Geneva (2003)

United Nations Conference on Trade and Development (UNCTAD), Capacity Building for Academia in Trade for Development: A Study on Contributions to the Development of Human Resources And to Policy Support for Developing Countries (2010)

United Nations Conference on Trade and Development (UNCTAD), Least Developed Countries Report 2007 128–29 (UNCTAD/LDC/2007 2007)

United Nations Conference on Trade and Development (UNCTAD), Trade and Development Report 2011, at: http://unctad.org/en/docs/tdr2011_en.pdf

United Nations Conference on Trade and Development (UNCTAD), World Investment Report 2013 – Global Value Chains: Investment and Trade for Development 2013

United Nations Conference on Trade and Development (UNCTAD), World Investment Report 2005, New York and Geneva, United Nations (2005)

United Nations Development Programme (UNDP), Human Development Indices, http://hdr.undp.org/en/content/human-development-index-hdi

United Nations Development Programme (UNDP), The Human Development concept, at: http://hdr.undp.org/en/humandev

United Nations Educational, Scientific and Cultural Organization (UNESCO), Institute for Statistics, Measuring R&D: Challenges Faced by Developing Countries, UIS/TD/10–08 (2008), available at http://www.uis.unesco.org/Library/Documents/tech%205-eng.pdf

United Nations Educational, Scientific and Cultural Organization (UNESCO), S&T glossary, at: http://uis.unesco.org/en/glossary

United Nations Educational, Scientific and Cultural Organization (UNESCO), Report on the Experts' Meeting on the Right to Enjoy the Benefits of Scientific

Progress and Its Applications (UNESCO Pub. SHS-2007/WS/13, June 7–8, 2007)

United Nations Educational, Scientific and Cultural Organization (UNESCO) Science & Technology (S&T) database at: http://data.uis.unesco.org/

United Nations Educational, Scientific and Cultural Organization (UNESCO) (2010), Technical Paper No. 5, Measuring R&D: Challenges Faced by Developing Countries

United Nations Educational, Scientific and Cultural organization (UNESCO) Science and Technology (S&T) Statistical report, at: http://data.uis.unesco.org/.

United Nations Industrial Development Organization (UNIDO), *Industrial Development Report 2013, Sustaining Employment Growth: The Role of Manufacturing and Structural Change*, United Nations Industrial Development Organization, Vienna (2013)

United Nations Industrial Development Organization (UNIDO), Strategic Long-Term Vision Statement, GC11/8/Add.1 (Oct. 14, 2005)

United Nations Industrial Development Organization (UNIDO), Valentina De Marchi, Elisa Giuliani and Roberta Rabellotti, Local Innovation and Global Value Chains in Developing Countries – Inclusive and Sustainable Industrial Development, Working Paper Series WP 05 (2015)

United Nations Millennium Project, Calestous Juma and Lee Yee-Cheong, *Innovation: Applying Knowledge in Development, London: Task Force on Science*, Technology and Innovation, Earthscan (2005)

United Nations Statistics Division (UNSTATS), Composition of Macro Geographical (continental) Regions, Geographical Sub-Regions, and Selected Economic and other Groupings (17 February 2011)

United Nations, The Millennium Development Goals Report 2007 (2007)

United States Congressional Research Service, Patent Ownership and Federal Research and Development (R&D): A Discussion on the Bayh-Dole Act and the Stevenson-Wydler Act (December 11, 2000)

United States Int'l Trade Comm'n (USITC), Pub. No. 4199, China: IP Infringement, Indigenous Innovation Policies and Framework for Measuring the Effects on the US Economy (2010)

United States Patent and Trademark Office (USPTO), Patent Full-Text and Image Database – Tips on Fielded Searching (Inventor Country (ICN)), at: http://www.uspto.gov/patft/help/helpflds.htm#Inventor_Country

United States Patent and Trademark Office, Enhanced Patent Quality Initiative pillars, at: http://www.uspto.gov/patent/enhanced-patent-quality-initiative-pillars

Vaitsos, Constantine, Patent Revisited: Their Function in Developing Countries, 9 *Journal of Development Studies*, 71, 89–90 (1972)

van Pottelsberghe, Bruno and Frank R. Lichtenberg, Does Foreign Direct Investment Transfer Technology Across Borders? *Review of Economics and Statistics*, 83(3), 490 (2001)

van Pottelsberghe, Bruno de la Potterie, The Quality Factor in Patent Systems, ECARES working paper 2010–027 (2010)

Vandenbussche, Jérôme, Philippe Aghion and Costas Meghir, Growth, Distance to Frontier and Composition of Human Capital, *Journal of Economic Growth*, 11(2) 97 (2006)

Vanhoudt, Patrick, Thomas Mathä and Bert Smid, How Productive Are Capital Investments in Europe?, *EIB papers*, 5(2) (2000)

Varga, Attila, Local Academic Knowledge Transfers and the Concentration of Economic Activity, *Journal of Regional Science*, vol. 40(2) 289 (2000)

Varsakelis, Nikos C., The Impact of Patent Protection, Economy Openness and National Culture on R&D Investment: A Cross-Country Empirical Investigation, 30 *Research Policy*, 1059 (2001)

Velho, Lea, *Science and Technology in Latin America and the Caribbean: An Overview*, Discussion Paper, 2004–4 (Maastricht: UNU-INTECH) (2004)

Verspagen, Bart and Wilfred Schoenmakers, The Spatial Dimension of Knowledge Spillovers in Europe: Evidence from Patenting Data, paper presented at the AEA Conference on Intellectual Property Econometrics, Alicante, 19–20 April (2000))

Visser, Coenraad, The Policy-Making Dynamics in Intergovernmental Organizations, 82 *Chicago-Kent Law Review*, 1457 (2007)

Vivarelli, Marco, Innovation, Employment, and Skills in Advanced and Developing Countries: A Survey of the Economic Literature, *Journal of Economic Issues*, 48(1) 123 (2014)

von Hagen, Jürgen and George Hammond, *Industrial localization: An Empirical Test for Marshallian Localization Economies*, Discussion paper 917, London: Centre for Economic Policy Research (1994)

Von Hipple, Eric, Sticky Information and the Locus of Problem Solving: Implications for Innovation, *Management Science*, 40, 429 (1994)

von Stadler, Manfred, Engines of Growth: Education and Innovation, *Jahrbuch für Wirtschaftswissenschaften/Review of Economics*, Vol. 63(2), 113 (2012)

Walsh, Vivien M., J.F. Townsend, *B.G. Achilladelis and C. Freeman, Trends in Invention and Innovation in the Chemical Industry*, Report to SSRC, SPRU, University of Sussex, mimeo (1979)

Wang, Qing, Zongxian Feng and Xiaohui Hou, A Comparative Study on the Impact of Indigenous Innovation vs. Technology Imports for Domestic Technological Innovation, Science of Science and Management of S & T (2010)

Watal, Jayashree, *Intellectual Property Rights in the WTO and Developing Countries* (Kluwer) (2001)

Waugh, David, *Geography, An Integrated Approach*, Nelson Thornes Ltd., 3rd ed., (2000)

Weeks, John and Howard Stein, Washington Consensus, in *The Elgar Companion to Development Studies* 676 (David Alexander Clark, ed.) (Edward Elgar Publishing, 2006)

Wei, Yehua H. Dennis, Ingo Liefner and Changhong Miao, Network Configurations and R&D Activities of the ICT Industry in Suzhou Municipality, *China, Geoforum*, 42(4) 484 (2011)

Weissman, Robert, A Long, Strange TRIPS: The Pharmaceutical Industry Drive to Harmonize Global Intellectual Property Rules, and the Remaining WTO Legal Alternatives Available to Third World Countries, 17 *University of Pennsylvania Journal of International Law*, 1079 (1996)

Whalley, John (ed.), *Developing Countries and the World Trading System*, Vols. 1 and 2, London: Macmillan, 1989

Whittington, Kjersten Bunker and Laurel Smith-Doerr, Gender and Commercial Science: Women's Patenting in the Life Sciences, *Journal of Technology Transfer*, 30, 355 (2005)

Whittington, Kjersten Bunker and Laurel Smith-Doerr, Women Inventors in Context: Disparities in Patenting across Academia and Industry, *Gender and Society*, 22(2), 194 (2008)

Whittington, Kjersten Bunker, *Gender and Scientific Dissemination in Public and Private Science: A Multivariate and Network Approach*, Department of Sociology, Stanford University; Kjersten Bunker Whittington, Laurel Smith-Doerr, Women and Commercial Science: Women's Patenting in the Life Sciences, *Journal of Technology Transfer*, 30, 355 (2005)

William N. Venables and Brian D. Ripley, *Modern Applied Statistics with S* (4th edition), Springer, New York (2002)

Williamson, Jeffrey G., Regional inequalities and the Process of National Development, *Economic Development and Cultural Change Quarterly Journal of Economics*, 13(1) 84 (1965)

Williamson, John, What Washington Means by Policy Reform, in *Latin American Adjustment: How Much Has Happened?* 7, 7, John Williamson, ed. (1990)

Williamson, Oliver, *Markets and Hierarchies, Analysis and Anti-Trust Implications: A Study in the Economics of Industrial Organizations*, New York: Free Press (1975)

Williamson, Oliver, *The Economic Institutions of Capitalism*, New York: Free Press (1985)

Wolfe, David, *Clusters Old and New: The Transition to a Knowledge Economy in Canada's Regions*, Kingston: Queen's School of Policy Studies (2003)

Wolff, Hendrik, Howard Chong and Maximilian Auffhammer, Classification, Detection and Consequences of Data Error: Evidence from the Human Development Index, *Economic Journal*, 121 (553): 843 (2011)

Woo, Wing Thye, Some Fundamental Inadequacies of the Washington Consensus: Misunderstanding the Poor by the Brightest, at 1 available at http://papers.ssrn.com/sol3/papers.cfm?abstract_id=622322

Woodward, Douglas, Octávio Figueiredo and Paulo Guimarães, Beyond the Silicon Valley: University R&D and High-Technology Location, *Journal of Urban Economics*, Vol. 60(1) 15 (2006)

World Bank Comm'n on Growth & Dev., The Growth Report Strategies for Sustained Growth and Inclusive Development (2008), available at https://open knowledge.worldbank.org/bitstream/handle/10986/6507/449860PUB0Bo x3101OFFICIAL0USE0ONLY1.pdf?sequence=1&isAllowed=y

World Bank Data IBRD-IDA—Glossary, at: http://data.worldbank.org

World Bank, Beyond Economic Growth Student Book (2004), at: http://www.worldbank.org/depweb/english/beyond/global/glossary.html

World Bank, Country and Lending Groups, at https://datahelpdesk.world bank.org/knowledgebase/articles/906519-world-bank-country-and-lending-groups (July 2016)

World Bank, Global Economic Prospects and the Developing Countries (vol. 12, 2002)

World Bank, How We Classify Countries, at: https://datahelpdesk.worldbank.or g/knowledgebase/topics/19280-country-classification

World Bank, Innovation Policy: A Guide for Developing Countries (2010), at 43, available at https://openknowledge.worldbank.org/bitstream/handle/10986/24 60/548930PUB0EPI11C10Dislosed061312010.pdf

World Bank, The Growth Report Strategies for Sustained Growth and Inclusive Development, Commission on Growth and Development, Conference Edition (2008), at: http://www.ycsg.yale.edu/center/forms/growthReport.pdf

World Bank, The Trading System and Developing Countries, available at http://siteresources.worldbank.org/INTTRADERESEARCH/Resources/Pa rt_7.pdf

World Health Organization (WHO), Exec. Bd. 124th Session, Public health, inno- vation and intellectual property: global strategy and plan of action: Proposed time frames and estimated funding needs, at 1, EB124/16 Add.2 (Jan. 21, 2009), at: www.who.int/gb/ebwha/pdf_files/EB124/B124_16Add2-en.pdf

World Health Organization (WHO), Public Health, Innovation and Intellectual Property and Trade—Expert Working Group on R&D Financing, http:// www.who.int/phi/R_Dfinancing/en

World Health Organization (WHO)-WIPO-WTO, Promoting Access to Medical Technologies and Innovation: Intersections between Public Health, Intellectual Property and Trade (5 February 2013)

World Health Organization (WHO)-WIPO-WTO, Public Health, Intellectual Property, and TRIPS at 20: Innovation and Access to Medicines; Learning from the Past, Illuminating the Future—Joint Symposium by WHO, WIPO, WTO (October 28, 2015)

World Health Organization WHO, WIPO, WTO Joint Technical Workshop on Patentability Criteria (October 27, 2015)

World Health Organization, Globalization, TRIPs and Access to Pharmaceuticals, World Health Organization Policy Perspectives on Medicines, No. 3, WHO/ EDM/2001.2 (Mar. 2001)

World Intellectual Property Organization (WIPO) Report 2011

World Intellectual Property Organization (WIPO), Academy Education and Training Programs Portfolio (2013), available at: http://www.wipo.int/export/ sites/www/freepublications/en/training/467/wipo_pub_467_2013.pdf

World Intellectual Property Organization (WIPO), Copyright Treaty, Dec. 20, 1996, 2186 U.N.T.S. 121 WIPO Doc. CRNR/DC/94 (23 December 1996), WIPO Doc. CRNR/DC/95 (23 December 1996), available at http://www.wip o.int/treaties/en/ip/wct/summary_wct.html

World Intellectual Property Organization (WIPO), Director General Report on Implementation of the Development Agenda, Committee on Development and Intellectual Property (CDIP) Thirteenth Session, CDIP/13/2, Geneva, May 19 to 23, 2014 (March 3, 2014)

World Intellectual Property Organization (WIPO), Economic Aspects of Intellectual Property in Countries with Economies in Transition, Ver. 1, the Division for Certain Countries in Europe and Asia, WIPO (2012)

World Intellectual Property Organization (WIPO), Economics & Statistics Series, World Intellectual Property Report – The Changing Face of Innovation 26 (2011)

World Intellectual Property Organization (WIPO), *Global Innovation Index Rankings 2015*, Geneva (2015)

World Intellectual Property Organization (WIPO), Press Release, Member States Adopt a Development Agenda for WIPO, WIPO/PR/2007/521, at www.wipo.int/pressroom/en/articles/2007/article_0071.html (Oct. 1, 2007)

World Intellectual Property Organization (WIPO), Press Release, Member States Agree to Further Examine Proposal on Development, WIPO/PR/2004/396 (Oct. 4, 2004), available at www.wipo.int/pressroom/en/prdocs/2004/wipo_pr_2004_396.html

World Intellectual Property Organization (WIPO), Protecting Innovations by Utility Models, at: http://www.wipo.int/sme/en/ip_business/utility_models/utility_models.htm

World Intellectual Property Organization (WIPO), Shahid Alikhan, Socioeconomic Benefits of Intellectual Property Protection in Developing Countries 1–9 (2000)

World Intellectual Property Organization (WIPO), Standing Committee on the Law of Patents, Tenth Session, Draft Substantive Patent Law Treaty, SCP/10/2, available at http://www.wipo.int/edocs/mdocs/scp/en/scp_10/scp_10_2.pdf

World Intellectual Property Organization (WIPO), Standing Committee on the Law of Patents, Quality of Patents: Comments Received from Members and Observers of the Standing Committee on the Law of Patents (SCP), Seventeenth Session, Geneva, December 5 to 9, 2011, SCP/17/INF/2, October 20, 2011

World Intellectual Property Organization (WIPO), The 45 Adopted Recommendations under the WIPO Development Agenda, at: http://www.wipo.int/ip-development/en/agenda/recommendations.html

World Intellectual Property Organization (WIPO), The Development Agenda, Cluster E: Institutional Matters including Mandate and Governance

World Intellectual Property Organization (WIPO), The Economics of Intellectual Property: Suggestions for Further Research in Developing Countries and Countries with Economies in Transition 22 (2009)

World Intellectual Property Organization (WIPO), The Global Innovation Index 2014: The Human Factor in Innovation, Soumitra Dutta, Bruno Lanvin, and Sacha Wunsch-Vincent (eds.) (2014)

World Intellectual Property Organization (WIPO), Update on the Management Response to the External Review of WIPO Technical Assistance in the Area of Cooperation for Development, CDIP/16/6 (September 2015)

World Intellectual Property Organization (WIPO), WIPO Intellectual Property Handbook: Policy, Law and Use, WIPO Publication No. 489, 2d ed., at http://www.wipo.int/about-ip/en/iprm (2004)

World Intellectual Property Organization (WIPO), Working Document for the Provisional Committee on Proposals Related to a WIPO Development Agenda (PCDA), WIPO Doc. PCDA/3/2, Annex B, ¶ 28, pp. 14–15, Feb. 20, 2007

World Intellectual Property Organization (WIPO), World Intellectual Property Indicators (2013)

World Intellectual Property Organization (WIPO), World Intellectual Property Indicators 2015 (2015)

World Intellectual Property Organization (WIPO), World Intellectual Property Report 2015 – Breakthrough Innovation and Economic Growth (2015)

World Intellectual Property Organization (WIPO), World Patent Report: A Statistical Review – 2008 edition, at: http://www.wipo.int/ipstats/en/statistics/patents/wipo_pub_931.html; WIPO, WIPO Patent Report: Statistics on Worldwide Patent Activities (2007)

World Intellectual Property Organization(WIPO), The 54th Session of the WIPO Assemblies of 22–30 September 2014, at: http://www.wipo.int/meetings/en/details.jsp?meeting_id=32482

World Intellectual Property Organization, Annual Statistical Report of the WIPO Academy for 2013, at: http://www.wipo.int/export/sites/www/academy/en/about/pdf/academy_statistics_2013.pdf

World Intellectual Property Organization, Standing Committee on the Law of Patents, at http://www.wipo.int/patent-law/en/scp.htm

World Trade Organization (WTO) Comm. on Trade & Dev., Note by the Secretariat, Report on Technical Assistance 2000, WT/COMTD/W/83 (May 2, 2001)

World Trade Organization (WTO), Comm. on Trade & Dev., Note by Secretariat, A New Strategy For WTO Technical Cooperation: Technical Cooperation for Capacity-building, Growth and Integration, WT/COMTD/W/90 (Sept. 21, 2001)

World Trade Organization (WTO), Comm. on Trade & Dev., Technical Assistance and Training Plan 2004, WT/COMTD/W/119/Rev.3 (Feb. 18, 2004)

World Trade Organization (WTO), Comm. on Trade and Dev, Note by Secretariat, Coordinated WTO Secretariat Annual Technical Assistance Plan 2003, WT/COMTD/W/104 (Oct. 3, 2002)

World Trade Organization (WTO), Declaration of the Group of 77 and China on the Fourth WTO Ministerial Conference at Doha, Qatar, WT/L/424 (Oct. 22, 2001), available at http://www.wto.org/english/thewto_e/minist_e/min01_e/proposals_e/wt_l_424.pdf

World Trade Organization (WTO), Least-developed Countries (LDCs) Members, http://www.wto.org/english/tratop_e/trips_e/trips_groups_e.htm

World Trade Organization (WTO), Ministerial Declaration of 20 November 2001 WT/MIN(01)/DEC/1, 41 I.L.M. 746 (2002)

World Trade Organization (WTO), World Trade Report 2003 (2003)

World Trade Organization (WTO)-WIPO Cooperation Agreement, entered into force Jan. 1, 1996

World Trade Organization (WTO), Who are the Developing Countries in the WTO?, at: https://www.wto.org/english/tratop_e/devel_e/d1who_e.htm

Wyer, Mary, Mary Barbercheck, Donna Cookmeyer, Hatice Ozturk, Marta Wayne, eds., *Women, Science, and Technology: A Reader in Feminist Science Studies* (Routledge, 3rd ed., 2014)

Wyllys, R.E., Overview of the Open-Source Movement, The University of Texas at Austin Graduate School of Library & Information Science (2000)

Xie, Yu and Kimberlee A. Shauman, Sex Differences in Research Productivity: New Evidence about an Old Puzzle, *American Sociological Review*, 63 (6), 847 (1998)

Xie, Yu and Kimberlee A. Shauman, *Women in Science: Career Processes and Outcomes*, Cambridge, MA: Harvard University Press (2003)

Yu III, Vicente Paolo B., Unity in Diversity: Governance Adaptation in Multilateral Trade Institutions Through South-South Coalition-building, Research papers 17, South Centre (July 2008)

Yu, Geoffrey, The Structure and Process of Negotiations at the World Intellectual Property Organization, 82 *Chicago-Kent Law Review*, 1445 (2007)

Yu, Peter K., Building Intellectual Property Coalitions for Development, in *Implementing the World Intellectual Property Organization's Development Agenda* 79, Wilfrid Laurier University Press, CIGI, IDRC (Jeremy de Beer, ed.) (2009)

Yu, Peter K., Currents and Crosscurrents in the International Intellectual Property Regime, 38 *Loyola of Los Angeles Law Review*, 323 (2004)

Yu, Peter K., The Middle Intellectual Property Powers, Drake University Legal Studies, Research Paper Series, Research Paper No. 12–28 (2012)

Yu, Peter, Déjà Vu in the International Intellectual Property Regime 113, in *Sage Handbook on Intellectual Property*, SAGE Publications Ltd, (Matthew David and Debora Halbert, eds.) (2014)

Yu, Peter, Five Oft-Repeated Questions About China's Recent Rise as a Patent Power, 2013 *Cardozo Law Review De Novo*, 78 (2013)

Yu, Peter, The ACTA/TPP Country Clubs 258, in Dana Beldiman (ed.), *Access to Information and Knowledge: 21st Century Challenges in Intellectual property and Knowledge Governance*, Edward Elgar (2014)

Yu, Peter, Toward a Nonzero-Sum Approach to Resolving Global Intellectual Property Disputes: What We Can Learn from Mediators, Business Strategists, and International Relations Theorists, 70 *University of Cincinnati Law Review*, 569 (2001)

Zellner, Christian, The Economic Effects of Basic Research: Evidence for Embodied Knowledge Transfer via Scientists' Migration, *Research Policy*, 32 1881 (2003)

Zhao, Minyuan, Conducting R&D in Countries with Weak Intellectual Property Rights Protection, *Management Science*, 5, 1185 (2006)

Zhou, Yu and Tong Xin, An Innovative Region in China: Interaction Between Multinational Corporations and Local Firms in a High-Tech Cluster in Beijing, *Economic Geography*, 79(2) 129 (2003)

Ziman, John, *Prometheus Bound*, Cambridge University Press (1994)

Zoellick, Robert B., America will not wait (September 21 2003), at http://www.fordschool.umich.edu/rsie/acit/TopicsDocuments/Zoellick030921.pdf

Zucker, Lynne G. and Michael R. Darby, *Costly Information in Firm Transformation, Exit, or Persistent Failure*, NBER Working Papers 5577, National Bureau of Economic Research, Inc. (1996)

Zucker, Lynne G., Michael R. Darby and Jeff Armstrong, Geographically Localized Knowledge: Spillovers or Markets?, Economic Inquiry, *Western Economic Association International*, vol. 36(1) 65 (1998)

Zucker, Lynne G., Michael R. Darby, Marilynn B. Brewer, Intellectual Human Capital and the Birth of US Biotechnology Enterprises, *American Economic Review*, 87(1) (1997)

Zucker, Lynne G. and Michael R. Darby, Star Scientists and Institutional Transformation: Patterns of Invention and Innovation in the Formation of the Biotechnology Industry – Proceedings of the National Academy of Science 93 (November): 12709 (1996)

Index of Subjects

A2 K. *See* Access to Knowledge
abroad sector, GERD data analysis of,
    157–64, 173
Access to Knowledge (A2 K), 46
    patent club convergence coalitions and, 58
ACTA. *See* Anti-Counterfeiting Trade
    Agreement
Adams, James, 132, 187
advanced economies, patent intensity in,
    114–17
Agreement on Textiles and Clothing
    (ATC), 34, 128
Agreement on the Trade Related Aspects of
    Intellectual Property (TRIPS)
    in business sector, 127, 128
    developing countries' promotion of, 35
    economic growth policies and, 15–17
    economic growth theory and, 62
    economic reform through, 1
    FTAs and, 28
    innovation-led economic growth and, 51
    IPRs under, 15–17
    LDCs and, 15–17
    MFN principle of, 32
    patent club convergence and, 62
    patent regulatory tradeoffs, 32–36
    technology transfer policies, 120
    trade innovation policies, 32–36
    TRIPS plus, 31–32
agricultural trade, 33
Amon, Christian, 187
Amsden, Alice, 131–32, 141–42
Annan, Kofi, 24
Anti-Counterfeiting Trade Agreement
    (ACTA), 59
applied research, 185–86
    basic research compared to, 187–89
    hierarchy in, 187–89
    regression models for, 208
    in universities and colleges, patents as
        result of, 186
        under Bayh-Dole Act, 186, 247

professor's privilege and, 186
Argentina, patent intensity in, 126
Arnold, Lutz, 218
Arrow, Kenneth, 181, 262–63
ATC. *See* Agreement on Textiles and
    Clothing
Audretsch, David, 264
Averch, Harvey, 181

Bahamas, patent intensity in, 126
Bailén, Jose, 175
Baloch, Irfan, 23
Barro, Robert, 222
Bartelsman, Eric, 133
basic R&D
    applied research compared to, 187–89
    defined, 184–85
    economic benefits of, 188
    global coordination of, 203
    hierarchy in, 187–89
    non-basic, 185
    Pasteur's Quadrant and, 205
    regression models for, 208
Baumol, William, 240–41
Bayh-Dole Act, 186, 247
Becker, Gary, 218–19
Beijing Consensus, 278
Beijing Treaty on Audiovisual Performances
    (BTAP), 22
Benhabib, Jess, 224–25, 239–40
Bermuda, patent intensity in, 126
Bernard, Jean-Thomas, 132–33
Blackburn, Keith, 218
bloc-type coalitions, 61
Bonin, Bernard, 241
Botswana, economic growth in, 18
Braithwaite, John, 20
Brazil
    economic growth in, 18
    indigenous patenting in, 273
    patent intensity in, 126
    PVP in, 200–1

Index of Persons